Yanliang Du · Baochen Sun
Jianzhi Li · Wentao Zhang

Optical Fiber Sensing and Structural Health Monitoring Technology

图书在版编目(CIP)数据

光纤传感与结构健康监测技术=Optical Fiber Sensing and Structural Health Monitoring Technology:英文/杜彦良等著. —武汉:华中科技大学出版社,2019.3
ISBN 978-7-5680-4729-6

Ⅰ.①光… Ⅱ.①杜… Ⅲ.①光纤传感器-智能材料-结构-监测-研究-英文
Ⅳ.①TP212.4

中国版本图书馆 CIP 数据核字(2019)第 038465 号

Sales in Mainland China Only
本书仅限在中国大陆地区发行销售

光纤传感与结构健康监测技术
GUANGXIAN CHUANGAN YU JIEGOU JIANKANG JIANCE JISHU

杜彦良　等著

策划编辑：俞道凯	
责任监印：周治超	
出版发行：华中科技大学出版社(中国·武汉)	电话：(027)81321913
武汉市东湖新技术开发区华工科技园	邮编：430223
录　　排：武汉三月禾文化传播有限公司	
印　　刷：湖北恒泰印务有限公司	
开　　本：710mm×1000mm　1/16	
印　　张：23	
字　　数：490 千字	
版　　次：2019 年 3 月第 1 版第 1 次印刷	
定　　价：198.00 元	

本书若有印装质量问题，请向出版社营销中心调换
全国免费服务热线：400-6679-118　竭诚为您服务
版权所有　侵权必究

Website: http://www.hustp.com/
Book Title: Optical Fiber Sensing and Structural Health Monitoring Technology
Copyright@2019 by Huazhong University of Science & Technology Press. All rights reserved. No part of this publication may be reproduced, stored in a database or retrieval system, or transmitted in any form or by any electronic, mechanical, photocopy, or other recording means, without the prior written permission of the publisher.

Contact address: No. 6 Huagongyuan Rd, Huagong Tech Park, Donghu High-tech Development Zone, Wuhan City 430223, Hubei Province, P.R. China.
Phone/fax: 8627—81339688; **E-mail**: service@hustp.com

Disclaimer

This book is for educational and reference purposes only. The authors, editors, publishers and any other parties involved in the publication of this work do not guarantee that the information contained herein is in any respect accurateor complete. It is the responsibility of the readers to understand and adhere to local laws and regulations concerning the practice of these techniques and methods. The authors, editors and publishers disclaim all responsibility for any liability, loss, injury, or damage incurred as a consequence, directly or indirectly, of the use and application of any of the contents of this book.

First published: 2019
ISBN: 978-7-5680-4729-6

Cataloguing in publication data: A catalogue record for this book is available from the CIP-Database China.

ISBN 978-7-5680-4729-6

Printed in The People's Republic of China

Preface

The international academic and engineering community have reached a broad consensus to monitor the health of civil engineering structure, study the health state of structure in real time, discover and eliminate the potential safety hazard in time, and guarantee the long-term safe service of various civil engineering structures. The health monitoring technology based on optical fiber sensing has unique advantages such as small volume, lightweight, high sensitivity, anti-electromagnetic interference capability, integrated transmission and sensing functions, easy networking, and distributed measurement. Therefore, it made rapid development and considerable progress in the recent 20 years. Its role in the health monitoring of engineering structure is increasingly important.

The authors and their research team have dedicated themselves to the research on the development and application of health monitoring technology based on optical fiber sensing for a long time. They have rich research achievements and engineering practice experience. The book introduces the basic theory and method of optical fiber sensing technology as well as the structural design, manufacture, and practical application of sensors from the point of structural health monitoring. The book is divided into 10 chapters. Chapter 1 is Introduction. It mainly introduces the basic concept, type, and feature of optical fiber sensing technology as well as its application and development trend in the structural health monitoring field. Chapters 2–5 mainly focus on the demand for engineering structure health monitoring. They describe various optical fiber sensors, including F-P optical fiber sensor, optical fiber laser sensor, fiber Bragg grating sensor, and fully distributed optical fiber sensor, from the aspects of sensor working principle, structure design, manufacture process, and sensing characteristics. Chapters 6–9 emphasize the health monitoring technologies of different engineering structures such as prestressed tendon, cable, and concrete. Chapter 10 focuses on the engineering application of optical fiber sensing technology. It mainly introduces the long-term health monitoring, security evaluation, and alarm system of several projects with optical fiber sensors, including Wuhu Yangtze River Bridge, Liaohe Grand Bridge on Qinhuangdao-Shenyang Passenger Dedicated Line, Hemaxi Grand Bridge, Xiaogou Grand Bridge on Shanxi Xinyuan Highway, and the high and steep slope of Shuohuang Railway.

The book summarizes and integrates the research achievements and engineering applications of the author and his research team in over 20 years. Sincerely thank the research team members for their dedicated cooperation and hard work for many years, including Sun Baochen, Su Mubiao, Wang Xinmin, Zhao Weigang, Liu Yongqian, Chen Baoping, Li Jianzhi, Chen Shuli, Li Yiqiang, etc. In addition, Zhang Xushe, Jin Xiumei, Wei Bin, Zhang Wentao, Li Feng, Xu Hongbin, Li Xiaoyang, Liu Chenxi, Shao Lin, Yang Yaoen, Yang Liping, Dai Jingyun, Xu Hua, Hao Gengjie, Han Jing, and Hou Yuemin devoted their wisdom and painstaking efforts to the research contents in the book when they were studying for doctoral or master's degree. When the book was drafted and compiled, Liu Bo, Li Feng, Xu Hongbin, Sun Xu, Zheng Xinyu, Ren Zexu, Li Zhendong, Wang Qingyou, Sun Haokai, Wang Haiyong, and other doctoral or master students made great efforts to the literature arrangement, translation, and review. The authors are hereby deeply grateful to all of them.

The book was funded by China High Technology Research and Development Projects "Inspection and Reinforcement Technology of the Bridge and Subgrade for Heavy-haul Railway" (2009AA11Z102) and "Monitoring Technology of Heavy-haul Train Operation Danger State Based on Optical Fiber Sensing" (2009AA11Z212); National Natural Science Foundation of China "Critical State Evaluation and Critical Technology Research on Cable-stayed Bridge Structure Based on Smart Stay Cable" (50778116), "Monitoring Technology and Method Research on Critical Bearing Component State of Prestressed Reinforced Concrete Structure" (50278058), "New Method of Long-term Composite Monitoring for Prestressed Anchor Cable Corrosion Damage" (51778379) and "New Method of Long-term Monitoring for Geotechnical Anchor Cable with Fully Distributed Stress and Damage Positioning Function" (51508349); Natural Science Foundation of Hebei Province "Research on the Real-time Monitoring and Security Evaluation System of Stayed Cable Based on Fiber Bragg Grating" (E2004000417), "Critical Technology Research on Carbon Fiber Composite Reinforcement with Automatic Monitoring Function" (E2006000389), "Dynamic Stress Analysis and Property of Intelligent Composite Material for Fiber Bragg Grating" (E2015210094); Scientific Research Project of Hebei Province "Research and Development of Optical Fiber Strain Sensor and Its Signal Acquisition System" (03213539D); Science and Technology Research and Development Projects of Ministry of Railways "Long-term Monitoring and Security Evaluation System of Wuhu Yangtze River Bridge" (2000G19-B) and "Application Research on Optical Fiber Testing Technology of Liaohe Grand Bridge on Qinhuangdao-Shenyang Passenger Dedicated Line" (2001G018-B). The authors hereby express their sincere thanks to the Ministry of Science and Technology, National Natural Science Foundation of China, Hebei Province, Ministry of Railways and other science and technology administration departments for their full support and project grant.

It is expected that the book is helpful for scientific researchers and engineering technicians in the field of structural health monitoring and optical fiber sensing technology, as well as the teachers and students in related majors of colleges and universities. The research contents in this book involve civil engineering, mechanics, photology, machinery, material, measurement and control, featuring multidisciplinary interpenetration, large span, and high difficulty. Given the limited level of the author, it is unavoidable to have omissions and improperness in the book. Please kindly offer advice and correct them.

Shijiazhuang, China
Spring 2018

Yanliang Du

Contents

1	**Introduction**		1
	1.1 Optical Fiber and Optical Fiber Sensor		1
		1.1.1 Optical Fiber	2
		1.1.2 Optical Fiber Sensor	3
	1.2 Classification and Characteristics of Optical Fiber Sensor		4
		1.2.1 Classification of Optical Fiber Sensor	4
		1.2.2 Characteristics of Optical Fiber Sensor	11
	1.3 Current Status and Development Trends of Optical Fiber Sensing Technology		13
		1.3.1 Current Status of Optical Fiber Sensing Technology	13
		1.3.2 Development Trends of Optical Fiber Sensing Technology	17
	1.4 Structural Health Monitoring Based on Optical Fiber Sensing Technology		24
	References		28
2	**Optical Fiber Interferometer Based on F-P Cavity**		31
	2.1 White Light Interferometric F-P Optical Fiber Sensor		31
		2.1.1 Principle of White Light Interferometric F-P Optical Fiber Sensor	31
		2.1.2 White Light Interferometric Sensor Head	35
		2.1.3 Embedded White Light Interferometric Optical Fiber Temperature Sensor	44
		2.1.4 Embedded White Light Interferometric Optical Fiber Strain Sensor	55
	2.2 Optical Accelerometers Based on F-P Cavity		66
		2.2.1 Preliminary Test for Encapsulation	66
		2.2.2 Structure of Accelerometer	69

i

		2.2.3	Principle of Accelerometer	70
		2.2.4	Test Results and Discussions	72
	2.3	Summary		73
	References			74
3	**Fiber Bragg Grating Sensor**			**77**
	3.1	Basic Principle of Fiber Bragg Grating		77
		3.1.1	Coupled Mode Theory for Fiber Bragg Grating	77
		3.1.2	Principle and Sensitivity of Fiber Bragg Grating Temperature Sensor	81
		3.1.3	Principle and Sensitivity of FBG Strain Sensor	83
		3.1.4	Theoretical Analysis of FBG Temperature–Strain Cross-Sensitivity	84
	3.2	Temperature Self-Compensated FBG Sensor Based on Thermal Stress		85
		3.2.1	Principle of Temperature Self-Compensation	86
		3.2.2	Structural Design	87
		3.2.3	Theoretical Analysis of Strain Sensing Characteristics	91
		3.2.4	Parameter Analysis of Temperature Compensation Structure Design	94
		3.2.5	FBG Strain Sensor with Integral Temperature Compensation Structure	99
		3.2.6	Small FBG Strain Sensor	114
	3.3	FBG Soil-Pressure Sensor Based on Dual L-Shaped Levers		120
		3.3.1	Structure and Principle of the Soil-Pressure Sensor	121
		3.3.2	Design and Strength Check of Soil-Pressure Sensor	123
		3.3.3	Laboratory Calibration Tests	125
		3.3.4	Field Tests	126
	3.4	Fiber Bragg Grating Displacement Sensor		128
		3.4.1	Sensor Design	128
		3.4.2	Tests and Results	130
	3.5	Fiber Bragg Grating Tilt Sensor		131
		3.5.1	Structure Design of Fiber Bragg Grating Tilt Sensor	132
		3.5.2	Sensing Performance of Fiber Bragg Grating Tilt Sensor	139
		3.5.3	Indoor Simulation Experiment of Fiber Bragg Grating Tilt Sensor	143
	3.6	Summary		145
	References			147
4	**Fiber Laser Sensor**			**149**
	4.1	Acoustic Emission Receiver Based on DFB		149
		4.1.1	Operation Principles	150

		4.1.2	Investigation of AE Directional Sensitivity of DFB Fiber Laser	151

        4.1.2 Investigation of AE Directional Sensitivity of DFB Fiber Laser 151
        4.1.3 Location Algorithm 158
        4.1.4 Tests and Results 160
  4.2 DFB Fiber Laser Accelerometers 162
        4.2.1 Principles..................................... 163
        4.2.2 Wavelet Denoising 169
        4.2.3 Inertial Algorithm.............................. 169
        4.2.4 Test Scheme 170
        4.2.5 Test Results 171
  4.3 Summary 172
  References 174

5 Fully Distributed Optical Fiber Sensor 177
  5.1 Spontaneous Scattering Spectrum in Optical Fiber 177
  5.2 Application of Spontaneous Scattering in Fully Distributed Optical Fiber Sensing Technology 178
  5.3 Winding Optical Fiber Strain Sensor 179
        5.3.1 Theoretical Basis and Analysis 179
        5.3.2 Structure and Parameters of Winding Optical Fiber Strain Sensor 186
        5.3.3 Measurement System of Winding Optical Fiber Strain Sensor 190
        5.3.4 Sensing Characteristic of Winding Optical Fiber Strain Sensor 198
        5.3.5 Distributed Sensing Characteristics of Winding Optical Fiber Strain Sensor 204
  5.4 Large Displacement Sensor Based on Fully Distributed Optical Fiber Sensor 206
        5.4.1 Principle of Fully Distributed Displacement Sensing Based on Fiber Bragg Grating................... 206
        5.4.2 Displacement Loading Test..................... 208
        5.4.3 Analysis of the Displacement Sensing Characteristics 210
  5.5 Summary 213
  References 216

6 Monitoring Technology for Prestressing Tendons Using Fiber Bragg Grating 219
  6.1 Theoretical Analysis on Prestress Loss of Concrete Structure... 219
        6.1.1 Calculation of Prestress Loss.................... 219
        6.1.2 Calculation of Effective Prestress................. 229
  6.2 Design of FBG Prestress Sensor at Anchor Head 230
        6.2.1 Prestress Monitoring Principle at Anchor Head of Prestressed Concrete Structure................. 231

		6.2.2	Structure Design and Principle of FBG Prestress Sensor at Anchor Head	233
		6.2.3	Design of FBG Prestress Sensor at Anchor Head	234
		6.2.4	Calibration Experiment for FBG Prestress Sensor at the Anchor Head	237
	6.3	Prestress Monitoring Technology Using Fiber Bragg Grating Sensor Arrays		238
		6.3.1	Structure and Performance Parameters of Steel Strands	238
		6.3.2	Combination of Fiber Bragg Grating and Steel Strand and Stress Measurement Principle for Steel Strand	240
		6.3.3	Quasi-Distributed Stress Monitoring of Prestressing Steel Strand Based on Fiber Bragg Grating	241
	6.4	Summary		247
	References			247
7	**Cable Stress Monitoring Technology Based on Fiber Bragg Grating**			249
	7.1	Current Status		249
	7.2	Cable Tension Monitoring System Based on FBG		251
		7.2.1	Composition and Working Principle of Cable Tension Monitoring System	251
		7.2.2	Characteristics of Cable Tension Monitoring System	252
		7.2.3	FBG Pressure Sensor	253
		7.2.4	Hardware Design	257
		7.2.5	Software Design	257
	7.3	Distributed Stress Monitoring System for Cable Based on FBG		260
		7.3.1	Composition and Working Principle of Distributed Stress Monitoring System for Cable	260
		7.3.2	Characteristics of Cable Tension Monitoring System	261
		7.3.3	System Design of Signal Acquisition Processing and Analysis	261
		7.3.4	Realization of Remote Monitoring for Smart Structure in Cable	261
	7.4	Test for Condition Monitoring of Cable Structure		263
		7.4.1	Cable Model and Test System	263
		7.4.2	Test of Cable Tension	264
		7.4.3	Test of Cable Stress Distribution	265
		7.4.4	Test of Cable Modal Parameter	266
	7.5	Summary		268
	References			269

8 Intelligent Monitoring Technology for Fiber Reinforced Polymer Composites Based on Fiber Bragg Grating ... 271

- 8.1 Preparation and Properties of Fiber Reinforced Polymer Composites ... 271
  - 8.1.1 Selection and Proportioning of Component Materials ... 271
  - 8.1.2 Performance Test and Analysis of Fiber Reinforced Polymer Bar ... 274
  - 8.1.3 Experiment Study on Anchorage System for Fiber Reinforced Polymer Bar ... 282
- 8.2 Interface Bonding Analysis of Fiber Bragg Grating Sensors and Composite Materials ... 284
- 8.3 Sensing Characteristics of Smart FRP Rod ... 285
  - 8.3.1 Preparation of Smart FRP Rod ... 287
  - 8.3.2 Test and Analysis on Sensing Characteristics of the Smart FPR Rod ... 289
- 8.4 Summary ... 291
- References ... 291

9 Concrete Crack Monitoring Using Fully Distributed Optical Fiber Sensor ... 293

- 9.1 Main Parameters of Fully Distributed Optical Fiber Sensing Technology ... 293
- 9.2 Brillouin Scattering Principle and Sensing Mechanism in Optical Fiber ... 295
  - 9.2.1 Brillouin Scattering in Optical Fiber ... 295
  - 9.2.2 Sensing Mechanism Based on Brillouin Scattering ... 303
- 9.3 FBG-Based Positioning Method for BOTDA Sensing ... 306
  - 9.3.1 Traditional Positioning Method for Fully Distributed Optical Fiber Sensing ... 306
  - 9.3.2 Description of FBG-Based Positioning Method ... 307
  - 9.3.3 Results and Discussion ... 309
- 9.4 Concrete Crack Monitoring Using Fully Distributed Optical Fiber Sensing Technology ... 315
  - 9.4.1 Tests ... 315
  - 9.4.2 Results and Discussion ... 318
- 9.5 Summary ... 325
- References ... 325

10 Engineering Applications of Optical Fiber Sensing Technology 327

10.1 Long-Term Health Monitoring and Alarm System for Wuhu Yangtze River Bridge 327
    10.1.1 Brief Introduction to Wuhu Yangtze River Bridge 328
    10.1.2 General Overview of the Long-Term Health Monitoring and Alarm System 329
    10.1.3 Strain Monitoring System Based on Optical Fiber Sensing 330

10.2 Monitoring System of Liaohe Bridge on Qinhuangdao-Shenyang Passenger Dedicated Line..................... 334
    10.2.1 Brief Introduction to Liaohe Bridge............... 334
    10.2.2 Application of Optical Fiber Sensor in Concrete Hydration Heat Testing........................ 335
    10.2.3 Application of Optical Fiber Sensor in Construction Quality Monitoring of Concrete Bridge 337
    10.2.4 Application of Optical Fiber Strain Sensor in Dynamic Monitoring of Concrete Box Beam 337

10.3 Long-Term Health Monitoring System for Xinyuan Highway Xiaogou Grand Bridge 342
    10.3.1 Brief Introduction to Xiaogou Grand Bridge 342
    10.3.2 Composition of Long-Term Health Monitoring System................................... 342
    10.3.3 Strain Monitoring System Based on Fiber Bragg Grating 345
    10.3.4 Effect Analysis of Strain Monitoring 347

10.4 Long-Term Monitoring System of Shuohuang Railway High-Steep Slope 348
    10.4.1 Brief Introduction to Monitoring Section 348
    10.4.2 Monitoring Scheme for Slope Deformation 349
    10.4.3 Monitoring Points Layout and Monitoring Equipments Installation....................... 351
    10.4.4 System Operation and Monitoring Results 352

10.5 Summary 353
References 354

Chapter 1
Introduction

1.1 Optical Fiber and Optical Fiber Sensor

Sensor is a tool to measure the parameters of various structural systems and obtain the information of structure. It is not only one of the three pillars in the information technology field (sensing and control technology, communication technology, and computer technology) but also an indispensable component of production automation, scientific testing, calculation and accounting, monitoring and diagnosing, and other systems. Therefore, the research and development of advanced sensing and testing technology have great significance to increase the safety and reliability of structure and equipment, promote the scientific and technological progress, and improve the economic and social benefits.

In recent years, with the rapid development of science and technology, the common sensing technology is also brewing new changes. In current research and development boom of sensing technology, the optical fiber sensing and testing technology has drawn high attention from various countries in the world because of its unique advantages. The optical fiber sensing technology can integrate optical waveguide technology and optical fiber technology with distinct characteristic of unifying "transmission" and "sensing" into one. The light that works for sensing and transmission has dominant advantages such as great transmission capacity and strong anti-electromagnetic interference capability. The optical fiber that works as lightwave carrier features chemical inertness and flexibility. Thus, the optical fiber sensing technology has incomparable competitive advantages in the aspects of intelligent material and structure, large structural safety monitoring, high voltage, and intense magnetic field.

This chapter mainly introduces the basic concept, type, and feature of optical fiber sensing technology as well as its application and development trend in the structural health monitoring field.

1.1.1 Optical Fiber

Optical fiber, also called as light-guide fiber, is a light transmission structure with the total reflection principle of light in the glass or plastic fiber. Gao Kun and G.A. Hockham first proposed the scientific idea of applying optical fiber in the communication transformation field. Hence, Gao Kun was granted with Nobel Prize in Physics in 2009. In general, optical fiber is the long and thin fibrous cylinder where the high-purity silica glass is mixed with a little germanium, boron, phosphorus, and other impurities.

The main structure of ordinary communication optical fiber is composed of fiber core and cladding. The cladding is covered by a coating layer, as shown in Fig. 1.1. The cladding is used to seal the optical wave in the optical fiber to realize transmission.

To transmit the optical wave, the refractive index n_1 of fiber core must be slightly greater than refractive index n_2 of cladding. The maximum relative refractive index difference Δ of optical fiber is shown in Formula (1.1) (Sun et al. 2000).

$$\Delta = \frac{n_1^2 - n_2^2}{n_1^2} \approx \frac{n_1 - n_2}{n_1} \tag{1.1}$$

The fiber core and cladding must be made of different materials in order to make their refractive indexes different. Currently, for the common quartz fiber, the refractive index will be increased if the fiber core is provided with GeO_2, TiO_2, Al_2O_3, or P_2O_5, while it will be decreased if the quartz is provided with B_2O_3. Therefore, these materials can be used as dopants of cladding. In addition, plastic optical fiber, photonic crystal fiber, and other types of optical fibers emerge in recent years.

Optical fiber is mainly used in the communication field first and develops rapidly in the sensing field in recent years. Various optical fiber sensors now are widely used in many fields. Particularly in 1978, Hill et al. (1978), Kawasaki et al. (1978) discovered the light sensitivity of optical fiber and fabricated the world's first fiber grating, which causes a new revolution in the optical fiber sensing field.

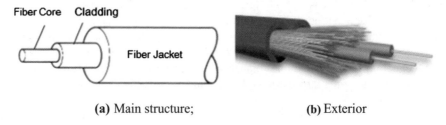

(a) Main structure;   (b) Exterior

Fig. 1.1 Fiber structure

1.1.2 Optical Fiber Sensor

With the continuous development of optical communication technology, optical fiber sensor (Fig. 1.2) has become a new type of sensor since 1970s. Basic working principle of optical fiber sensor: transmit the light from light source to modulator through optical fiber, interact the parameters to be tested and the light in the modulation area, change the optical property (e.g., intensity, wavelength, frequency, phase, and polarization state) and call as modulated signal light, transmit it to optical detector through optical fiber, and obtain the parameters to be tested after modulation. According to the working principle, the optical fiber sensing system is mainly composed of light source, transmission optical fiber, sensing element, photoelectric detector, and signal processing unit. With the development in the over forty years, the optical fiber sensor has been widely used in the national defense and military, space flight and aviation, industry, mining, agriculture, energy conservation, biomedicine, metrology testing, automatic control, domestic appliance, and many other fields because of its advantages such as high precision, corrosion resistance, small size, light weight, flexibility, and anti-electromagnetic interference.

In recent years, various countries have applied the optical fiber sensing technology in the civil engineering structures. For example, the optical fiber sensor is used to monitor several physical quantities such as strain, damage, crack, and displacement. It plays an important role in the health monitoring of dam, bridge, and other civil engineering structures. Moreover, it gradually introduces a new trend of combining advanced monitoring technology and traditional civil engineering structures.

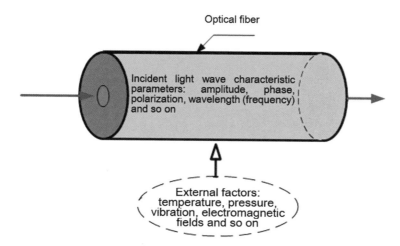

Fig. 1.2 Optical fiber sensor

1.2 Classification and Characteristics of Optical Fiber Sensor

1.2.1 Classification of Optical Fiber Sensor

There are various optical fiber sensors. Based on different classification standards, the common classification methods are shown below:

1. By the Principle of Light Modulation in the Optical Fiber

It is a key technology of optical fiber sensor to detect the light modulation by external parameters. According to the principle of light modulation in the optical fiber sensor, the optical fiber sensor is classified into five types: strength intensity, phase modulation, frequency modulation, wavelength modulation, and polarization modulation.

(1) Intensity-modulated optical fiber sensor

Intensity-modulated optical fiber sensor is to sense the parameters by measuring the change of light intensity in the optical fiber due to external factors. It is mainly composed of reflection type intensity-modulated optical fiber sensor, transmission type intensity-modulated optical fiber sensor, evanescent field coupling type intensity-modulated optical fiber sensor and physical effect type intensity-modulated optical fiber sensor.

(2) Phase-modulated optical fiber sensor

Phase-modulated optical fiber sensor changes the phase of optical wave transmitted in the optical fiber with the effect of energy field tested and then converts the phase change to light intensity change through interferometry technology. Thus, parameters are detected. At present, optical detectors cannot directly sense the phase change of optical wave. Therefore, the light interference technology must be applied to convert the phase change to light intensity change. Then, the external parameters are sensed. Common optical fiber interferometers include Michelson optical fiber interferometer, Mach–Zehnder optical fiber interferometer, Sagnac optical fiber interferometer, and Fabry–Perot optical fiber interferometer.

(3) Frequency-modulated optical fiber sensor

Frequency-modulated optical fiber sensor is to sense the external parameters by measuring the light frequency change due to external factors with Doppler effect.

(4) Wavelength-modulated optical fiber sensor

The wavelength or spectrum distribution of light energy in the optical fiber is changed due to external factors. Wavelength-modulated optical fiber sensor measures the parameters by monitoring the spectrum distribution. The wavelength modulation is also called color modulation because the wavelength is directly related to color.

1.2 Classification and Characteristics of Optical Fiber Sensor

(5) Polarization modulated optical fiber sensor

Polarization modulated optical fiber sensor detects various physical properties by measuring the change of light polarization state due to external factors. In the optical fiber sensor, the modulation of polarization state is mainly based on artificial optical rotation and artificial birefringence, such as Faraday rotation effect, Kerr electro-optical effect, and elasto-optical effect.

2. By the Role of Optical Fiber in the Sensor

According to the role of optical fiber in the sensor, the optical fiber sensor is classified into functional sensor and nonfunctional sensor.

Functional optical fiber sensor is also called sensing or detecting type sensor. Optical fiber is not only a carrier to transmit light but also a sensitive element, namely, featuring "transmission" and "sensing". However, this type of sensors also have disadvantages such as great technical difficulties, complex structures, and many difficulties in adjustment. Optical fiber voltage/current sensor and optical fiber liquid level sensor are subject to functional sensors.

Nonfunctional optical fiber sensor is also called as light transmitting type sensor. Optical fiber is not sensitive element in this type of sensor. Optical materials and mechanical or optical sensitive elements are placed at the end surface of optical fiber or in the middle of two optical fibers to sense the change of parameters. Thus, the optical characteristics of sensitive element change with them. In this process, optical fiber only works as the transmission circuit of light. To obtain great light and luminous power, the multimode optical fiber that has large numerical aperture and fiber core is mainly used. This type of sensor has advantages such as simple structure, reliability, and easily implemented technology. However, compared with functional optical fiber sensor, the nonfunctional optical fiber sensor has lower sensitivity and measurement precision. Optical fiber speed sensor and optical fiber radiation temperature sensor are subject to nonfunctional sensors.

3. By Physical Quantity Measured

By physical quantity measured, optical fiber sensor is classified into optical fiber pressure sensor, optical fiber temperature sensor, optical fiber image sensor, optical fiber liquid level sensor, optical fiber acceleration sensor, optical fiber strain sensor, optical fiber methane sensor, optical fiber hydrogen sensor, and optical fiber inclination sensor.

4. By Sensing Mechanism

By sensing mechanism, optical fiber sensor is classified into optical fiber grating sensor, interference type optical fiber sensor, polarization state modulated optical fiber sensor, Rayleigh optical fiber sensor, Brillouin optical fiber sensor, and Raman optical fiber sensor.

(1) Fiber grating sensor

Fiber grating is space phase grating that has the refractive index changing periodically in the fiber core because of the photosensitivity of optical fiber core material containing germanium plasma and the UV-light on optical fiber.

When the light beam with certain spectral width enters into optical grating, the fiber grating only reflects the light that meets Bragg diffraction in the incident light and the remaining light is transmitted out. As shown in Fig. 1.3.

The central wavelength of fiber grating reflected wave is affected by grating period Λ and refractive index n. Grating is very sensitive to axial strain and temperature changes of optical fiber. When the optical fiber is affected by external strain and temperature, the elasto-optical effect and thermo-optic effect affect the refractive index n of optical fiber. The length and thermal expansion of optical fiber affect the grating period Λ. Therefore, the basic principle of fiber grating sensor is to convert the change of parameters to the movement of central wavelength through the sensitivity of effective refractive index n and periodical Λ space change of fiber grating to external parameters, and then monitor the movement of central wavelength for sensing.

Fiber grating not only has very high reflection, frequency selection and dispersion characteristics, rapid response of wavelength shift, and linear output with a wide dynamic range but also can realize absolute measurement of parameters. Meanwhile, it is not affected by light intensity and not sensitive to the interference of backlight. It is so small and compact that can be embedded into material easily and it can be directly coupled with optical fiber system. Therefore, it significantly promotes the development of optical fiber sensing technology. The typical structure of fiber grating sensor is shown in Fig. 1.4.

(2) Interference type optical fiber sensor

Interference type optical fiber sensor is phase-modulated optical fiber sensor. The basic sensing mechanism is described as below: the phase of optical wave transmitted in the optical fiber changes with the effect of field energy to be measured, and then the phase change is converted to amplitude change through interferometry technology, thus detecting parameters.

Michelson optical fiber interferometer, Mach–Zehnder optical fiber interferometer, Fabry–Perot optical fiber interferometer and Sagnac optical fiber interferometer (Ferreira et al. 2016; Liu et al. 2016; Jasim and Ahmad 2017; Lopez-Dieguez et al.

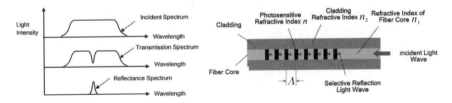

Fig. 1.3 Principle of fiber Bragg grating

1.2 Classification and Characteristics of Optical Fiber Sensor

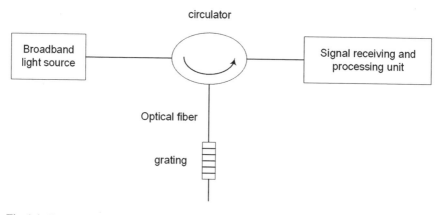

Fig. 1.4 Structure of fiber Bragg grating sensor

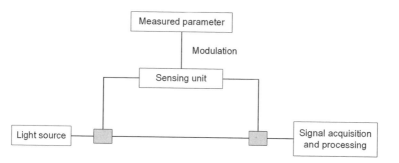

Fig. 1.5 Schematic of Mach–Zehnder fiber interferometer

2017; Xie et al. 2017) have been researched and developed according to the optical interference principle. Figure 1.5 is the schematic diagram of Mach–Zehnder optical fiber interferometer.

The optical phase in the optical fiber is very sensitive to external parameters so the phase-modulated optical fiber sensor usually has very high detection sensitivity. However, because of the high sensitivity of optical phase, the system is easily affected by external interference, thus increasing the random system noise and reducing stability.

(3) Polarization modulated optical fiber sensor

The polarization of optical wave plays an important role in many optical systems. Numerous physical effects will affect the polarization state of light. The polarization modulated optical fiber sensor usually adopts the following physical effects, including optical rotation effect, magneto-optic effect, Pockels effect, Kerr effect, and elasto-optical effect.

The typical examples are optical fiber current sensor and single-mode optical fiber polarization modulated temperature sensor. The basic structure of optical fiber current sensor is shown in Fig. 1.6.

5. By Working Mode

By working mode, optical fiber sensor is classified into point type optical fiber sensor and fully distributed optical fiber sensor, as shown in Fig. 1.7.

(1) Point type optical fiber sensor

Point type optical fiber sensor is also called as discrete optical fiber sensor. Due to different quantities of sensor units, the point type optical fiber sensor is further divided into single-point and multipoint optical fiber sensors (Bohnert et al. 2002).

The single-point optical fiber sensor works through a single unit to sense and measure the change of parameters in a small range near a point predetermined. The common point type sensing units are used to measure certain physical quantity, such as optical fiber Bragg grating and various interferometers. Point sensor can effectively complete the measurement in specific position (Blin et al. 2007).

Multipoint optical fiber sensor contains several sensing units to constitute sensor array and realize multipoint sensing (Ren et al. 2007). This type of optical fiber sensing system connects multipoint sensors in specific sequence to constitute sensor array or several multiplexed sensing units. A single channel or multiple information transmission channels are shared through time division multiplexing, frequency division multiplexing, and wavelength division multiplexing technologies to form distribution system. The system integrates the characteristics of point sensor and distributed sensor, so it is called as quasi-distributed optical fiber sensor.

Although the quasi-distributed optical fiber sensor can measure the information in several positions at the same time, it can only measure the information where sensor is preset. Other optical fibers are only used to transmit the wave rather than sense like point sensors. Meanwhile, when there are many sensing units, it is available to increase the degree of construction complexity and the difficulty of signal demodulation. For the point optical fiber sensing technology, optical fiber only works as the signal transmission medium, not sensing medium in most cases.

The multiplex function of sensor is a unique technology of optical fiber sensor, and the multiplex fiber grating sensor is typical representative (Liu et al. 2008). Fiber grating can easily realize multiplex function through wavelength encoding and

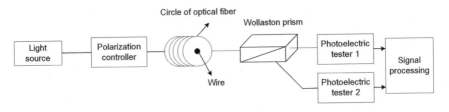

Fig. 1.6 Schematic diagram of polarization state modulation type fiber current sensor

1.2 Classification and Characteristics of Optical Fiber Sensor

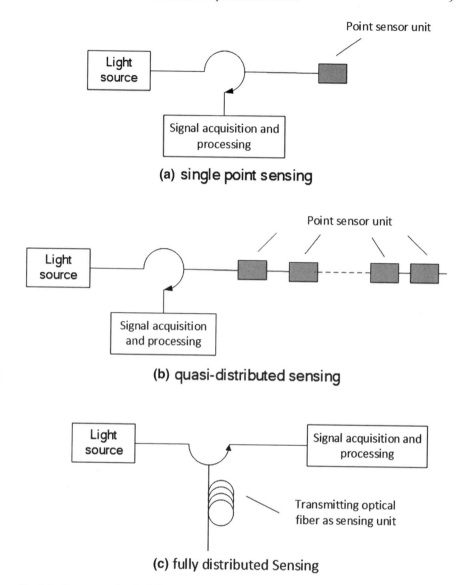

Fig. 1.7 Two types of optical fiber sensing systems

other technologies. The key technology of multiplex fiber grating is multiwavelength detection and demodulation. The common demodulation methods include scanning optical fiber F-P filter method, wavelength division multiplexing based on linear array CCD detection as well as frequency division multiplexing, time division multiplexing, and wavelength division multiplexing technologies based on mode-locked laser. The structure of quasi-distributed fiber grating sensor with scanning optical fiber F-P filter method is shown in Fig. 1.8.

(2) Fully distributed optical fiber sensor

Some objects detected are usually not a single point or several points, but spatial distributed field (Ozeki et al. 2006), such as temperature field and stress field. Such objects not only have long distance and large range but also are distributed continuously in 3D space. Thus, the point sensor or multipoint quasi-distributed sensor cannot realize sensor monitoring. Therefore, the fully distributed optical fiber sensor emerges. Optical fiber works as signal transmission medium and sensing unit in the fully distributed optical fiber sensor. The whole optical fiber can be used as sensing unit. Sensing points are distributed continuously. It is also called as mass sensing head. The fully distributed optical fiber sensor can measure the information in any position along the optical fiber (Wuilpart et al. 2002). With the continuous development of optical device and signal processing technology, the fully distributed optical fiber sensing system has expanded its maximum sensing range to tens and even hundreds of kilometers, even tens of thousands of kilometers. Thus, this sensing technology is increasingly concerned by people. It has become an important research field of optical fiber sensing technology.

The working mechanism of fully distributed optical fiber sensor is mainly based on the reflection and interference of light. The reflection method to sense the change of scattering light and nonlinear effect of optical fiber related to the external environment is widely used and most remarkable in recent years. The schematic diagram of its structure is shown in Fig. 1.9.

According to the difference of detected light signals, the fully distributed optical fiber sensor is divided into Rayleigh scattering, Raman scattering, and Brillouin scattering. According to the different signal analysis methods, it is divided into fully distributed optical fiber sensor based on time domain and frequency domain (Kim et al. 1982).

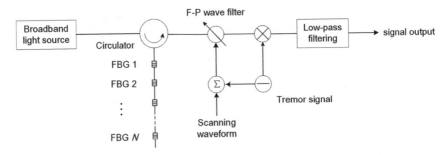

Fig. 1.8 The structure of quasi-distributed fiber grating sensors based on F-P filter scanning

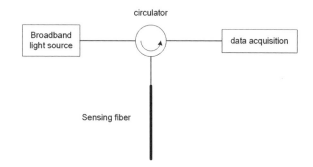

Fig. 1.9 Schematic diagram of distributed optical fiber sensor

1.2.2 Characteristics of Optical Fiber Sensor

Compared with traditional electric or mechanical sensors, the optical fiber sensor has the following advantages:

(1) Anti-electromagnetic interference, good insulating property, and corrosion resistance

The main component of optical fiber that works as sensing medium is silicon dioxide, so it is safe. Therefore, optical fiber sensor features anti-electromagnetic interference, lightning protection, better temperature resistance, and corrosion resistance. It can be used in severe environment (such as intense radiation, high corrosion, easy combustion, and explosion).

(2) Small volume, lightweight, and strong plasticity

Optical fiber is the main component of sensor. It not only has small volume and lightweight but also can be bent to a certain extent. Thus, its direction can be changed with the shape of the object detected to adapt the detection environment as far as possible. It can be either buried into composite material or pasted on the surface of material to realize good compatibility with material detected.

(3) Large bandwidth, low loss, and easy long-distance transmission

Optical fiber has wide working band and a small transmission loss in the optical fiber (the loss of 1,550 nm optical wave in standard single-mode optical fiber is only 0.2 dB/km). Therefore, it is appropriate to perform long-distance sensing and remote monitoring.

(4) Many measurable parameters and objects

Optical fiber sensor can sense various parameters through modulation and demodulation technologies. In addition to the traditional sensing fields such as stress, temperature, vibration, current, and voltage, it is also used in many new sensing fields such as speed, acceleration, revolving speed, corner, vibration, bending, twisting, displacement, refractive index, humidity, pH, solution concentration, and liquid leakage (Lee 2003). Therefore, optical fiber sensor can measure many objects and sense more than 100 types of parameters, including but not limited to sensing parameters in Fig. 1.10.

(5) High sensitivity

The optical fiber sensor that is designed effectively (for example, using optical fiber interference technology) can realize a very high sensitivity.

(6) Easy multiplex and large-scale array/array network observation

There is no mutual interference among optical waves, so the wavelength division multiplexing technology can be used to transmit multiwavelength light signals at the same time in one optical fiber in the communication field. It is convenient for optical fiber to network. It is benefit to combine with existing optical communication device to form telemetry net and optical fiber sensing network.

Thus, optical fiber sensor is deeply concerned by people and develops rapidly. Meanwhile, the optical fiber sensors that have new mechanisms and new application objects constantly emerge.

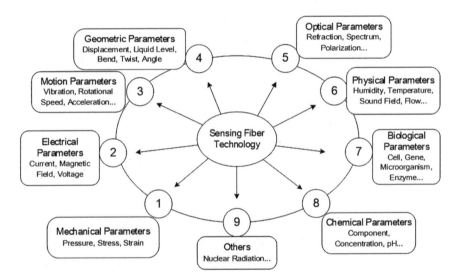

Fig. 1.10 Diagram of sensing parameters for optical fiber sensor technology

1.3 Current Status and Development Trends of Optical Fiber Sensing Technology

1.3.1 Current Status of Optical Fiber Sensing Technology

1. Development History and Present Situation of International Optical Fiber Sensing Technology

Since the first low-loss optical fiber was researched and developed by USA Corning Company in 1970s, the optical fiber communication technology has developed rapidly. Various new optical devices and photoelectric devices were researched and developed successively. Optical fiber sensing technology also emerges. In 1977, USA Naval Research Laboratory (NRL) started performing the Foss (optical fiber sensor system) plan hosted by Dr. Charles M. Davis (Doebling et al. 1996). It is usually thought as a hallmark event for the emerging of optical fiber sensor. Hence, the concept of optical fiber sensor becomes reality in many laboratories all over the world. The optical fiber sensor has wide application range and broad market prospect so its research and development have drawn high attention from various countries in the world.

Optical fiber sensor has been widely researched in the academic circle. For example, J. Bucaro realized the sense of acoustic wave with phase modulation method through the structure of Mach–Zehnder interferometer (Bucaro et al. 1977). M.K. Barnoski proposed the optical time domain reflection (OTDR) technology to measure the loss in each position along the optical fiber according to the intensity change of Rayleigh scattering light in the optical fiber (Barnoski et al. 1977). From then on, the research achievements on this subject are increased year by year (Rogers 1980; Dakin et al. 1985).

The most famous international conference in the optical fiber sensor field is International Conference on Optical Fiber Sensors (OFS). The conference was first held in UK in 1983. From then on, it was held every 18 months or so. Till now, 25 conferences have been held in USA, Europe, Asia, etc. It gradually becomes the hallmark conference in the international optical fiber sensor field.

In the industrial circle, 15 companies and research institutes, including Toshiba and Nippon Electric Company, have researched and developed 12 types of first-class civil optical fiber sensors in 1990s. Large companies in Western Europe also actively participated in the research, development and market competition of optical fiber sensors, including British Standard Telecom Company, French Thomson Company, and German Siemens Company. In twenty-first century, energy problems have become increasingly obvious. The internationally famous energy companies, such as Sercel, Halliburton, and Schlumberger, have invested heavily in the research and development of optical fiber sensor and the energy exploration field.

2. Development History and Present Situation of China Optical Fiber Sensing Technology

China started the research on optical fiber sensor at the end of 1970s, almost synchronously with the world. The development process of China's optical fiber sensing technology is generally divided into three stages (Liao 1996, 2003).

(1) From the end of 1970s to the middle of 1980s

China's optical fiber sensing technology developed rapidly in this period and forms a small upsurge. At the end of 1970s, the optical fiber sensing technology developed rapidly in the world. The new theory studies and application fields were opened up constantly. The academic activities were very frequent. Relevant industries increased at the speed of 30% every year. It revealed the vital force of new technology. These situations drew a high attention from China's academic circle and industrial sector. Many scientific research institutions, colleges, and universities, and industrial sectors took actions one after another. They researched and developed various optical fiber sensing technologies from different aspects, and tried to promote their applica-tions. On this basis, the New Technology Bureau of State Scientific and Technological Commission held the "Demonstration Meeting of Optical Fiber Current and Voltage Sensor Schemes" in Hangzhou in June 1983. Hence, State Scientific and Technological Commission,the State Economic and Trade Commission and other relevant departments successively held many optical fiber sensor conferences. Table 1.1 shows some main conferences.

(2) 1990s

In 1990s, China's optical fiber sensor enters into the second stage. With the rapid development of optical fiber communication industry and the sharp increase of optical communication market demand, the national planning and investment departments as well as the optical fiber technology research institutes turn to the research, development, and application of optical fiber communication field successively. By contrast, the optical fiber sensing is still developing in the initial stage. Its technologies, processes and component research and development meet with a cold acceptance and enter into a low ebb. The optical fiber sensing is researched slowly. The main reasons are as below:

① Immature technology. Due to technologies, processes, components and other reasons, the optical fiber sensing not only has a low rate of finished products but also meets with a lot of interference factors. Therefore, it cannot be recognized in the actual application.
② Expensive components. Component is a key factor of optical fiber sensing technology and system. The components used for optical fiber sensing usually should meet special fabrication requirements. In the initial period of research and development, the performance and quality of components cannot reach application standards. In addition, most components are researched and developed by researchers, so the small quantity and high cost restrict the increase of performance and quality and the reduction of price.

1.3 Current Status and Development Trends of Optical Fiber Sensing … 15

Table 1.1 Main domestic conferences on optical fiber sensing in early period

Conference date	Conference name	Conference place	Paper number	Conference sponsor
1983.06	Argumentation Conference of Optical Fiber Current and Voltage Sensor Scheme	Hangzhou	3	New Technology Bureau of State Scientific and Technological Commission
1983.09	Fiber Sensor and its Applications Development Forecast Symposium	Yangzhou	11	New Technology Bureau of State Scientific and Technological Commission
1984.05	National Conference on Sensor Academic Exchange	Wuhan	10	State Economic and Trade Commission, State Scientific and Technological Commission, China Instrument and Control Society
1984.08	Fiber Optic Sensor Planning Symposium	Beijing	–	Ministry of Electronics Industry
1984.10	China Optical Society Fiber Optic and Integrated Optical Professional Committee Establishment and Academic Exchange Conference	Xian	15	Chinese Optical Society
1984.10	Seminar on Optical Fiber Technology in Electrical Industry	Tianjin	5	Bureau of Electrical Engineering of the Ministry of Machinery Industry
1984.11	The Meeting of Optical Fiber Sensor Academic Exchange	Nanjing	11	Bureau of Industry Instrument of the Ministry of Machinery Industry, China Instrument and Control Society

(continued)

Table 1.1 (continued)

Conference date	Conference name	Conference place	Paper number	Conference sponsor
1984.12	The Meeting of Optical Fiber Sensor Academic Exchange	Hefei	20	Bureau of Industry Instrument of the Ministry of Machinery Industry, China Instrument and Control Society
1985.01	National Fiber Optic Sensing Technology "Seventh Five-Year Plan" Symposium	Beijing	27	New Technology Bureau of State Scientific and Technological Commission
1985.11	Fiber Optic Sensing Technology Information Network Establishment and Academic Exchange	Nanjing	–	Bureau of Industry Instrument of the Ministry of Machinery Industry China Instrument and Control Society

③ Inferior characteristics. The optical fiber sensing technology has unparalleled advantages compared with traditional sensing technology. However, it only focused on the estimate of performance in the early stage. Thus, it has no advantages over traditional sensing technology during actual application. Many advantages are not reflected. For example, the optical fiber sensing has high sensitivity but very low signal-to-noise ratio and stability; the optical fiber sensing has strong anti-interference capability but is limited due to immature technology; the application inconvenience and low-cost performance also affect the recognition for its advantages.

④ No urgent need in market. The domestic production technology level is limited and the automation level is very low at that time. Therefore, there is no market demand for the optical fiber sensor, a new high-precision safe monitoring system. For example, due to the low domestic oil depot management level, low manual detection cost and other social factors, it is difficult to promote the application of optical fiber oil tank monitoring system though it has met technical indexes.

(3) Twenty-first century

In the twenty-first century, the optical fiber communication rises from the valley bottom and China's optical fiber sensing technology develops rapidly. Many optical fiber and relevant component manufacturers turn to optical fiber sensing. Numerous investment organizations are optimistic about the market. In the academic circle, the quantity of published papers on China's optical fiber sensing technology ranks steadily first in the world. Academic Conference on China's Optical Fiber Sensing (OFS-C) is held once every 18 months at certain interval with International Conference on Optical Fiber Sensors. It has been held 10 times till now. In the industrial circle, with the rapid development of optical devices and electronic technologies, the companies engaged in optical fiber sensing technology emerge in large numbers. Many optical fiber sensing systems can meet the market application requirements. The market share of China's optical fiber sensing has been up to billions of RMB. The sharp increase in market demand has become an active force to promote the rapid development of optical fiber sensing technology.

Currently, the development of optical fiber sensing is full of opportunities and challenges. Over 40 years' development, the optical fiber sensing technology has made significant progress. Many practical products appear, and an independent system is also formed basically. However, with the social development and technology progress, the demand for optical fiber sensor not only promotes the rapid increase in quantity but also leads to the diversified and high-standard performance parameters. The present development situation of optical fiber sensing technology is still far from the actual demand. Therefore, there is large research and development space.

In a word, with the rapid development of optical fiber sensing technology, the optical fiber sensing industry is gradually becoming another optical fiber technology application industry after optical fiber communication.

1.3.2 Development Trends of Optical Fiber Sensing Technology

The rapid development of optical fiber sensing and relevant technologies meets the higher requirements of various control devices and systems for information obtaining and transmission, which gradually increases the degree of automation in each field. However, there are still some problems for in-depth research. For example, it is necessary to research the performance of components in the optical fiber sensor, especially further improve the workmanship and structure of sensitive components explore new sensitivity principle, and develop the optical fiber sensor technology with high precision for special demands and in the long-distance monitoring application. At the same time, it will be a trend in the future to make full use of the micro-process technology and computer software to improve and complement the performance of optical fiber sensor. It's also essential to develop the new digital, integrated, automatic and engineering optical fiber sensor, research and develop the optical fiber sensor array which is appropriate to engineering structure and other special measurement requirements. The book discusses the research status and development trend of sensing technologies that are commonly used for engineering structure at home and abroad, mainly including the following aspects.

1.3.2.1 Development Trends of F-P Sensing Technology

Optical fiber Fabry–Perot (F-P) interferometric sensor is a kind of phase-modulated optical fiber sensor. It detects physical quantities with multibeam interference. When the light in the optical fiber meets two reflecting mirrors, two different beams of reflected light are generated. Then, they meet and interfere with each other (Ferreira et al. 2017). When the length of interferometric cavity changes with physical quantity, the phase difference of two beams of reflected light also changes. Thus, the electric signal output by photoelectric detector changes with phase difference, and the output signal of detector can be detected to sense the physical quantity.

Fiber grating is a reflecting mirror in the optical fiber with intrinsic wavelength selection characteristics. It provides a new technical scheme for intrinsic F-P optical fiber sensor. Two FBGs are used as reflecting mirror to constitute a Fabry–Perot sensor based on FBG. This type of dual-grating structure inherits the excellent wavelength division multiplexing capability of FBG. Compared with the traditional optical fiber F-P cavity, it has higher mechanical strength and sensitivity.

Over 20 years' development, F-P optical fiber sensor has been gradually applied in engineering. However, the technology still has the following main problems:

1. Extrinsic F-P optical fiber sensor has great difficulty in workmanship and poor repeatability.

(1) The technology of plating reflecting mirror on the end surface of optical fiber is to be improved. Now, the traditional glass coating process is used for optical fiber. Film is coated on the section at the level of μm, so the glass coating process and parameter control do not fit perfectly. Thus, the film coating technology of optical fiber is to be further researched.

(2) The optical fiber connection and fixing technology of F-P cavity are to be improved. Now, optical fibers are connected by two methods: first, fusion splicer. Its advantages are small loss and good repeatability. However, a fusion splicer costs hundreds of thousands of RMB. In addition, the high fusion temperature will damage the film of optical fiber. The second method is cold connection. Use optical cement that has optical refractive index near fiber core. It is not easy to damage the film, and the light transmission effect is good. However, it is difficult to control the collimation of two optical fibers.

2. Extrinsic F-P optical fiber sensor is easily interfered by environment.

Any change of environment, such as temperature, humidity, vibration, and even air flow, will affect F-P optical fiber sensor.

3. Difficult multiplexing.

Either for intrinsic F-P optical fiber sensor or extrinsic F-P optical fiber sensor, a large scale of multiplexing is still a problem that has not been solved for a long time. In many application situations, such as the health monitoring of large structure, hundreds or thousands of sensors are required in general. Increasing the multiplexing capability is an important subject to reduce the system cost. Although the multiplexing technology has been explored persistently and many multiplexing methods have been put forward, such as time division multiplexing, coherence multiplexing, wavelength division multiplexing, and spatial division multiplexing. However, the effect is not very ideal, especially for the time division multiplexing and coherence multiplexing, the signal-to-noise ratios are degraded sharply with the increase of sensor quantity. In a word, the above methods have the following problems in general: as the quantity of sensors is increased, the system is more complex, the cost increases, and the requirement for sensor is raised gradually. Therefore, the exploration of more advanced, simpler, and more economical multiplexing methods has a great significance to promote the research and application of F-P optical fiber sensor.

1.3.2.2 Development Trends of Fiber Grating Sensing Technology

The application of fiber grating sensor is an ascendant field and has very broad development prospect. However, there are many problems to be solved urgently to apply the optical fiber Bragg grating sensor to real works in a better way. For example, the wavelength displacement monitoring of fiber grating requires very complicated technologies and devices, large-power broadband light source, tunable light source, etc. In current technical conditions, its detention resolution and dynamic range will be limited to some extent. Meanwhile, the problem of cross-sensitivity between temperature and strain of fiber grating is also a major obstacle that affects its practical application.

1. Cost

The writing equipment and demodulation system of fiber grating are very expensive now, which seriously restricts the general application of fiber grating sensor.

2. Signal Demodulation

The sensing mechanism of fiber grating needs to be further researched, and improving the sensitivity coefficient of fiber grating sensor or the resolution of demodulation system is also an important problem to be solved urgently. In addition, the fiber grating sensor is usually applied in dynamic system. Therefore, it is important to research its dynamic response characteristics. However, it is not researched deeply now.

3. Packaging

Optical fiber is so weak that it should be packaged for protection. Therefore, it is necessary to deeply research the sensing mechanism of fiber grating sensor packaged so as to maximize its sensitivity without damaging optical fiber.

4. Layout of Sensing Network

Fiber grating is distributed by point. It needs further research to realize the optimized layout of sensor, namely, to obtain as much information as possible through minimum fiber grating sensors in the large engineering structure.

5. Development of Information Processing System

Because of the importance of the health monitoring system of large engineering structure, periodic measurement should be conducted. It is necessary to handle a large amount of information storage and calculate the parameters that are involved in the structure safety performance indexes through signal measurement.

6. Cross-Sensitivity

Because the fiber grating is sensitive to both strain and temperature, it is difficult to distinguish the wavelength change caused by strain and temperature when the grating wavelength changes. When fiber grating is used as the optical communication device, the wavelength stability of fiber grating is required to have high stability. The smaller the influence of temperature on the wavelength, the smaller the test error gets. The central wavelength drift of fiber grating will seriously affect its application to laser frequency stabilization and wavelength division multiplexing. When the fiber grating is used as strain sensing element, the wavelength drift caused by temperature change and mutual strain interference will make it difficult to recognize the accurate response of temperature and physical quantities to be measured.

Currently, the research on fiber Bragg grating sensor is mainly concentrated on the following aspects:

(1) Research of New Fiber Grating

Some new gratings appear successively on the basis of traditional FBG, such as tilted grating, long period grating, photonic crystal, and plastic grating. These gratings have different sensing characteristics from FBG, which are research hotspots in the world in recent years.

(2) Design and Research of Fiber Grating Sensor

According to the specific requirements of physical quantity type, property, and distribution in engineering, the structure of optical fiber grating sensor needs to be designed specially to ensure the accuracy and repeatability of test results. The existing fiber grating sensors can detect the following physical quantities: temperature, strain, pressure, displacement, intensity of pressure, torsional angle, torque (torsional stress), acceleration, current, voltage, magnetic field, frequency, concentration, coefficient of thermal expansion, vibration, etc. Therefore, developing a series of sensors with high sensitivity and resolution has become the research hotspot and development trend of fiber grating sensing technology.

(3) Research of Fiber Grating Sensor Packaging Technology

The naked grating is weak so it cannot be practically used in most cases. On the one hand, the sensitivity varies with measured parameters, so the fiber grating needs to be packaged with enhanced sensitivity. On the other hand, the environment in the construction site is severe and complex so the fiber grating should be packaged with advanced technology. Therefore, since the fiber grating sensor appears, the packaging technology is always an important research field.

(4) Cross Sensitivity of Strain and Temperature

The wavelength of Bragg grating is sensitive to strain and temperature, so it is difficult to distinguish the respective variables of strain and temperature from the variables of a single Bragg wavelength. The cross sensitivity of strain and temperature of fiber grating sensor has become an important obstacle that directly affects its practical application. Meanwhile, multiparameter test has become a new research field of fiber grating. A large number of theories and experiments indicate that the sensing schemes that are based on dual-wavelength matrix manipulation method, dual-parameter matrix manipulation method or strain (temperature) compensation method can solve above problems effectively.

(5) Stability and Durability of Optical Fiber Grating Sensor

The designed service life of large engineering structure is tens of years and even longer. Thus, the stability and durability of optical fiber grating directly affect the application of optical fiber grating sensor to the large structure health monitoring. There are many factors that affect the stability and durability of optical fiber grating, such as ambient temperature, humidity and chemical corrosion, etc. It is found that the central wavelength, refractive index, and reflectivity of optical fiber grating change with time and temperature. Although they change a little, the stability and durability of sensor will be affected.

(6) Research of Signal Demodulation System

The wavelength detection technology that features high precision and low cost is the application basis of signal demodulation. Currently, the optical fiber grating demodulation equipment generally has disadvantages such as low resolution, low sampling frequency, poor multiplexing, and excessive cost. Therefore, it is necessary to research and develop the detection technologies that feature low cost, small scale, reliability, and sensitivity.

(7) Research of Optical Fiber Grating Sensing Network

The distributed sensing network system that has multiple parameters and functions is a basis of realizing real-time monitoring of large structure. Optical fiber grating can be flexibly connected in series or in parallel, and can also sense pressure, temperature, vibration, and other parameters in real time. Therefore, with reference to optical multiplex communication technology, optical wave multiplex, and space division, it is available to establish multidimensional (linear array, area array, volume array, and their combination), multiparameter (physical quantity, thermal quantity, geometric quantity, etc.), multifunctional, distributed (multipoint quasi-distribution, continuous distribution), and intelligent (alert, covering, etc.) sensing system.

(8) Damage Positioning and Assessment Technology Research

For the large scale of optical fiber grating sensing network, the processing efficiency of sensing signal must be increased so that the large structure health is monitored in real time, the measured signal and external interference signal are distinguished rapidly and effectively, and the damage position and extent are determined. The optical fiber sensing network is distributed in a broad range, so the output signal of sensing network may be a large area of distribution signal in the nonlinear relationship. The calculation and analysis loads are very heavy. Therefore, the rapid signal analysis and processing algorithm of large-scale distributed sensing network is a special development field of optical fiber grating sensing technology.

With the continuous development of photoelectronic technology, a new generation of sensor that has the active optical fiber grating laser (two main structures: distributed feedback optical fiber laser and reflection type optical fiber laser) as sensing element appears on the basis of traditional passive optical fiber grating sensor. This type of optical fiber laser writes optical fiber grating to active optical fiber to form the effective gain type laser resonant cavity. When the external physical quantity monitored changes, the outgoing laser wavelength, polarization or mode also changes. The change of physical quantity measured can be obtained by monitoring the change of laser parameters.

1.3.2.3 Development Trends of Optical Fiber Laser Sensing Technology

Both narrow bandwidth interference-demodulated laser sensor and polarization type beat frequency demodulated laser sensor must meet the requirement of laser operating in the single mode to obtain high signal-to-noise ratio. Therefore, the short Er-doped fiber laser cavity is adopted basically. The highly doped rare earth active fiber is required to ensure stable work in the short laser. However, it will reduce the pump conversion efficiency of laser. In addition, the inherent ion focusing effect of doped fiber reduces the quantum conversion efficiency of optical fiber grating laser and results in self-pulsing, thus leading to additional noise to laser sensor. Sanchez et al. carried out experiments and theoretical research. The feedback modulation can effectively conquer these disadvantages. However, this type of Er-doped fiber grating laser has very complex technology, so it is very difficult to obtain satisfactory effect. The root cause for this effect is the low conversion efficiency of erbium ion. Double-doped optical fiber is adopted in order to solve the problem fundamentally. Yttrium ion has very wide absorption bandwidth and large stimulated emission interface. Thus, the erbium–yttrium-doped fiber has a pump absorption strength that is two magnitudes higher than ordinary Er-doped fiber. It can also effectively avoid the laser self-pulsing effect due to low dope concentration. However, the erbium–yttrium-doped fiber that has higher concentration will result in many ion noises. In addition, the extra small laser structure will lead to instable lasing, thus resulting in many noises.

In current situation, although the phase type optical fiber lasing sensor has achieved great development, there are many critical technologies to be further researched and discussed, such as low-frequency noise suppression of optical fiber laser, power equilibrium of optical fiber laser array, Rayleigh scattering noise suppression in the long-distance sensing network, anti-interference technology of demodulation system, large-scale networking technology of optical fiber lasing sensor, etc. Polarization type optical fiber laser sensor has relatively small coupling noise during demodulation and relatively simple demodulation mode. However, current research is concentrated on the measurement of beat frequency signal of extra-short optical fiber grating laser sensor in the orthogonal mode. This type of laser requires additional optical fiber polarization analyzer to couple the cross polarization signal. It not only brings more additional noises to sensing signal but also reduces the strength of signal. In addition, it is not easy to control the polarization of short-cavity optical fiber laser so its long-term stability is to be verified. It is expected to develop a kind of stable and simple optical fiber grating laser sensor that is based on beat frequency demodulation technology.

1.3.2.4 Development Trends of Fully Distributed Optical Fiber Sensing Technology

With the vigorous development of optical fiber sensing technology, the fully distributed optical fiber sensing technology has become an increasingly strong and important power in the field. The distributed optical fiber sensing technology was put forward at the end of 1970s. It is gradually developed on the basis of optical time domain reflection (OTDR) technology which is still widely used in the optical fiber engineering now. The distributed optical fiber sensing technology can extract the information distributed in a large range of measurement field, so it can solve many difficult problems in the measurement field. A series of distributed optical fiber sensing mechanisms and measurement systems appear in a short time and are widely used in many fields. Currently, the technology has become one of the most promising optical fiber sensing technologies.

The fully distributed optical fiber sensor has the unique feature with continuous distributed sensing so it has become the most potential development direction in the optical fiber sensing technology. The main development directions are shown below:

(1) Increase the signal reception and processing capability. For example, to further increase the space resolution, measuring precision, sensitivity and measuring range of sensing system, and to reduce measuring time and cost.
(2) Explore the new fully distributed optical fiber sensing mechanism, and research and develop the new fully distributed optical fiber sensor.

(3) Design the special optical fiber materials and devices that are applicable to engineering structure. Especially, research and develop the new optical fiber and optical cable that can be used for fully distributed optical fiber sensing in urgent need.
(4) Develop array and network of fully distributed optical fiber sensing technology. The fully distributed optical fiber sensing technology can only detect the unidimensional sensing information on a single axial optical fiber. With the enlargement of detection range and increase of information, the limitation will be increasingly prominent. Therefore, only multidimensional fully distributed optical fiber sensing network can sense a large range of information rapidly and accurately and realize the monitoring from point to area and from area to volume.
(5) Develop the multiuse and multifunctional fully distributed optical fiber. Many physical parameters need to be measured for a large structure, such as temperature, strain, acceleration, and displacement. The research and development of an optical fiber sensor that can measure many physical parameters can effectively save the cost of sensor and reduce other problems caused by sensor setting to large structure, such as effect of mechanical property.
(6) Put the fully distributed optical fiber sensor into use and engineering. On the basis of existing research level, the application research should be conducted to great extent, including real-time dynamic sensing, layout mode of sensing fiber and cable, and the effect of natural environment change on sensing system.
(7) Solve the problem of position drift and cross-sensitivity between temperature and strain of fully distributed optical fiber sensor. The fully distributed sensor has unique advantages such as large sensing range and full-scale continuity. However, the long range of measurement will cause the length of optical fiber to vary with external temperature and strain, resulting in spatial position drift. Therefore, how to solve the position drift of optical fiber is a difficult problem to be solved urgently. Meanwhile, the cross-sensitivity of temperature and strain in the fully distributed optical fiber sensing is another difficult problem to be solved in the engineering application. The synchronous measurement of multiple physical quantities or the coupling of different sensing technologies provides ideas to solve the cross sensitivity of temperature and strain in the fully distributed optical fiber sensing.

1.4 Structural Health Monitoring Based on Optical Fiber Sensing Technology

It is not long for the world to apply optical fiber sensor in the large engineering structure health monitoring. Now is the transitional period from bud to development. In 1989, Mendez et al. first proposed to apply the optical fiber sensor to concrete structure monitoring. From then on, scientific researchers from Japan, British, America,

Germany, and other countries successively research the application of optical fiber sensing system to civil works. The application of optical fiber sensors from Japan, America, and Switzerland is relatively wide in the civil works from measurement or monitoring of concrete pouring to pile, foundation, bridge, dam, tunnel, building, earthquake, landslide, and other complex systems.

The optical fiber intelligent health monitoring system is mainly composed of three parts: optical fiber sensor system, signal transmission, and collection system as well as data processing and early warning system. The optical fiber sensor system includes the type selection of optical fiber sensor, namely, selecting the specific modulation mode and the optical fiber sensor that meets performance requirement. It also involves the topology and installation modes of optical fiber sensor, either pasting on the surface or embedding. The signal transmission and collection system includes the optical fiber sensor calibration, sampling module as well as real-time mass data storage structure and mode. The data processing and monitoring mechanism is the core part of health monitoring system, including mass data validity analysis, structure health performance index selection, structure operating state visualization system, and corresponding early warning function. The three parts of optical fiber intelligent health monitoring system supplement each other. Each is the organic component of the whole system, and none is dispensable. Each sensor requires a specific demodulation system. Once the sensor is determined, the corresponding signal collection and processing system will be determined. Therefore, the optimized layout method of optical fiber sensor and the analysis and monitoring of real-time signal become the key problems of optical fiber intelligent health monitoring system applied to large structure. During real-time monitoring, the real-time signal measured by optical fiber sensor is sent to monitoring center through signal transmission and collection system. Then, the health state of structure is evaluated through corresponding processing and judgment. If the key health parameters monitored exceed set threshold values, relevant management organizations will be notified through instant message (SMS), E-mail or other modes to take corresponding emergency measures, thus avoiding serious loss of personnel and property.

1. Application to Bridge Structure Health Monitoring

The optical fiber grating sensor is pasted on the surface of composite materials or embedded into composite materials during manufacture. The bridge structure is monitored for a long time by measuring the stress, strain, and other mechanical parameters of composite cable, rib, etc. Take Beddington Trail Bridge near Calgary, Canada, for example. It first used optical fiber grating sensor to monitor the bridge structure by pasting it on the steel reinforced bar and carbon fiber composite rib under the support of prestressed concrete so as to measure the stress. For another example, Storck Bridge in Winterthur, Switzerland is the world's first cable-stayed bridge that replaces steel cable with CFRP inhaul cable. Optical fiber grating sensors are installed in the carbon fiber reinforced polymer cables to monitor the bridge for a long term. In 2004, Ou (2005) monitored the health of Hulan River Bridge in Heilongjiang Province. Optical fiber grating sensors were laid out to monitor the rebar strain process during tensioning of prestressed box girder and the rebar strain increment and distribution in the static load test of box girder. In 2007, they successfully laid FBG strain and temperature sensors in the key suspenders of Dadu River Arch Bridge in Ebian, Sichuan Province, and monitored the suspender strain and temperature change process under the vehicle load as well as the influence of the same vehicle load on suspenders at different lengths. In 2003, Magne et al. laid 11 optical fiber grating sensors in the concrete box girder to monitor the reaction of bridge under dynamic load and evaluate the health of bridge.

2. Application to Bridge Structure Construction Monitoring

Optical fiber grating sensors are laid in the important parts of bridge structure to monitor the geometric position and strained condition of bridge structure in each construction stage. Based on test results, the next stage of control variables is predicted and regulation schemes are made to control the construction of structure. A prestressed concrete bridge at the length of 147 m was built over the ring canal in Ghent, Belgium. 18 optical fiber grating sensors were buried in the prestressed bridge during pouring to monitor the construction process and the structure health for a long term. Yang Jianchun from Chongqing University pasted the optical fiber grating on a thin copper sheet and protected it properly to form a fiber grating dynamic strain sensor to monitor the strain of main structure of Lupu Bridge, obtaining excellent measurement results. Harbin Institute of Technology, Huazhong University of Science and Technology, Wuhan University of Technology, and South China University of Technology, respectively, apply the optical fiber grating sensors to monitor the bridge construction, prestressed tendon tensile deformation, completed bridge, and bridge tie bar cable replacement.

3. Application to Bridge Structure Damage Monitoring

Optical fiber grating sensors are pasted or embedded into the key parts of bridge structure to form the optical fiber measuring network to monitor and identify the damages of bridge structure, especially bridge reinforcing and repair works. For example, a steel structural bridge that was built on the Interstate Highway 10 in Las Cruces, New Mexico, USA in 1970 has many fatigue cracks. Optical fiber grating sensing system installed in the bridge can not only detect and count the quantity of standard vehicles but also measure the vehicle speed and weight. The system can monitor the structure response, degradation, and damage caused by dynamic load in real time to know about the long-term change of bridge responding to traffic volume. For another example, United States Naval Research Laboratory buried 60 optical fiber grating sensing systems in a bridge model at the scale of 1/4 to carry out damage test on model and obtain more details about damaged bridge strain distribution. Jiang Desheng from Wuhan University of Technology deeply studied the application of optical fiber grating to the strengthening monitoring of building structure.

4. Application to the Monitoring of Concrete Crack, Temperature Field, Durability, and Other Performances

Optical fiber grating sensors that are buried in the concrete can be used to monitor the internal parameter change of concrete during pouring, curing and application and predict the health state of concrete. Researchers from Switzerland buried optical fiber grating sensors in the concrete to measure the width of concrete elongation zone at rupture. Wuhan University of Technology buried optical fiber grating sensors into concrete to monitor the girder body poured by high-performance concrete in Badong Yangtze Bridge for a long term, hoping to evaluate the durability of high-performance concrete in the practical project objectively and provide reference data for the durability design of concrete structure. Tongji University monitors the temperature of large concrete foundation with optical fiber grating sensors to enhance the thermal insulation measure during construction and avoid temperature crack in the foundation. Roctest Company in Canada monitors the stress, vibration, damage, crack, and other internal states of bridge structure with white-light extrinsic F-P optical fiber sensor. The scientific research team under the leadership of Rao Yunjiang from University of Electronic Science and Technology of Chengdu monitored the state of Chongqing Yangtze River Bridge with extrinsic FFP stress sensor system.

5. Application to the Monitoring of Geotechnique, Tunnel, Oil Field, Mine, and Dam Projects

Optical fiber grating sensors monitor the geological data and structural parameters to control the structure health. For example, GFZ Potsdam in Germany buried optical fiber grating in a GFRP rock anchor bolt and developed FBX anchor bolt of optical fiber grating sensor to detect the static and dynamic strains in the rock component and rock engineering structure. For another example, Europe STABILOS plan applies the optical fiber grating sensing system to the long-term static displacement measurement of mine girder and the monitoring of Mont-Terri tunnel in Switzerland. Material Testing and Research Laboratory of Swiss Confederation installed the optical fiber grating sensor in Luzzone Dam for safety monitoring. Wuhan University of Technology researched and developed a new optical fiber Bragg grating sensor that monitors the porewater pressure and also researched the application of optical fiber grating sensing technology to the monitoring of anchor cable and drilling derrick stress. In 2009, optical fiber grating strain sensors were applied to Jiulong Tunnel Project of Yunan Xiaomo Highway (Wang et al. 2009) and Xiang'an Seabed Project in Xiamen, Fujian (Hu 2009). In 2009 and 2010, optical fiber gratings were, respectively, used to monitor the settlement of permafrost subgrade on Qinghai–Tibet Railway and the settlement of constructed ballastless track on Zhengzhou-Xi'an Railway (Hao and Zhu 2010). In 2009, Ge (2009) applied the optical fiber grating technology to monitor Xincheng Seawall, Lingang, and Shanghai. In 2013, Duan et al. (2013) applied the optical fiber grating displacement sensor to the model test of underground rock salt gas reservoir group. In 2014, Sun (2014) measured the slope displacement in Majiata open pit mine 2 of Shendong Tianlong Group. Shi Bin et al. (Tong et al. 2014) conducted distributed measurement based on PHC pile deflection of BOTDA.

6. Common Optical Fiber Sensors in Structural Health Monitoring

The main optical fiber sensors commonly used in the structural health monitoring include intensity modulation, Fabry–Perot interference, optical fiber grating, and distributed types. They sense the change of certain characteristic parameter of light in the optical fiber caused by the disturbance of physical quantity for measurement. The intensity-modulated optical fiber sensor has simple structure and demodulation mode, but low sensitivity. Fabry–Perot optical fiber sensor has high sensitivity and small probe, but only measures the relative change amount of physical quantity. The intensity-modulated, Fabry–Perot interference and optical fiber grating sensors are connected and multiplexed in certain mode for distributed measurement. The main multiplexing technologies adopted by FBG sensing network include wavelength division multiplexing (WDM), time division multiplexing (TDM), spatial division multiplexing (SDM), and their combination (Zhang and Guo 2014). The engineering structure usually applies distributed optical fiber sensor. The reason is that the sensor multiplex quantity of intensity-modulated, Fabry–Perot interference, optical fiber grating, and other point type sensing technologies is limited while the distributed optical fiber sensing technology can extract the distribution information in a large range of measurement field so it is applicable to the health monitoring of large structure (Zhong 2017).

References

Barnoski MK, Rourke MD, Jensen SM et al (1977) Optical time domain reflectometer. Appl Opt 16(9):2375–2379. https://doi.org/10.1364/AO.16.002375
Blin S, Digonnet JFM, Kino SG (2007) Noise analysis of an air-core fiber optic gyroscope. IEEE Photonics Technol Lett 19(19):1520–1522. https://doi.org/10.1109/LPT.2007.903878
Bohnert K, Gabus P, Nehring J et al (2002) Temperature and vibration insensitive fiber-optic current sensor. J Lightwave Technol 20(2):267–276
Bucaro JA, Dardy HD, Carome EF (1977) Fiber-optic hydrophone. J Acoust Soc Am 62(5):1302–1304. https://doi.org/10.1121/1.381624
Dakin JP, Pratt DJ, Bibby GW et al (1985) Distributed optical fibre Raman temperature sensor using a semiconductor light source and detector. Electron Lett 21(13):569–570. https://doi.org/10.1049/el:19850402
Doebling SW, Farrar CR, Prime MB et al (1996) Damage identification and health monitoring of structural and mechanical systems from changes in their vibration characteristics: a literature review. United States. https://doi.org/10.2172/249299
Duan K, Zhang Q, Zhu H et al (2013) Application of fiber Bragg grating displacement sensors to geotechnical model test of underground salt rock gas storages. Rock Soil Mech 34(10):471–476. https://doi.org/10.16285/j.rsm.2013.s2.034
Ferreira AC, Coêlho AG, Sousa JRR et al (2016) PAM–ASK optical logic gates in an optical fiber Sagnac interferometer. Opt Laser Technol 77(Supplement C):116–125. https://doi.org/10.1016/j.optlastec.2015.09.002
Ferreira MFS, Statkiewicz-Barabach G, Kowal D et al (2017) Fabry–Perot cavity based on polymer FBG as refractive index sensor. Opt Commun 394(Supplement C):37–40. https://doi.org/10.1016/j.optcom.2017.03.011
Ge J (2009) Application of BOTDR to monitoring sea dyke subsidence. Rock Soil Mech 30(6):1856–1860. https://doi.org/10.16285/j.rsm.2009.06.059
Hao JY, Zhu SJ (2010) Discussion on schemes for monitoring post-construction subsidence of subgrade of Zhengzhou-Xi'an passenger dedicated line. J Railway Eng Soc 27(03):33–36

References

Hill KO, Fujii Y, Johnson DC et al (1978) Photosensitivity in optical fiber waveguides: application to reflection filter fabrication. Appl Phys Lett 32(10):647–649. https://doi.org/10.1063/1.89881

Hu N (2009) Application of optical FBG sensors to long term health monitoring for tunnel. Transp Sci Technol 3:91–94. https://doi.org/10.3963/j.issn.1671-7570.2009.03.031

Jasim AA, Ahmad H (2017) A highly stable and switchable dual-wavelength laser using coupled microfiber Mach–Zehnder interferometer as an optical filter. Opt Laser Technol 97(Supplement C):12–19. https://doi.org/10.1016/j.optlastec.2017.06.004

Kawasaki BS, Hill KO, Johnson DC et al (1978) Narrow-band Bragg reflectors in optical fibers. Opt Lett 3(2):66–68. https://doi.org/10.1364/OL.3.000066

Kim B, Park D, Choi S (1982) Use of polarization-optical time domain reflectometry for observation of the Faraday effect in single-mode fibers. IEEE J Quantum Electron 18(4):455–456. https://doi.org/10.1109/JQE.1982.1071597

Lee B (2003) Review of the present status of optical fiber sensors. Opt Fiber Technol 9(2):57–79. https://doi.org/10.1016/S1068-5200(02)00527-8

Liao Y (1996) Current status of optical fiber sensor technology in China. Laser Infrared 26(4):244–246

Liao Y (2003) The promotion of OFS to the development of industry. Optoelectron Technol Inf 16(5):1–6

Liu W, Guan Z-G, Liu G et al (2008) Optical low-coherence reflectometry for a distributed sensor array of fiber Bragg gratings. Sens Actuators A 144(1):64–68. https://doi.org/10.1016/j.sna.2008.01.002

Liu Y-C, Lo Y-L, Liao C-C (2016) Compensation of non-ideal beam splitter polarization distortion effect in Michelson interferometer. Opt Commun 361(Supplement C):153–161. https://doi.org/10.1016/j.optcom.2015.09.099

Lopez-Dieguez Y, Estudillo-Ayala JM, Jauregui-Vazquez D et al (2017) Multi-mode all fiber interferometer based on Fabry–Perot multi-cavity and its temperature response. Opt Int J Light Electron Opt 147(Supplement C):232–239. https://doi.org/10.1016/j.ijleo.2017.08.091

Ou J (2005) Research and practice of smart sensor networks and health monitoring systems for civil infrastructures in mainland China. Bull Natl Sci Found China 1:8–12. https://doi.org/10.16262/j.cnki.1000-8217.2005.01.003

Ozeki T, Seki S, Iwasaki K (2006) PMD distribution measurement by an OTDR with polarimetry considering depolarization of backscattered waves. J Lightwave Technol 24(11):3882–3888

Ren L, Song G, Conditt M et al (2007) Fiber Bragg grating displacement sensor for movement measurement of tendons and ligaments. Appl Opt 46(28):6867–6871. https://doi.org/10.1364/AO.46.006867

Rogers AJ (1980) Polarisation optical time domain reflectometry. Electron Lett 16(13):489–490. https://doi.org/10.1049/el:19800341

Sun J (2014) Application research of fiber grating displacement sensor in slope monitoring. Ind Mine Autom 40(2):95–98. https://doi.org/10.13272/j.issn.1671-251x.2014.02.025

Sun S, Wang T, Xu Y (2000) Optical fiber measurement and sensing technology. Harbin Institute of Technology Press, Harbin

Tong H, Shi B, Wei G et al (2014) A study on distributed measurement of PHC pile deflection based on BOTDA. J Disaster Prev Mitig Eng 34(6):693–699. https://doi.org/10.13409/j.cnki.jdpme.2014.06.005

Wang X, Zhang H, Hu Y (2009) Application of fiber-optic Bragg grating in test tunnel geological disaster. Qual Test 27(12):11–13

Wuilpart M, Ravet G, Megret P et al (2002) Polarization mode dispersion mapping in optical fibers with a polarization-OTDR. IEEE Photonics Technol Lett 14(12):1716–1718. https://doi.org/10.1109/LPT.2002.804653

Xie D, Xu C, Wang AM (2017) Michelson interferometer for measuring temperature. Phys Lett A 381(36):3038–3042. https://doi.org/10.1016/j.physleta.2017.07.036

Zhang B, Guo Y (2014) Survey of aerospace structural health monitoring research based on optic fiber sensor networks. J Shanghai Univ (Nat Sci) 20(1):33–42

Zhong D (2017) Application of distributed optical fiber sensing technology in structural health monitoring system of large-scale tunnel. Mod Trans Technol 14(3):67–71

Chapter 2
Optical Fiber Interferometer Based on F-P Cavity

The interferometric optical fiber sensor is classified into Michelson, Mach–Zehnder, Fabry–Perot, and Sagnac types. Fabry–Perot interferometric optical fiber sensor senses the external parameters through the change of optical cavity length. It gains popularity among domestic and overseas researchers because of advantages such as flexible design, high space resolution, small scale, and high sensitivity, especially in the structural health monitoring field. This chapter mainly introduces the white light interferometric F-P sensor and the acceleration sensor based on F-P cavity, and deeply studies the working principle, structural design, manufacturing process, and performance of the sensor as well as the software and hardware design of the signal acquisition system.

2.1 White Light Interferometric F-P Optical Fiber Sensor

2.1.1 Principle of White Light Interferometric F-P Optical Fiber Sensor

F-P optical fiber sensor has an artificial cavity that has two reflectors in the optical fiber (Fig. 2.1a). The cavity is called optical fiber Fabry–Perot (F-P) cavity. When the light beam enters into cavity through sensing optical fiber, it reflects on two cavity surfaces and returns by the original way. Then, the two light beams meet and interfere with each other to generate a new beam of interference light. Obviously, when the cavity length or refractive index changes, some characteristic parameters of interference light (e.g., optical intensity, wavelength, etc.) also change (Yang et al. 2005b).

When the optical fiber (F-P) cavity, light source, and photoelectric detector are simply connected through optical fiber, it forms F-P optical fiber sensor system (EFPI), as shown in Fig. 2.1b. When the optical fiber sensor is installed on structure, the internal strain of structure will result in synchronous change of vacuum cavity length t of F-P optical fiber sensor, and thus changes output optical intensity I_R. It can be seen from Formula (2.1) that the relationship between output light optical I_R of F-P optical fiber sensor and vacuum cavity length is shown in Fig. 2.1. Obviously, F-P cavity length and even the structural deformation can be acquired only when the output wavelength λ of F-P optical fiber sensor is detected. The strain of structure can be obtained directly by subtracting the measured F-P cavity length from the standard length a of F-P optical fiber cavity. It can be seen from Formula (2.1) that F-P optical fiber strain sensor based on the principle above has the wavelength of light beam as minimum unit of measurement so it has high measurement sensitivity (Yang et al. 2005a, 2006). What's more important, optical wavelength has no relation with power fluctuation, circuit drift, light path change, and other environmental factors. Therefore, the disadvantages of conventional strain measuring instruments, such as temperature drift and zero drift, can be eliminated easily. In addition, F-P cavity is made of full sealed silicon, so it has advantages such as waterproof, corrosion prevention resistance, good reliability, and long service life. It can be stably used for various structures (e.g., steel structure, concrete structure, etc.) in the severe environment for a long term.

F-P cavity can be seen to be composed of two surface plates in parallel, as shown in Fig. 2.2. The light in the cavity continuously reflects and transmits between two surface plates. Therefore, multiple beams 1, 2, 3… are generated in the reflecting direction of plate, and multiple beams of light $1'$, $2'$, $3'$… are generated in the transmitted light direction. Assume $\theta_1 \approx 0$, the incident light $I_0 = 1$, and multi-beam interference is generated at reflection and transmission ends of the cavity.

If the two reflection ends of cavity are made of glass without coating, which has a reflectivity of 0.04, the reflected and transmitted beams are distributed as shown in Table 2.1. 1 and 2 are near to reflected beam. Compared with 1 and 2, the optical intensities of 3 and the beams followed can be ignored. For transmitted beam, $1'$ is much greater than the light intensity of beams followed. Except $1'$, the light intensities of all other beams are weakened rapidly so no clear stripe is formed. Therefore, in case of very low reflectivities on the two surfaces of plate, the sufficient degree of

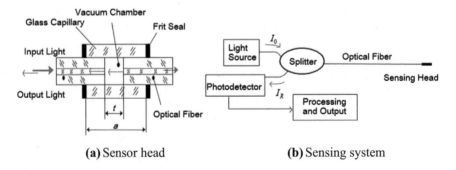

(a) Sensor head (b) Sensing system

Fig. 2.1 Schematic diagram of optical fiber interferometer

2.1 White Light Interferometric F-P Optical Fiber Sensor

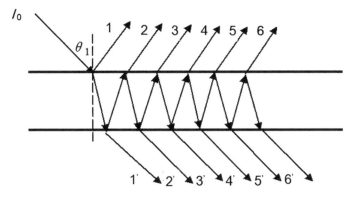

Fig. 2.2 Multiple reflections and transmissions of light in the cavity

Table 2.1 Light intensity distribution in the case of low reflectivity ($R = 0.04$)

Number	Reflected beam	Number	Transmitted beam
1	0.04	1'	0.92
2	3.7×10^{-2}	2'	1.5×10^{-3}
3	6×10^{-4}	3'	2.4×10^{-6}
4	9.49×10^{-8}	4'	3.8×10^{-9}
5	1.5×10^{-10}	5'	6.0×10^{-12}
6	2.4×10^{-13}	6'	9.7×10^{-15}

approximation can be achieved only in consideration of the two-beam interference of reflected light.

For the two-beam interference, if the intensity of incident light is I_0, the output optical intensity after coherent reflection is I, and the length of air cavity is L,

$$\frac{I}{I_0} = 2R\left[1 + \cos\left(\frac{4\pi nL}{\lambda}\right)\right] \quad (2.1)$$

where R is interface reflectivity.

It is assumed that L is $L + \Delta L = L + \varepsilon L$ and output optical intensity is changed to $I + \Delta I$ from I when the optical fiber sensor suffers from longitudinal stress and generates strain ε; the formula is

$$\frac{I + \Delta I}{I_0} = 2R\left\{1 + \cos\left[\frac{4\pi n}{\lambda}L(1+\varepsilon)\right]\right\} \quad (2.2)$$

It can be known from the above formula that the change of physical quantity (e.g., strain, temperature, etc.) to be measured can be obtained through the change of optical intensity when the monochromatic light is used as light source. However, the change of physical quantity to be measured with light intensity is easily affected by optical fiber bending degree, circuit drift, and other factors. Therefore, it is difficult to measure the absolute value of measured physical quantity.

The interferometric optical fiber sensor that applies low-coherence light source is called as "white-light interferometric optical fiber sensor" (Maaskant et al. 1997). According to spectral distribution, the low-coherence light source used to measure the white light interferometry is classified into three types: broadband continuous spectroscopic light source, narrowband continuous spectroscopic light source, and comb-type continuous spectroscopic light source.

When narrowband continuous spectroscopic light source light-emitting diode (LED) that have different frequencies is used as the white light adopts, it can be regarded as the light source overlapped by several monochromatic waves (assume the lengths of these monochromatic waves are $\lambda_1, \lambda_2, \ldots, \lambda_n$). These monochromatic waves are not coherent. Therefore, the total output light intensity I after the light emitted from LED to F-P cavity should be equal to the sum of output light intensity of each monochromatic wave, namely

$$\int_0^{+\infty} G(\omega) I_R d\omega = \int_0^{+\infty} G(\omega) 2 I_0 R (1 + \cos \Phi) d\omega \qquad (2.3)$$

where $G(\omega)$ is spectral power density of LED.

The spectral power density is subject to Gaussian distribution.

$$G(\omega) = \frac{1}{\sqrt{2\pi}\sigma} e^{-\frac{(\omega-\omega_0)^2}{2\sigma^2}} \qquad (2.4)$$

where σ is the spectral half-width of LED; ω_0 is the center circle frequency of LED. With white light, it is available to measure the strain of stressed F-P optical fiber sensor through wavelength changes. It can be known from Formula (2.3) that the reflected light intensity I_R is the greatest when $\cos \Phi = 1$. At this time, $\Phi = 2k\pi$ (k is arbitrary integer). For $\Phi = \frac{4\pi}{\lambda} L$,

$$L = \frac{\Phi}{4\pi} \lambda = \frac{2k\pi}{4\pi} \lambda = \frac{k}{2} \lambda \quad (k \text{ is arbitrary integer}) \qquad (2.5)$$

Namely, when the length of F-P cavity is integral multiple of half wavelength of transmission light, the reflected light intensity I_R is the greatest. The light emitted from LED enters into F-P cavity. Namely, when the complex wavelength of incident light is $\lambda_1, \lambda_2, \ldots, \lambda_n$, the length of F-P cavity is $L = \frac{k}{2} \lambda_i$ ($i = 1, 2, \ldots, n$). When L changes due to external factors, it can be known from Formula (2.5) that there is always a monochromatic wave λ_i ($i = 1, 2, \ldots, n$) meeting $L = \frac{k}{2} \lambda_i$. The interference stripe that corresponds to the wavelength is the brightest. This is the wavelength modulation mechanism.

For the wavelength modulation mechanism, the optical fiber sensor has the wavelength of light beam as minimum unit of measurement, so it has high measurement sensitivity. What's more important, the optical wavelength has no relation with power fluctuation, circuit drift, light path change, and other environmental factors. Thus, the problems of conventional strain measuring instruments, such as temperature drift and zero drift, can be solved easily. It is able to measure the absolute physical quantity. Therefore, the white light interferometry principle and wavelength modulation mechanism are ideal optical fiber sensing and testing technologies for the long-term health monitoring of concrete structure.

2.1.2 White Light Interferometric Sensor Head

According to the sensing principle of the interference fiber-optic sensor, the F-P interference cavity is the key element for the fiber-optic sensor to sense external signals. It consists of three parts: optical fiber, reflective fiber, and glass capillary. A complete F-P interference cavity can be formed by inserting the optical transmission fiber and the reflective fiber into the glass capillary, adjusting the cavity length, and fixing the optical transmission fiber and the reflective fiber at both ends of the glass capillary. However, the F-P interference cavity has not formed a complete sensor, because it cannot be directly embedded in the concrete structure; only after encapsulation protection, it can be embedded for testing various parameters, and it can only be used as a sensitive element of the optical fiber sensor. To facilitate the discussion, the F-P interference cavity can be referred to as a white light interference sensor head, and the complete structure, which can be directly used after protected through encapsulation, and can be referred to as a white light interference-type optical fiber sensor (according to the measuring type, it can also be divided into the white light interference optical fiber strain sensor and temperature sensor). Therefore, the white light interference sensor head is a key element in the successful development of various white light interference optical fiber sensors.

2.1.2.1 Structural Design and Parameter Analysis of White Light Interferometric Sensor

1. Structural Design of White Light Interference Sensor Head

According to the sensing principle of interference-type fiber-optic sensor, the structural design of white light interference sensor head is shown in Fig. 2.3. An F-P interference cavity is formed by inserting the incident fiber and the reflecting fiber into the glass capillary, and fixing them at both ends of the glass capillary, respectively. The optical fiber sensor with this structure has simple structure and low cost, and controllable gauge length (the distance between two bonding points) and cavity length. It can be made into various optical fiber sensors with different specifications according to the test requirements for the measured physical quantity.

The light entering from the incident optical fiber passes through the optical transmission fiber end face M1 to form the first reflected light and the first transmitted light. The transmitted light impinges on the end face M2 of reflective fiber through the F-P cavity and is reflected by the M2 to form the second reflected light. The reflected light is then returned to the incident optical fiber via M1 to form interference light with the first reflected light. As the cavity length changes, the interference light wavelength will be modulated according to the cavity length to form modulated light (Escober et al. 1992; Rao 1999). In Fig. 2.3, L_c refers to the cavity length of F-P cavity; the distance L_g between the two bonding points is referred to as the gauge distance of optical fiber sensor, which defines the operating range and sensitivity of the sensor. If the sensor is stuck on the structure to be tested, the structural strain will be converted into the change of cavity length, and the strain can be calculated according to the following formula:

$$\varepsilon = \frac{\Delta L}{L_g} \qquad (2.6)$$

where ΔL refers to the variation of F-P cavity length; L_g refers to the initial length (gauge length) of the sensor.

2. Analysis of the Influence of Structural Parameters on Sensing Characteristics

The cavity length and gauge length are the most important structural parameters of white light interference optical fiber sensor. The variation range of initial cavity length and cavity length and the size of sensor head distance determine the testing range of optical fiber sensor. The variation range of cavity length is determined by the test range of cavity length tester and the measured physical quantity. At present, the common measurement range of cavity length tester is 9,000–24,000 nm. Therefore, the allowable variation range of cavity length cannot exceed 9,000–24,000 nm. For the initial cavity length, it must be determined by the allowable variation range of cavity length and the maximum positive and negative value of the measured physical quantity.

(1) Determination of Initial Cavity Length

Taking the strain sensor as an example, if the maximum positive and negative strain of the measured structure is equal, the initial cavity length should be taken as

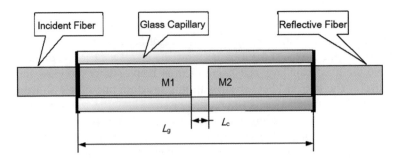

Fig. 2.3 Structural diagram of white light interference sensor head

2.1 White Light Interferometric F-P Optical Fiber Sensor

the middle value of the allowable variation range of cavity length. If the structure is mainly subjected to the positive (negative) strain, the initial cavity length can be set as a smaller (greater) value.

If the known measure range is $x^-\sim x^+$ (x refers to the temperature or strain), the scale factor of positive and negative measured physical quantity is i,

$$\left|\frac{x^-}{x^+}\right| = i \tag{2.7}$$

Then, the optimum initial cavity length L_c should meet

$$\frac{L_c - y_1}{y_2 - L_c} = i \tag{2.8}$$

i.e.,

$$L_c = \frac{y_2 i + y_1}{1 + i} \tag{2.9}$$

In the above analysis, the determination of initial cavity length is on the condition that its corresponding strain or temperature is zero. Although this is the principle, cavity length L_c is influenced by the specific circumstances when the optical fiber sensor head is manufactured. This is because that the white light interference sensor head applied in engineering practice still must be encapsulated for protection. If the temperature is measured, the ambient temperature when the sensor head is packaged will have an influence on the measured range of the sensor, and it must therefore be taken into account. If the ambient temperature of manufacture is T_0, the relative zero point should be T_0, and the scale factor i' will be

$$i' = \left|\frac{x^- - T_0}{x^+ - T_0}\right| \tag{2.10}$$

Then, the relational expression that the initial cavity length L_c should meet will become

$$L_c = \frac{24,000\, i' + 9000}{1 + i'} \tag{2.11}$$

The initial cavity length calculated according to the above formula can truly meet the requirements of the test range.

(2) Determination of the Gauge Distance

According to the sensing principle, the sensing characteristic relationship of the F-P optical fiber sensor can be expressed as

$$y = kx + L_c \tag{2.12}$$

the sensitivity (nm/με or nm/°C); x refers to the measured quantity (strain (με) or temperature (°C)); and y refers to the cavity length (nm). Depending on the requirements of the measuring range and the limitations of the test instrument, the sensitivity k can be expressed as

$$k = \frac{y_2 - L_c}{x^+} = \frac{L_c - y_1}{x^-} \tag{2.13}$$

If the measured physical quantity is temperature,

$$\Delta L_c = L_g \alpha_{matrix} \Delta T \tag{2.14}$$

The variation of cavity length per unit temperature, that is, the sensor sensitivity, is as follows:

$$k = \frac{\Delta L_c}{\Delta T} = \alpha_{matrix} L_g \quad (nm/°C) \tag{2.15}$$

In turn, the gauge distance can be deduced as

$$L_g = \frac{k}{\alpha_{matrix}} \quad (mm) \tag{2.16}$$

where k refers to the temperature sensitivity(nm/°C); α_{matrix} refers to the coefficient of thermal expansion(10^{-6}/°C).

When the measured physical quantity is strain, it can be known from the sensor head structure that the strain of structure under test will cause a change in the cavity length L_c, the strain can be expressed as

$$\varepsilon = \frac{\Delta L_c}{L_g} \tag{2.17}$$

Thereby,

$$\Delta L_c = L_g \varepsilon \tag{2.18}$$

The change in cavity length corresponding to the unit micro-strain is the sensitivity of the F-P strain sensor, i.e.,

$$k = \frac{\Delta L_c}{1 \times 10^{-6}} = L_g \quad (nm/\mu\varepsilon) \tag{2.19}$$

Therefore, when the strain value is constant, the greater the gauge L_g is, the greater the ΔL_c is, namely, the greater the amplitude of cavity length variation caused by the same amount of strain variation is, the higher the sensor sensitivity to the structural strain.

2.1 White Light Interferometric F-P Optical Fiber Sensor

By determining the structural parameters L_c and L_g of the F-P sensor head according to the above method, the highest sensing sensitivity can be obtained when the range requirement is satisfied.

(3) Analysis of Specific Parameters of the Sensor

Assuming that the deformation carrier is made of copper alloy, the coefficient of thermal expansion is $19 \times 10^{-6}/°C$, the ratio factor of positive and negative measured physical quantities I is 1, the initial cavity length L_c is 16,500 nm, and the sensor fabrication temperature is zero degrees Celsius; then, the corresponding relation among sensitivity, gauge distance, and test range is as shown in Table 2.2.

For the sensor embedded in concrete, it is mainly subjected to shrinkage stress, load stress, and temperature stress during pouring and curing, using various large concrete structures. The strain range of concrete structure is generally from -500 to $+100$ με due to the combined effect of various stresses, so the test range of strain sensor should be within -800 to $+300$ με. In addition, that environmental operating temperature of concrete structure is mainly between -30 and 45 °C, and the maximum temperature of concrete during pouring and curing can reach 60 °C. Therefore, the optical fiber strain sensor suitable for monitoring concrete structural strain should work normally in the temperature range of -40 to 70 °C.

For the sensor manufacturing, the temperature is generally about 20 °C; so for the temperature sensor, the ratio factor $i' = 1.2$, the initial cavity length L_c is 17,000 nm, and in order to ensure that the sensor works normally in the temperature range of -40 to 70 °C, then $k = (24{,}000 - 17{,}000)/50 \sim (17{,}000 - 9000)/60 = 140 \sim 133$. If the copper alloy is used as the deformation carrier and the thermal expansion coefficient is $19 \times 10^{-6}/°C$, then the gauge distance is $7.3 \sim 7$ mm; if the stainless steel is used as the deformation carrier and the thermal expansion coefficient is $12 \times 10^{-6}/°C$, then the gauge distance is $11.6 \sim 11$ mm. The corresponding relation among sensitivity gauge length, and test range is as shown in Table 2.3.

Table 2.2 Parameters of sensor with copper alloy deformation element

Sensitivity	Gauge length (mm)	Temperature (°C)	Strain (με)	Condition
19	1	±390	±7500	Using copper alloy, thermal expansion coefficient of $19 \times 10^{-6}/°C$, positive and negative measured physical quantity scale factor $I = 1$, the initial cavity length L_c should be 16,500 nm, the sensor production temperature is zero degrees Celsius
38	2	±195	±3750	
76	4	±97	±1675	
152	8	±48	±837	
304	16	±24	±418	
380	20	±19	±209	

Table 2.3 Parameters of sensor with stainless steel deformation element

Sensitivity	Gauge length (mm)	Temperature (°C)	Strain (με)	Condition
12	1	±625	±7500	Using stainless steel, thermal expansion coefficient of $12 \times 10^{-6}/°C$, positive and negative measured physical quantity scale factor $I = 1$, the initial cavity length L_c should be 16,500 nm, the sensor production temperature of zero degrees Celsius
24	2	±312	±3750	
36	4	±156	±1675	
72	8	±78	±837	
144	16	±39	±418	
240	20	±31	±209	

For the strain sensor, the influence of maximum strain needs to be taken into account in addition to the temperature sensor. If the copper alloy is used as the deformation carrier, -800 με is equivalent to $-800/19 = -42$ °C, and $+300$ με is equivalent to $300/19 = 16$ °C; therefore, it is equivalent to temperature sensor of $-82 \sim 86$ °C.

Therefore, the scale factor $i' = 102/66 = 1.55$, and the initial cavity length L_c is 18,000 nm.

$k = (24{,}000 - 18{,}000)/66 \sim (18{,}000 - 9000)/102 = 91 \sim 88$

$L_g = 91/19 \sim 88/19 = 4.79 \sim 4.6$ mm.

If the stainless steel is used as deformation carrier, -800 με is equivalent to $-800/12 = -67$ °C, and $+300$ με is equivalent to $300/12 = 25$ °C; therefore, it is equivalent to temperature sensor of $-107 \sim 95$ °C; therefore, the scale factor $i' = 127/75 = 1.69$, and the initial cavity length L_c is 18,400 nm. Then, $k = (24{,}000 - 18{,}400)/75 = (18{,}400 - 9000)/127 = 74$. Therefore, the gauge distance is $74/12 = 6$ mm.

To sum up, the following conclusions are drawn:

For the temperature sensor, if the copper alloy is used as the deformation carrier, the initial cavity length L_c is 18,000 nm and the gauge distance is <7 mm; if the stainless steel is used as the deformation carrier, the initial cavity length L_c is 18,400 nm and the gauge distance is <11 mm.

For the strain sensor, if the copper alloy is used as the deformation carrier, the initial cavity length L_c is 18,000 nm and the gauge distance is <4.5 mm; if the stainless steel is used as the deformation carrier, the initial cavity length L_c is 17,000 nm and the gauge distance is <6 mm.

According to this method, the initial cavity length and gauge distance of the white light interference sensor head not only can meet the test range requirements for the measured physical quantity but also ensure that the developed optical fiber sensor has a high sensitivity.

2.1.2.2 Materials of White Light Interferometric Sensor

1. Selection of Optical Fibers and Glass Capillaries

Optical fibers can be divided into four types according to their composition materials. The silica glass is composed of silica (SiO_2) as a main material, and a small amount (about a few percent) of oxides such as B_2O_3, CeO_2, P_2O_5, Ae_2O_3, etc. is added to change the refractive index distribution. This kind of optical fiber has become the most widely used by virtue of its high fire resistance (melting temperature above 1,500 °C) and low loss. In addition, this kind of optical fiber fully meets the technical requirements of white light interference sensor head for lightwave transmission and reflection. Hence, both the incident and the reflective optical fibers of white light interference sensor head are made of quartz glass optical fiber.

According to the number of transmission modes in optical fiber, the optical fiber can be divided into single mode and multimode. At present, the step-type multimode fiber gradually varied multimode fiber and single-mode fiber, i.e., multimode 50/125, 62.5/125, and single mode 9/125 are most commonly used. Single-mode fibers can transmit only one mode, while multimode fibers can transmit multiple modes, up to hundreds to thousands. Because of the different transmission modes, the transmission characteristics of single-mode fiber and multimode fiber are obviously different, and the main difference is that the multimode fiber is more complicated in attenuation and dispersion.

In the optical fiber sensing system using laser as light source, a single-mode optical fiber is usually selected to ensure the consistency of polarization state. However, the optical fiber sensing system in this design is based on the principle of white light interference, and the F-P cavity length is very small; the influence of polarization state is not large, and the high coupling efficiency can be obtained by adopting multimode optical fiber. Therefore, both the incident fiber and the reflective fiber are selected to be 50/125 multimode. To ensure the better reflection effect of light, the end face of reflective fiber should be larger than that of the incident fiber. Thus, the reflective fiber can be selected to have a core diameter of 62.5 μm, i.e., 62.5/125 multimode, while the incident fiber is selected to have a 50/125 multimode.

To match the 50/125 multimode optical fiber and ensure that the sensor head has a high tensile strength, the glass capillary should be YN-type standard polyamide-coated capillary, with inner diameter of 138 μm, outer diameter of 250 μm, and tensile force >300 kpsi.

2. Selection of Adhesive Agents

The designed white light interference sensor mainly relies on adhesive to bond and fix the glass capillary, the incident optical fiber, and the reflective optical fiber together. Therefore, the performance of adhesive directly determines the sensing characteristics and the service life of sensor (Li 1992; Luo 2002).

According to the requirements for strain and temperature test of concrete structure, the selected adhesive should meet the following rules:

① The temperature is between −50 and 80 °C;

② The bond strength is high;
③ The curing time is short;
④ It has a better thermal stability and long-term stability.

The Norland UV curable optical adhesive NOA83H is selected according to the properties of various adhesives. NOA83H is a high-performance adhesive produced in the United States. It is a single-component liquid photopolymer that can be cured under ultraviolet light. Its outstanding characteristic is the fast curing speed. By using the ultraviolet light source, the curing time for thin and thick film cure can be shorter than 10 and 20 s, respectively. In the absence of UV light, the glue has strong stability and will not coagulate or block the glue injection nozzle. NOA83H is very sensitive to UV light at 320~380 nm, and has the peak sensitivity at 365 nm. It can be cured by UV curing lamp. After curing, it becomes a hard film, but the secondary film is not brittle. It has a small elasticity and can release stress caused by vibration or extreme ambient temperature variation. This toughness ensures the long-term properties of glue after curing. Within the range of operating temperature of −60~120 °C, the shear strength is >32 MPa, the impact strength is >4.8 MPa, the tensile strength is >35 MPa, and it has an excellent chemical aging resistance as well as the resistance to various oils and solvents.

2.1.2.3 Manufacturing Device and Process of White Light Interferometric Sensor

In the field of optical fiber sensor technology at home and abroad, the F-P cavity optical fiber sensor technology is a high-tech and is still in the research and development stage. To reach the advanced level in the world in this field, it is more important to study it. The core element of F-P cavity optical fiber sensor is the optical fiber sensor head. Its specific structure is a 1-cm-long capillary glass tube inserted with two bare optical fibers. The capillary glass tube has an inner diameter of 138 μm, an outer diameter of 250 μm, a bare fiber core of 50 μm, and a cladding outer diameter of 125 μm, i.e., an optical fiber having an outer diameter of 125 μm is inserted into the capillary glass tube having an inner diameter of 138 μm. In the light of pre-photo sensor head, it is required not only to ensure the quality of fiber end face and the bond fixing but also to precisely control the initial cavity length (about 17 μm). Such high-precision technical requirements cannot be accomplished by hand. In order to manufacture the white light interference sensor head with high precision, a manufacturing device for the white light interference sensor head is designed and developed. This device is composed of two sets of three-dimensional precision regulators and a microscope (see Fig. 2.4). The movement stroke of regulator is 12.5 mm, the adjustment accuracy is 0.1 μm, and the magnification of microscope is 100~500 times.

The incident optical fiber is placed on the optical fiber fixing bracket of the right side adjuster, the reflective optical fiber is placed on the optical fiber fixing bracket of the left side adjuster, the glass capillary is placed on the holder on the microscope stand, and the adjusting knobs on Y and Z directions are adjusted to align the incident

2.1 White Light Interferometric F-P Optical Fiber Sensor

optical fiber; the reflective optical fiber and the glass capillary complete and keep them on the same axis. The adjusting knob on X direction is adjusted to insert the incident optical fiber and the glass capillary into the middle of glass capillary, and the cavity length test instrument is used for the precise control of the initial cavity length. After the initial cavity length is adjusted properly, the white light interference sensor head can be manufactured by glue curing at both ends of the glass capillary tube.

The structure of white light interference sensor head is as shown in Figs. 2.5 and 2.6.

In the white light interference sensor, the core part is the F-P interference cavity inserting a 50/125 multimode optical fiber coated by two end faces into a glass capillary. Of them, the multimode optical fiber has an outer diameter of is 125 μm; the hollow fiber has an inner diameter of 138 μm, an outer diameter of 250 mm, and a length of 5~10 mm. The optical fiber lead-out line of the core part of sensor head is protected by polyethylene acid grease coating layer on the basis of bare optical fiber. The length of this section is 20 mm. Despite the protection of coating layer, it is relatively fragile, and protection should be taken in use. After coating the protective layer with 20-mm-long polyvinyl ester, there is a 16-mm-long protective layer of polyimide plastic having a diameter of 0.9 mm, which is relatively strong and generally not easily broken. The main part of the transmission line is provided with a polyimide plastic protective layer having a diameter of 0.9 mm based on the protection of bare optical fiber by polyethylene acid grease coating, and then protected by polyvinyl chloride having a diameter of 2.9 mm.

Fig. 2.4 Physical photo of white light interference sensor head production device

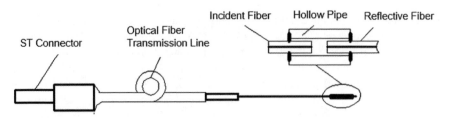

Fig. 2.5 White light interference-type sensor head

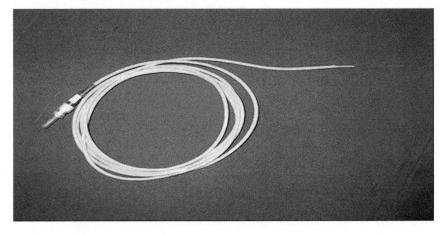

Fig. 2.6 Physical photo of white light interference sensor head

2.1.3 Embedded White Light Interferometric Optical Fiber Temperature Sensor

To avoid the damage to optical fiber sensor head during construction and installation and ensure the fast and effective transmission of structural temperature to the F-P optical fiber interference sensor head and the durability of temperature sensor, it is necessary to provide special package protection for F-P optical fiber interference sensor to meet the requirements for long-term monitoring of concrete temperature (Dai et al. 2008).

2.1.3.1 Performance Requirements of Sensors for Concrete Temperature Monitoring

(1) Testing Range. The ambient temperature in China is mainly between −30 and 45 °C, while the maximum temperature of concrete can reach 60 °C during pouring and curing. Therefore, the optical fiber temperature sensor should be able to work normally within the temperature range of −40~70 °C.

(2) Achievable Absolute Measurement of Temperature. Once embedded in the concrete structure, the temperature sensor is required to enable long-term monitoring of the process from concrete pouring to using. Therefore, the sensor is required to be free of zero drift.
(3) Good Performance of Anti-electromagnetic Interference and Long-Distance Transmission for Sensing Signals. Large concrete structures tend to work under strong electromagnetic fields (such as bridges in electrified railways, dams in hydropower facilities, etc.), and the actual measuring points tend to be far away from the acquisition system, which requires a longer signal transmission line. Because of the large electromagnetic interference and the limited signal transmission distance, the common electric measuring method cannot meet the requirements for abovementioned tests. The optical fiber has the natural anti-electromagnetic interference performance and can realize the long-distance transmission.
(4) Conveniently Embedded in the Concrete and No Impact on the Overall Performance of Concrete. To ensure the integrity of concrete structure, it is required that the sensor size should be as small as possible, but too small sensor will lead to insufficient strength. At the same time, it is very easy to deform, bend, or damage the sensor under the impact and shock during concrete pouring, resulting in inaccurate test data or abnormal operation of the sensor.
(5) Good Long-Term Stability and Durability. The long-term monitoring of concrete structure requires the sensor to have good long-term stability and durability; otherwise, the long-term monitoring of concrete structure cannot be achieved. The optical fiber itself is made of quartz glass materials and has good antiaging performance and corrosion resistance. The long-term stability and durability can be guaranteed as long as the stainless steel or copper alloy with better corrosion resistance is selected as the outer protective materials.

2.1.3.2 Structural Design of White Light Interferometric Optical Fiber Temperature Sensor

1. Material Selection for Temperature-Sensitive Element and Encapsulated Protective Sleeve

Because the multimode optical fiber and the glass capillary forming the fiber-optic sensor head are made of quartz glass, and the thermal expansion coefficient of quartz glass is 0.5×10^{-6} °C^{-1}. Thus, when the temperature is increased or decreased, the multimode optical fiber and the glass capillary will be shrunk or expanded simultaneously due to the temperature influence. Therefore, the F-P cavity length of the optical fiber sensor head is not affected by temperature itself, that is, the sensor head itself is not sensitive to temperature. If the optical fiber sensor head is fixed on the temperature-sensitive element, when the temperature changes, the expansion and contraction of the temperature-sensitive element will cause the extension or shortening of the glass capillary, which will lead to the expansion or shortening of the

F-P cavity length. Due to the variation of cavity length, the variation of temperature can be reversed. Therefore, the selection of temperature sensor material with stable and large-value coefficient of thermal expansion is the key to realize optical fiber temperature sensing.

In terms of material selection, the temperature-sensitive element requires not only a relatively large coefficient of thermal expansion and the stability within the operating temperature range but also a good corrosion resistance. Therefore, in combination with the physical and chemical properties of various materials, it is suitable to select stainless steel as the temperature-sensitive element of sensor (Fig. 2.7).

The thermal expansion coefficient of 3Gr13 stainless steel is 11.8×10^{-6}, and its performance is stable in the temperature range of $-40{\sim}70\ °C$. Due to the small size, the optical fiber sensor head can deform synchronously with the temperature variation of stainless when it is cemented to the stainless substrate, and the relationship between the deformation and the temperature variation is linear. Therefore, the requirements of temperature sensing can be satisfied.

The encapsulated protective sleeve should have high mechanical strength and good corrosion resistance. It is also suitable to use 3Gr13 stainless steel as the material of encapsulated protection sleeve according to the mechanical and chemical properties as well as processing cost of existing materials.

2. Structural Design for Temperature-Sensitive Element and Encapsulated Protective Sleeve

The structure of temperature-sensitive element is shown in Fig. 2.7. The structure of encapsulated protective sleeve is shown in Fig. 2.8. During the manufacturing of sensors, the optical fiber sensor head is fixed in the stainless pipe for the temperature-sensitive element with glue, and then the stainless steel pipe end is screwed into the encapsulated protective sleeve to form a complete set of optical fiber temperature sensor. The structural design of stainless steel tube type optical fiber temperature sensor is shown in Fig. 2.9.

3. Selection of Temperature Transfer Medium in Encapsulated Protective Sleeve

According to the structural characteristics of designed optical fiber temperature sensor, one end of the temperature-sensitive element is fixed, and the other end is

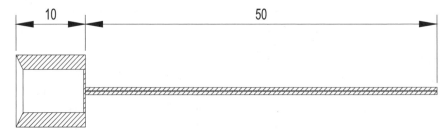

Fig. 2.7 Stainless steel tube for F-P temperature sensor

free to expand and contract in the protective sleeve. Since the temperature-sensitive element is not in direct contact with the metal protective sleeve, if the temperature transfer medium is not filled therein, the thermal conductivity will be poor, which will directly affect the reaction sensitivity and speed of the sensor. Therefore, it is necessary to fill in with a certain amount of temperature transfer medium in order to improve the reaction sensitivity and speed of the sensor.

Of all the liquid materials, oil has the best thermal conductivity, so the oil material can be selected as the filling medium. Considering the test range of sensor, the selected filling oil should be able to work stably at −40~70 °C and have no corrosive effect on the sensor head, temperature-sensitive element, and encapsulated protective material.

According to characteristics of all kinds of oil, the environment-friendly hydraulic oil (EAF hydraulic oil for short) is the most suitable filling material for the sensor. Compared with traditional hydraulic oil, this kind of hydraulic oil has the following advantages: ① good thermal conductivity, nontoxic; ② within the range of −30~82 °C, it has excellent bearing capacity and stability; ③ EAF is similar to traditional mineral oil, and the elastic seal of ordinary mineral oil and the hose material can also be dissolved in EAF; ④ it has no corrosive effect on metal materials such as steel and copper, and has strong antifouling ability; ⑤ the hydraulic oil can be used for a long time without replacement. Various performance indexes of EAF hydraulic

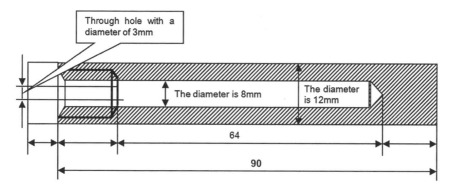

Fig. 2.8 Encapsulated protective sleeve

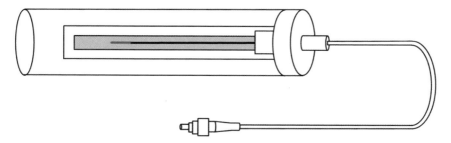

Fig. 2.9 Schematic diagram of F-P optical fiber temperature sensor

oil can completely meet the requirements of good thermal conductivity, long-term use, and no corrosion effect on encapsulating materials.

2.1.3.3 Manufacturing Process of White Light Interferometric Optical Fiber Temperature Sensor

The manufacturing process of embedded white light interference-type optical fiber strain sensor is mainly divided into three steps: equipment preparation, specific manufacturing, and sensor checking. The main process and parameters are as follows.

Step I: Equipment Preparation;

The finished encapsulating steel pipe, encapsulating glue, glue injection machine, acetone, cleaning container, sensor head, blower, electrothermostat, FIT-10 optical fiber signal analyzer, and other testing devices will be well prepared.

Step II: Specific Manufacturing;

(1) Clean the protective steel pipe of processed sensor (mainly with grease, or use acetone), and blow the fine holes in the pipe with a blower to dry.
(2) Take a certain amount of Epo-TEK353ND adhesive according to the required proportion to mix evenly and then load it into the glue injection machine, and adjust the glue injection time and pressure.
(3) Inject 15 mL of uniformly mixed encapsulating adhesive into the encapsulating tube, and irradiate the hole at the right end with an ultraviolet lamp for 5 s to cure when the adhesive overflows from the right end of encapsulating tube.
(4) Wipe the sensor head with alcohol cotton ball, cut off the excess length of the reflective fiber (leaving the reflective fiber extending about 1 cm out of the hollow glass tube), carefully insert the treated sensor into the encapsulating tube, and irradiate the left end hole of the encapsulating tube for 5 s with an ultraviolet lamp to cure it.
(5) Encapsulate the sensor in accordance with the above procedure and place the sensor in an electrothermostat (for 1.5 h at 120 °C) to cure it.
(6) After curing, use FIT-10 optical fiber signal analyzer to check the sensor for signal output, and if so, transfer to the next step for thermal cycle test; otherwise, it will be eliminated.

Step III: Sensor Checking;

(1) Conduct the thermal cycle test at -40 to $+80$ °C on the sensor, and if the linearity is good and the hysteresis is small, the sensor will be qualified.
(2) Conduct the calibration test on the sensor, establish the mathematical relation between the temperature and the wavelength variation, and determine the sensitivity coefficient of sensor.
(3) The sensor manufacturing is completed.

The embedded white light interferometric optical fiber temperature sensor manufactured is as shown in Fig. 2.10.

2.1 White Light Interferometric F-P Optical Fiber Sensor

2.1.3.4 Performance Evaluation of White Light Interferometric Optical Fiber Temperature Sensor

1. Calibration of White Light Interferometric Optical Fiber Temperature Sensor

(1) Test Apparatus and Methods

The heating and cooling equipment and the cavity length tester used in this test are, respectively, high-precision cryostat and interference-type optical fiber sensor cavity length tester FTI-10. The temperature control of high-precision cryostat can be accurate to 0.01 °C, and its temperature range is −30~100 °C; the measurement accuracy of cavity length tester FTI-10 for interference-type fiber-optic sensor is 1 nm.

First, the F-P optical fiber sensor is put into a high-precision cryostat, fixed on the shelf, and then the door is closed. The thermostat is set to a temperature value before heating, and the temperature is tested every half hour. The temperature is set every time based on the high-precision cryostat temperature, with a temperature interval of 10 °C. When the temperature rises to 80 °C, cool it down and record the data until the temperature reaches −30 °C, and then repeat the test procedure.

(2) Analysis of Sensing Characteristics

Temperature changes will cause the sensor itself thermal expansion and contraction, resulting in temperature drift. During the test, five sensors are selected randomly to carry out the temperature test. The test data are shown in Fig. 2.11.

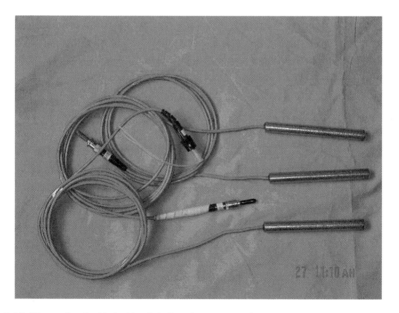

Fig. 2.10 Photo of embedded white light interference-type fiber-optic temperature sensor

It can be seen from Fig. 2.11 that the fitting curves for temperature repeat tests of five sensors are substantially coincident and tend to be straight line. The sensor will change slightly when it is heated. As can be seen from the linear equation of the sensor, the experimental test yields a temperature sensitivity coefficient of between 0.0073 and 0.0079, with an average of 0.0076. The experimental results are in good agreement with the theoretical ones. The sensor changes slightly with the temperature change, and it is not sensitive to temperature. When the sensor is encapsulated in a temperature sensor, analyze the change of sensor with temperature, and details are as follows.

(1) Relationship Between Cavity Length and Temperature

It can be called temperature sensor if the sensor is encapsulated in a certain deformation body and made to deform synchronously with the deformation body. Assuming that stainless steel is used for encapsulation, the thermal expansion coefficient of stainless steel is 11.8×10^{-6} °C^{-1}, the gauge distance of sensor is 7 mm, and the sensitivity coefficient of sensor should be about 82.6 nm/°C. That is to say, the temperature sensitivity coefficient of sensor encapsulated into stainless steel is 82.6 nm/°C. Therefore, the variation of 0.0076 nm/°C of sensor only accounts for 0.009% of the variation of 82.6 nm/°C of temperature sensor. Accordingly, with respect to the temperature sensor, the cavity length of sensor is negligible due to the subtle change in temperature.

(2) Analysis of the Repeatability of Sensor

Heat the sensor from −30 °C, measure it every 10 °C, stop heating at 80 °C, cool down the temperature, measure it every 10 °C, and then test for several times. The results of multiple tests show that the absolute cavity length variation of fiber-optic sensor is negligible, so the repeatability of sensor is good.

Therefore, the F-P fiber-optic sensor can meet the whole test range of temperature, and it is not sensitive to temperature variation and has a good repeatability. When it

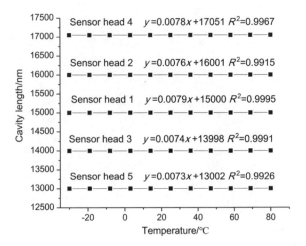

Fig. 2.11 Cavity length versus temperature

2.1 White Light Interferometric F-P Optical Fiber Sensor

is encapsulated as a temperature sensor, the change of sensor can be ignored, mainly considering the influence on temperature change sensitive elements.

2. Durability Assessment of White Light Interferometric Optical Fiber Temperature Sensor

In engineering, the working conditions of sensors cannot be compared with the ideal conditions in laboratory. When being embedded into the concrete, the sensor will go through a series of processes from concrete pouring to vibration, etc. At the same time, a lot of hydration heat will be released when the concrete is cured, resulting in a very severe test for the sensor. In the application process of large structure, the sensor must be subjected to the constantly changing static and dynamic loads of wind, sunshine, rain, bridge, and other structures for a long period of time, which requires higher service life of the sensor. Therefore, the sensor must be subjected to various examinations and tests to verify whether it can meet the requirements for large-scale structural health monitoring before it is applied to practical projects.

Processed sensors always accept the harsh test from external environment in the process of use. For instance, these sensors will be embedded in concrete for a long time, or installed on the structural surface, or subjected to long-term exposure to sunlight or in the atmosphere, or will withstand high- and low-temperature impact with temperature changes. Because the sensitive elements adopt the bonding technology, these environmental factors will undoubtedly influence the strength change of bonding. Therefore, it is necessary to test the aging resistance of the sensor, which mainly includes weather resistance test, heat resistance test, and water resistance test.

(1) Weather Resistance Test of the Sensor

In the natural environment, the mechanical properties of bonding joints will be affected by the combined action of corrosive chemicals such as oxygen, carbon dioxide, sulfide, etc., and substances such as heat, light, water, etc. Therefore, the weather resistance test is very necessary. The weather resistance test mainly includes air exposure test, accelerated aging test, natural storage test, mold test, etc.

(2) Heat Resistance Test of the Sensor

There are many occasions where the sensors are used in high-temperature environment, for instance, when being embedded in the concrete structure, the sensors are subjected to high temperature caused by hydration heat during concrete curing; and when the concrete structure is used, the temperature of surface exposed to the sun will be very high. Therefore, the sensor performance change under high-temperature and long-term heat is a very important issue. The heat resistance test mainly includes the thermal aging test, the high–low-temperature alternating test, and the low-temperature performance test.

The purpose of thermal aging test is to understand the changing rule and service life of the sensor under thermal action. During the test, samples should be placed on the thermostat frame and in a free state; and to ensure the air circulation, samples should be kept at a certain distance. The thermal aging test (see Fig. 2.12) was carried out using the 102-type thermostat. The test temperature is 200 °C and the duration is

48 h. Immediately after the test, the samples are taken out and the performance test is carried out after placing for 24 h in the standard state.

The high–low-temperature alternating test is mainly to test the ability of anti-cracking for the bonding part of the sensor. Under the action of heating and cooling impact, the adhesive and the adherend will contract and expand. Because of the difference of thermal expansion coefficient between the adhesive and the adherend, the bonding part will produce great stress, so this method has very important practical significance (Jackson 1994). The GDJS-0250 high–low-temperature alternating humidity test chamber is used for high–low-temperature alternating humidity test (see Figs. 2.13 and 2.14). The test is carried out with high temperature (120 °C, 12 h) → indoor temperature placing (24 h) → low temperature (−20 °C, 12 h) → indoor temperature placing (24 h) as a cycle.

The test results show that the temperature fluctuation of optical fiber temperature sensor is less than 0.07 °C after 30 times of temperature and humidity cycling test at −50~80 °C. Therefore, the optical fiber temperature sensor still has high testing accuracy and excellent performance stability under the harsh high–low-temperature alternating cycle impact.

(3) Water Resistance Test of the Sensor

In practical use, the sensor is always in contact with water, such as in the process of embedding into the concrete, exposure to the rain, etc. Water molecules will desorb the adhesive at the interface and reducing its strength. If water intrudes into the sensor interior, it will cause corrosion of the internal components, making the

Fig. 2.12 Sensors in a thermostat

2.1 White Light Interferometric F-P Optical Fiber Sensor

sensor to lose precision and even fail. Therefore, the water resistance test of sensor is also necessary. The water resistance test mainly includes the damp-heat test, the constant temperature immersion test, the dry–wet alternating test, etc. Among them, the damp-heat test and the constant temperature immersion test are commonly used.

The combination of high temperature and high humidity is the main cause of aging of the sensor and its internal adhesive. The GDJS-0250 high–low-temperature alternating damp-heat test chamber is used to carry out the damp-heat test. The temperature is controlled at 80 °C, humidity at 90% RH, and time at 48 h.

The constant temperature immersion test is to immerse the sensor in the constant temperature testing water tank, so that the waterproof performance of sensor joints can be well checked by the action of water. The test method is simple and low cost. The test is carried out by immersing the sensor directly at indoor temperature for 4 weeks and then removing it for test. The test results show that the maximum temperature fluctuation of the optical fiber temperature sensor is less than ± 0.08 °C after more than one month's antiaging test. Thus, it has an excellent impact resistance, heat resistance, water resistance, and accuracy stability.

After carrying out heat resistance test and water resistance test on the sensor, it is also necessary to carry out repeated examination test to ensure the good stability of temperature sensor. The test uses a Beckmann thermometer (GW-2 type with an

Fig. 2.13 High and low-temperature test chamber

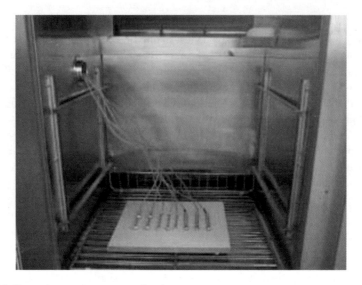

Fig. 2.14 Sensor in a temperature test chamber

accuracy of 0.01 °C) for at least three cycles. Taking the first sensor as an example, the results of three measurements are shown in Fig. 2.15. The sensor linearity is above 0.999, the temperature precision is above 0.5 °C, the stability of the sensor is good, and the error of its stability is not greater than 1%.

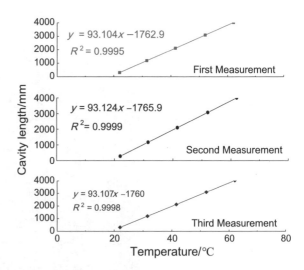

Fig. 2.15 Relationship between temperature and cavity length of the sensor in repeatability test

2.1.4 Embedded White Light Interferometric Optical Fiber Strain Sensor

Due to the vulnerability of F-P optical fiber interference sensor head and the extensive construction technology of concrete, the F-P optical fiber interference sensor needs special encapsulation protection to avoid damage of sensor head during construction and installation, and to ensure that the structural strain can be faithfully transmitted to F-P optical fiber interference sensor head (Zhang et al. 2008).

2.1.4.1 Performance Requirements of Sensors for Strain Monitoring of Concrete Structures

(1) Testing Range. It is mainly subjected to shrinkage stress, load stress, and temperature stress and other factors during cast-cured and using various large concrete structures. The strain range of concrete structure is generally -500 to $+100$ $\mu\varepsilon$ due to the combined effect of various stresses, so the test range of strain sensor should be within $-800 \sim +300$ $\mu\varepsilon$.
(2) Measuring Accuracy. The concrete can only bear compressive stress rather than the tensile stress. Therefore, any slight tensile stress of concrete structure will lead to cracking. This requires that the strain sensor should achieve the test accuracy of 0.1 $\mu\varepsilon$.
(3) With Achievable Absolute Measurement of Strain. Once embedded in the concrete structure, the strain sensor is required to realize the long-term monitoring from concrete pouring to using. Therefore, it is required that the sensor should perform an absolute measurement of the strain (i.e., the displayed strain value at each measurement should be the absolute strain at that time).
(4) In addition, the requirements for resistance to electromagnetic interference, implantable capacity, long-term stability, and durability of the sensor are the same as those for the F-P interference temperature sensor in Sect. 2.1.3.1.

2.1.4.2 Structural Design of White Light Interferometric Optical Fiber Strain Sensor

1. Material Selection for Encapsulated Protective Casing

To ensure the integrity of concrete structure, the sensor size is required to be as small as possible, but it will lead to the insufficient strength for too small sensor. It is very easy to deform, bend, or damage the sensor under the impact and shock during concrete pouring, resulting in inaccurate test data or abnormal operation of the sensor. The optical fiber sensor head itself is very slim and is very easy to be destroyed if directly embedded into the concrete. Therefore, it must be protected by metal material with higher strength. In particular, an optical fiber strain-sensitive element can be encapsulated in a metal tube or plate equivalent to the outer diameter of a steel bar, and the optical fiber strain-sensitive element and the metal tube or plate can be kept synchronously deformed along with the concrete deformation. This not only can guarantee the testing accuracy but also avoid the deformation, bending, or destruction of the sensor under the impact and shock during concrete pouring. At the same time, because the outer encapsulating material and size of the sensor are equivalent to the steel bar, the embedding of sensor will not affect the overall performance of the concrete structure.

In fact, the encapsulating protective material acts as a strain elastic element; in addition to the requirements of similar elastic modulus and thermal expansion coefficient to the concrete reinforcement, it is also required to have good resistance performance and high elastic limit, and can carry out stable works in the range of −40~70 °C. Therefore, combining with the mechanical, physical, and chemical properties of various materials, it is suitable to select stainless steel as the encapsulated protective material of sensor.

For the 3Gr13 stainless steel, the tensile strength is >740 MPa, the elastic limit is >380 MPa, the elasticity modulus is 2.11×10^5 GPa, the thermal expansion coefficient is 11.8×10^{-6} °C^{-1}, and its performance is stable in the temperature range of −40~70 °C. Due to the small size, the optical fiber sensor head can deform synchronously with the temperature variation of stainless steel when it is cemented to the stainless substrate, and the relationship between the deformation and the strain variation is linear. Therefore, the selection of 3Gr13 stainless steel as the encapsulating protective material can meet the technical requirements for strain sensing.

2. Structural Design for Encapsulated Protective Casing

Only when the deformation of optical fiber strain sensor is consistent with concrete, the test signal of the sensor can truly reflect the deformation of concrete, which requires the sensor must be tightly combined with concrete so as to keep the uniform deformation between them. However, the concrete is a kind of nonuniform structure, the size of sand aggregate is 10~20 mm, and it is difficult for the sensor to be closely combined with concrete. Meanwhile, the deformation of each point along the sensor is not uniform due to the action of aggregate. Therefore, ensuring the consistency of the deformation of the sensor and the concrete without affecting the concrete strength is an important problem to be discussed in depth.

Because of the effect of large size aggregate in concrete, the strain measured only at a certain point cannot reflect the concrete deformation; instead, the average strain of concrete should reflect the deformation within a certain length range. The statistical results from a large number of actual engineering tests in China and abroad show that when the sensor is longer than 70 mm and the average strain of this length is tested, the concrete deformation can be truly reflected; therefore, the effective length of the strain sensor should be greater than 60 mm.

To meet the test requirements for reflecting the entire concrete deformation with the average strain in a certain length range, it is considered to process the shape of sensor into a structure of thin middle and flanges at both ends. In particular, when the sensor is embedded in concrete, white gauze can be used to wrap the middle part of the sensor to isolate the cured concrete from the sensor, so that the middle sensing part does not directly contact with the concrete, and the effect of tightly bonding the flange parts at both ends with the concrete can be achieved. The deformation process of concrete structure drives the flange to deform synchronously, and the flange at both ends then drives the middle sensing part to deform uniformly. The resulting strain is the average strain of the middle sensing part.

To sum up, the structure shown in Fig. 2.16 can be selected as the encapsulation protection shell of the embedded strain sensor. When encapsulating sensor with the slender hole, the optical fiber sensor head is glued into the stainless steel hole (inner diameter of steel pipe is 0.4 mm), and the end of signal transmission line is fixed into the hole of stainless steel (inner diameter of steel pipe is 3 mm) to form a complete optical fiber strain sensor.

The advantage of this structure is that the encapsulation protection of optical fiber sensor head is good, the inner diameter of stainless steel tube is 0.4 mm, the outer diameter of optical fiber sensor head is 0.25 mm, and the middle is filled with glue, so that the synchronous deformation of the optical fiber sensor head and the stainless steel tube can be ensured.

3. Selection of Adhesive Agents

The designed optical fiber strain sensor is composed of the optical fiber interference sensor head and the encapsulation protection shell, and the connection between these two parts is mainly made of adhesives. Therefore, the performance of adhesive directly determines the sensing characteristics and the service life of sensor.

According to the structural characteristics and functional requirements of the sensor, the selected adhesives should meet the following requirements:

① The bonding between stainless steel and glass can be realized, and the bonding strength is higher in the temperature range of −40 ~120 °C;
② In a liquid state and easy to encapsulate;
③ Good aging resistance;
④ High-temperature curing and long initial hardening time (more than 5 min);
⑤ Low brittleness, good thermal stability, and able to withstand repeated high-temperature changes for a long time.
⑥ The adhesive shrinkage during curing is small so as to avoid shrinkage stress from influencing the sensing characteristics.

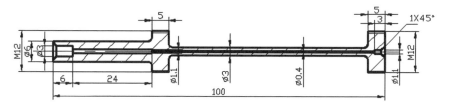

Fig. 2.16 Structure of embedded white light interference-type fiber strain sensor

The Epo-TEK 353ND-type adhesive is selected according to the properties of various adhesives. The Epo-TEK 353ND adhesive has the following characteristics: high tensile strength and excellent geometric stability within the wide temperature range. It can work continuously at 200 °C, but at the same time, it can work at 300~400 °C for several hours. The Epo-TEK 353ND adhesive is excellent in solubility and chemical resistance. It is an ideal adhesive for fixing optical fibers, optical products, metals, glass, ceramics, and most of plastics.

In application, the weight ratio of A and B glue is 10:1, the volume ratio is 8.5:1, and the recommended maximum mix quantity is not greater than 25 g. The adhesive is easy to penetrate into optical fiber bundles due to its longer curing time, and its color changes from amber to crimson during curing (whether the glue is fully cured or not can be judged by color change). The adhesive is cured by ultraviolet curing and high-temperature curing, and the specific curing process is as follows:

① First mix the glue A and B according to the abovementioned proportional relationship evenly, and then inject them into the encapsulation tube.
② Use ultraviolet lamp to cure both ends of the steel tube.
③ After curing, put it into the electrothermostat and heat it for curing (Cure for 1.5 h at 60 °C and 15 min at 80 °C).

2.1.4.3 Manufacturing Process of Embedded White Light Interferometric Optical Fiber Strain Sensor

The adhesive and the manufacturing process of the embedded white light interferometric optical fiber strain sensor are basically the same as those of the interference optical fiber temperature sensor. The photographs of interferometric-type optical fiber strain sensor are as shown in Figs. 2.17 and 2.18.

2.1.4.4 Calibration Tests of Embedded White Light Interferometric Optical Fiber Strain Sensor

1. Test Apparatus and Methods

The test beam is used for loading. The test beam is a simply supported beam (1,525 mm × 60 mm × 8 mm), and a steel plate (426 mm × 60 mm × 12 mm) is

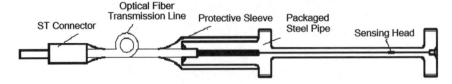

Fig. 2.17 Structure of embedded white light interference-type fiber strain sensor

2.1 White Light Interferometric F-P Optical Fiber Sensor

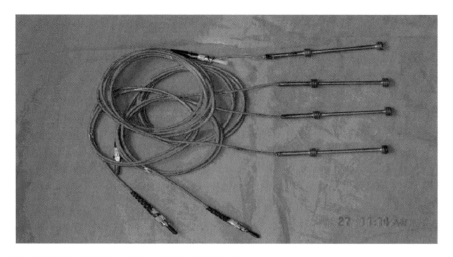

Fig. 2.18 Physical photo of embedded white light interference-type fiber strain sensor

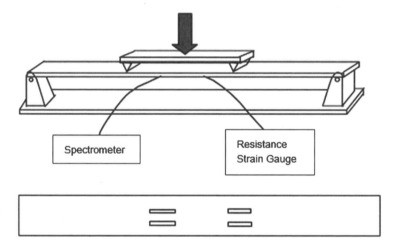

Fig. 2.19 F-P optical fiber sensor and strain gauges glued on the lower surface of simply supported beam

placed on the beam. As shown in Fig. 2.19, the optical fiber sensor and strain gauge are glued on the lower surface of the beam, respectively, and then the steel plate is tested.

The purpose of placing a steel plate in the middle of simply supported beam is to ensure that the strain is all the same on lower surface of the simply supported beam within the length of steel plate during loading. The purpose of sticking resistance strain gauge is to compare with the output signal of optical fiber strain sensor and use it as standard strain. The output signal of resistance strain gauge is measured by the DS3815-type static strain test system. The output signal of optical fiber strain sensor is measured by the FTI-10 optical fiber spectrometer. The FTI-10 optical fiber spectrometer can directly detect the cavity length of optical fiber strain sensor. By recording the standard strain and the cavity length of optical fiber strain sensor, the various sensing characteristics of optical fiber strain sensor can be analyzed using computer numerical calculation.

2. Analysis of Sensing Characteristics of the Sensor

The static loads of 2.55, 5.1, 7.65, 10.2, 12.75 and 15.3 kN are, respectively, applied to the steel beams during static loading. The test data are as shown in Table 2.4.

The linearity is an index reflecting the degree of coincidence (or deviation) between the sensor output–input calibration curve and the selected fitting straight line. Obviously, the values of calculated linearity are different for straight lines with different fitting. The principle of selecting the fitting straight line is to ensure that the minimum nonlinear error can be obtained (Rao et al. 2000), and the principle of least square method is selected to obtain the fitting straight line that can ensure the minimum residual sum of squares for sensor calibration data. The linearity can be analyzed by the curve of strain versus cavity length, as shown in Fig. 2.20. In the figure, only results from the third loading and unloading test are selected. The linear fitting of sensor is 0.9999 for the first loading, 0.9917 for the first unloading, 0.9999 for the second loading, and 0.9981 for the second unloading. The linear fitting of sensor is 0.9978 and 0.9998, respectively, when the sensor was loaded and unloaded for the third time. It can be seen that the cavity length of the sensor varies linearly with the strain.

The sensitivity refers to the ratio of the output increment of sensor to the input increment, i.e.,

$$k = \Delta y / \Delta x \qquad (2.20)$$

Table 2.4 Measured cavity length of F-P sensor applied with loads

Frequency		0	1	2	3	4	5	6
Strain (με)		0	83.5	167	250.5	334	417.5	501
Test cavity length (μm)	Load 1	18.395	18.867	19.338	19.838	20.338	20.822	21.306
	Unload 1	18.421	18.893	19.364	19.726	20.088	20.697	21.306
	Load 2	18.418	18.887	19.356	19.858	20.360	20.845	21.329
	Unload 2	18.433	18.857	19.281	19.839	20.397	20.863	21.329
	Load 3	18.427	18.896	19.365	19.797	20.229	20.783	21.336
	Unload 3	18.430	18.901	19.371	19.879	20.387	20.862	21.336

2.1 White Light Interferometric F-P Optical Fiber Sensor

Obviously, the sensitivity of linear sensor is the slope of its fitting line, which is expressed as $k = y/x$. The sensor sensitivity is very high, that is, its resolution is very high.

It can be seen from Fig. 2.20 that $k = 0.0058$ μm/με = 5.8 nm/με. The minimum resolution wavelength of optical fiber spectrometer is 0.3 nm, and the sensor sensitivity is the variation of micro-strain per actual strain, which makes the cavity length of sensor to generate a variation of 5.8 nm. If the sensor cavity length changes by 0.3 nm, the corresponding micro-strain is 0.05. It can be seen that the developed optical fiber-sensitive element has very high sensitivity.

The hysteresis is an index that reflects the coincidence degree of output–input curve in the process of forward and reverse strokes of sensor. It is usually expressed by the ratio percentage of the maximum deviation of output in the forward and reverse strokes to the full range output, i.e.,

$$e_H = \frac{\Delta H_{max}}{y_{FS}} \times 100\% = 0.14\% \quad (2.21)$$

Select a representative one-time loading and unloading test results, as shown in Fig. 2.21. By substituting the data into the hysteresis characteristic formula, it is known that the hysteresis is 0.14%, that is, when the adhesive bonding is used, the sensor characteristic has a certain hysteresis due to the presence of adhesive; however, the hysteresis is small.

The repeatability is an index to measure the sensor for the degree of consistency in the characteristics of curves obtained when the input amount changes for multiple times continuously with the full range under the same operating conditions. The closer each characteristic curve is, the better the repeatability is. The repeatability error reflects the dispersion degree of calibration data and is a random error, so it can be calculated according to the standard deviation.

$$e_R = \frac{a\sigma_{max}}{y_{FS}} \times 100\% \quad (2.22)$$

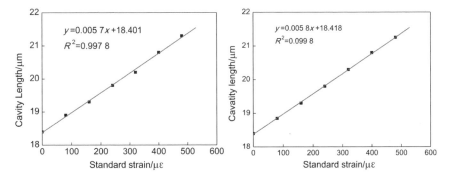

Fig. 2.20 Sensing characteristic of F-P sensor head

where σ_{max} refers to the maximum value in the standard deviation of the output values in the forward and reverse strokes at each calibration point; a refers to the confidence factor, usually is taken as 2 or 3; when $a = 2$, the confidence probability is 95.4%; and when $a = 3$, the confidence probability is 99.73%.

As the number of measurements in the test is generally less than 10, the range method can be used to calculate the deviation. The so-called range is the difference between the maximum value and the minimum value in the calibration data at a calibration point. The formula for calculating the deviation is

$$\sigma = \frac{W_n}{d_n} \tag{2.23}$$

where W_n refers to the range; d_n refers to the range coefficient related to the number of measurements n, and the value can be searched from Table 2.5.

By calculating the range, searching $d_n = 1.91$ on the basis of $n = 3$, taking $a = 2$, and substituting it into the Formulas (2.21) and (2.22), it can be obtained that the repeatability error $e_R = 14.3\%$. Several loading and unloading tests showed that the reproducibility is good, as shown in Fig. 2.22. Therefore, the sensor has a good linearity, a high sensitivity, a good repetition stability, and a low hysteresis.

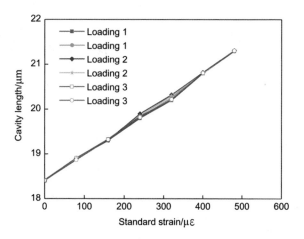

Fig. 2.21 Hysteresis characteristics

Table 2.5 Range coefficient

n	2	3	4	5	6	7	8	9	10
d_n	1.41	1.91	2.24	2.48	2.67	2.88	2.96	3.08	3.18

2.1.4.5 Performance Evaluation of Embedded White Light Interferometric Optical Fiber Strain Sensor

The stability of optical fiber strain sensor under long-term and changing load is an issue that should be paid more attention to. Therefore, it is very necessary to carry out fatigue test on the optical fiber strain sensor. The fatigue performance of strain sensor is evaluated by fatigue test, which includes fatigue stability, temperature drift, creep, fatigue life and composite process inspection of the sensor, etc.

In this test, two reinforced concrete rectangular beams were fabricated, and two comparative sections were, respectively, located within 1 m of the test beam span. Then, two white light interference-type optical fiber strain sensors and two differential-type strain sensors were embedded in the concrete rectangular beams, and eight steel string strain sensors (the steel string withstands no force in the fatigue test) and eight strain gauges were attached externally (see Figs. 2.23 and 2.24).

First, the static load test is to be carried out on the test beam with the maximum test load of 48 kN. At this time, the concrete of the beam body has reached the service state, and the cracking of beam body is completed. The test for load level is then repeated twice with the relevant parameters read as reference values for the fatigue test. The main experimental steps are as follows:

① Before the fatigue test, load the test beam to make it crack and reach the service state;
   Loading level: 0 → 4 kN → 8 kN → 16 kN → 20 kN → 22 kN → 24 kN → 26 kN → 28 kN → 32 kN → 40 kN → 44 kN.
② Load the cracked beam twice and obtain the basic data after cracking;
   Loading level: 0 → 8 kN → 16 kN → 24 kN → 32 kN → 40 kN → 44 kN.
③ Start the fatigue loading, and ensure that the fatigue amplitude is 8~40 kN. When the number of fatigue cycles reaches each measurement period, the fatigue loading will be stopped and unloaded;

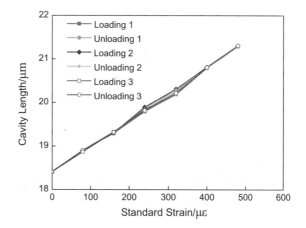

Fig. 2.22 Repeatability

Measurement period (number of fatigue cycles): 50,000 times, 100,000 times, 300,000 times, 500,000 times, 750,000 times, 1 million times, 1.25 million times, 1.5 million times, 1.75 million times, and 2 million times.

④ Carry out the static load test when the fatigue cycle times reach the measurement period;
Loading level: $0 \rightarrow 8$ kN $\rightarrow 16$ kN $\rightarrow 24$ kN $\rightarrow 32$ kN $\rightarrow 40$ kN $\rightarrow 44$ kN.

Fig. 2.23 Photo of optical fiber sensors embedded in F-P cavity

Fig. 2.24 Photo of steel string sensor glued on the beam

2.1 White Light Interferometric F-P Optical Fiber Sensor

⑤ Continue the fatigue loading and repeat step ③ to ④.
  The test result is as shown in Fig. 2.25.

The performance of white light interferometric optical fiber strain sensor is summarized as follows by analyzing the comparison chart and test data table of various strain sensors under various loads in different fatigue cycle periods:

① The linearity and stability are good under long-term working condition. The measured data show that after the number of fatigue cycle times reaches 2 million, and the measured strain values are in good consistent with those of other kinds of sensors. Besides, the comparison diagram shows that it has good convergence.
② High sensitivity. The sensitivity of white light interferometric optical fiber strain sensor used in this test is 6.0, which is much higher than that of other strain sensors (less than 3.0 in general), so it has higher resolution and can greatly improve the measurement accuracy.
③ Long fatigue life. It can be seen from the measured data and the comparison chart that, after 2 million fatigue cycles, the long-term stability of the white light interference-type optical fiber strain sensor and the differential-type strain sensor are similar, and both of them have the characteristics of small zero drift and small creep.
④ Less influence form mechanical hysteresis. Under repeated loadings, both white light interference optical fiber strain sensors and differential strain sensors are less affected by mechanical hysteresis and can be used for absolute measurement of long-term strain.
⑤ Easy to operate and high survival rate due to the hybrid process. Compared with other sensors, the white light interference optical fiber strain sensor is not affected by electromagnetic interference and insulation, and can meet the needs of monitoring and controlling as well as long-term monitoring under the complicated conditions of construction site.

In addition, the durability of white light interference optical fiber strain sensor is tested, and the contents and methods of aging resistant test are the same as those for

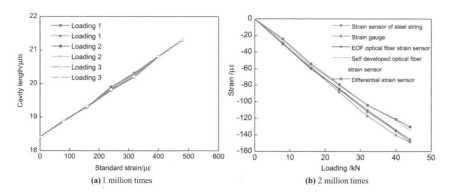

Fig. 2.25 Comparison of measured load–strain curves for various strain sensors

the optical fiber temperature sensor. The test results show that the strain measurement fluctuation of the sensor is less than one micro-strain.

2.2 Optical Accelerometers Based on F-P Cavity

Optical fiber sensors are widely used in structural health monitoring, offshore oil exploration, seismic wave detection, electricity, and other fields. These sensors have the superiority of small size, lightweight, electromagnetic immunity, high sensitive, high resolution, and easy integration (Baldwin 2014; Gillooly et al. 2014; Ko et al. 2014; Moslehi et al. 2014; Skinner and Maida 2014). The F-P interference accelerometer is one of the most important members in the optical fiber sensor because of its simple structure. It just uses two reflective films to make up an F-P cavity and one signal optical fiber to generate an interference pattern (Pocha et al. 2007; Yang et al. 2013). Test signal is only embodied in the F-P cavity, and thus it can reduce the external factors influencing the result of interference. Test signal is solved by changing the cavity length. But there are two common problems existing in the conventional fiber-optic F-P accelerometers: one is the problem between the sensitivity and the detection bandwidth, and the other one is the closed F-P cavity (Yang et al. 2005b).

This chapter proposes a novel F-P accelerometer structure, a 45° mirror fixed between two reflective films. The light of the mirror has a 45° change without closed cavity and it increases the optical path difference. The increased optical path difference has no effects on the resonance frequency of the system, but it can improve the sensitivity when the resonance frequency is kept unchanged.

2.2.1 Preliminary Test for Encapsulation

Before encapsulating the accelerometer, it is necessary to pick up the best structures and optical elements, as well as to measure the test length of F-P cavity, and to optimize the accelerometer performance. The main optimization includes the optical fiber core diameter (105/125 μm or 9/125 μm), the reflective film diameter (3 mm or 1.4 mm), and the accelerometer structure (the complete reflective inside or outside steel tube). The schematic diagram of test is as shown in Fig. 2.26 and the photograph of test is as shown in Fig. 2.27.

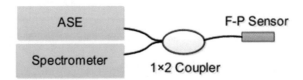

Fig. 2.26 Schematic diagram of sensor test system

2.2 Optical Accelerometers Based on F-P Cavity

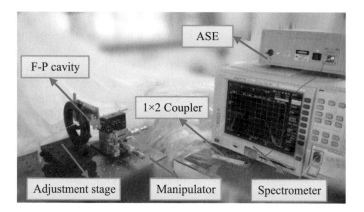

Fig. 2.27 Photograph of optical fiber test

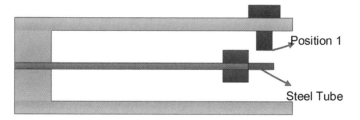

Fig. 2.28 Schematic diagram of accelerometer structure

2.2.1.1 Test of Optical Fiber Diameter

The test results showed that 105/125-μm-diameter fiber cannot form interference fringes, but the 9/125 μm one had good interference fringes. The reason is that the big diameter can reflect a stronger light intensity by the 90% reflective film and the reflected strength is greater than the reflected strength by the 4%. The intensity of two lights differs too much, so it cannot form any interference fringes.

The conclusion is a 9/125 μm diameter is better for the accelerometer, so the 9/125-μm-diameter fiber can be selected for encapsulation.

2.2.1.2 Test of Two Structures

The schematic diagram of two different structures is as shown in Fig. 2.28. For the first one, the 90% reflective film is fixed inside the steel tube and the G-lens is fixed at position 1; for the second one, the 90% reflective film is fixed outside of the steel tube (at position 1) and G-lens is inside the steel tube.

When the complete reflective film is fixed outside the steel tube, the G-lens end face fixed inside the steel tube has a bit long distance from the internal diameter of steel tube, and as a result, the optical losses increased. Thus, the reflected light coupled difficultly and the reflected light strength is not enough for forming good interference fringes.

When the complete reflective film is fixed inside the steel tube, the structure could make sure that the reflected light is coupled well and has good interference fringes. Because the steel tube's light hole and the G-lens end face diameter have almost the same diameter, the inside shorter distance makes a more reflected light.

So the structure of complete reflective film inside the steel tube is selected.

2.2.1.3 Test of Different Reflective Films

In contrast of two diameters films, small films of 1.4 mm are superior to those of 3 mm from test phenomenon, operation, stability, and test repeatability. The reason is that the big film has a longer cavity gap ranging from 2 to 8 mm; however, it is difficult to make sure that the influence of external interference and the stability of fine tuning with larger distance is poor. Then, the sensor encapsulation with 3 mm film has poor repeatability and it is hard to encapsulate.

Because the smaller film can guarantee the stability and repeatability, the 1.4 mm film is selected to encapsulate the accelerometer.

2.2.1.4 Test of Cavity Determination

Based on the above conclusion, the cavity length can be further determined. The structural test photograph is as shown in Fig. 2.29. By comparing with the length gap of test values and estimated values (only estimated the gap in the tube as the outside length can be taken from micro-displacement table), we can make sure that the estimated values are accurate enough to encapsulate. Because the distance between the complete reflective film and the 45° mirror cannot be accurately measured, we can calculate it according to the structure.

The optical cavity length formula is

Fig. 2.29 Photograph of cavity length determination

2.2 Optical Accelerometers Based on F-P Cavity

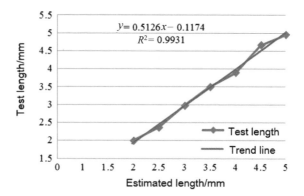

Fig. 2.30 Linear relationship between test length and estimated length

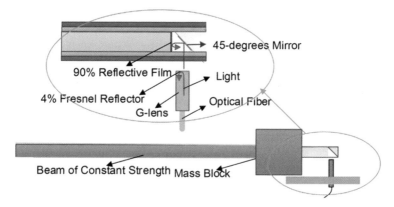

Fig. 2.31 Structural blueprint of 45° F-P accelerometer

$$d = \frac{\lambda_1 \lambda_2}{2(\lambda_1 - \lambda_2)} \quad (2.24)$$

where λ_1 and λ_2 are two adjacent resonance peak wavelengths in the free spectrum range.

According to the equation, we can obtain that the length of F-P cavity is from 2 to 5 mm.

Figure 2.30 shows the high consistency of estimated values and test values, and the correlative coefficient is 0.9965. This indicates that the estimated values are very accurate to encapsulate the sensor.

2.2.2 Structure of Accelerometer

The structural blueprint of 45° F-P accelerometer is as shown in Fig. 2.31.

The accelerometer uses the uniform intensity cantilever as the vibration elastic-sensitive element. The G-lens end face and complete reflective film make up the two reflective films of the F-P cavity and the reflectivity are 4 and 90%, respectively. The complete reflective film is fixed in the steel tube on the mass block, and the G-lens is fixed in the fine tuning bolt that can fine adjust the F-P cavity when the sensor is encapsulating. There is a 45° mirror fixed in the steel tube between two reflective films, and this mirror is used for changing the transmission of the light. The inner diameter of steel tube is 3 mm, and the distance between the reflective film and the middle of mirror is 1.5 mm. Because of the fixed 45°, the cavity has added as well as the optical path difference. The length of F-P cavity will change when the accelerometer is vibrating. Using the amount of change, we can obtain the accelerated velocity at that time.

2.2.3 Principle of Accelerometer

When the sensor is subject to vibration, the beam will vibrate; this can make a micro-distance change, and the maximum bending deflection is equal to the displacement of the F-P cavity. According to characteristics of the uniform intensity cantilever, we use the bending deflection formula to calculate the maximum displacement and obtain the displacement of F-P cavity. The bending defection ω at the mass block is related to the acceleration $a_{(t)}$ of the vibration by

$$\omega_{(t)} = \left| \frac{-6L^3 M a_{(t)}}{E b_0 h_1^3} \right| \tag{2.25}$$

where M is the equivalent weight of the mass block, beam, and the steel tube. It can be shown as

$$M = \left[l_1^2 h_1 + 2\left(l_1^2 h_2 - \frac{1}{2} l_3^2 l_1 \right) \right] \rho_{Cu} + \pi r^2 l_4 \rho_F \tag{2.26}$$

So the change of displacement is

$$\Delta L_{(t)} = \omega_{(t)} \tag{2.27}$$

According to Eqs. (2.25) and (2.27), we can obtain the acceleration:

$$a_{(t)} = \frac{E b_0 h_1^3}{6 L^3 M \Delta L_{(t)}} \tag{2.28}$$

where L is the effective length; l_1, h_1, and E are the mass effective length, thickness, and elastic modulus of the beam, respectively; h_2 is the thickness of the mass block; l_3 is the side length of the triangle fixture; l_4 is the diameter and length of the steel

2.2 Optical Accelerometers Based on F-P Cavity

tube; b_0 is the beam width of the stiff end; and ρ_{Cu} and ρ_F are the density of beryllium bronze and steel, respectively.

As shown in Fig. 2.32, the laser will be reflected and refracted on the two films, and the interference occurs when the laser is reflected back into the fiber. The change in the cavity length leads to the change in optical path difference, and then it alters the phase shifts. The phase shifts caused by acceleration excitation can be expressed as

$$\Delta \Phi_{(t)} = 2k \Delta d_{(t)} = 2 \times \frac{2\pi n \Delta d_{(t)}}{\lambda} \qquad (2.29)$$

where k is the phase velocity in the air; λ is the center wavelength of the accelerometer; and n is the index of the air.

According to Eqs. (2.28) and (2.29), we have the accelerometer sensitivity S_e:

$$S_e = \frac{\Delta \Phi_{(t)}}{\Delta a_{(t)}} = \frac{24\pi n L^3 M}{\lambda E b_0 h 3} = 0.697 \text{ rad/g} \qquad (2.30)$$

The stiffness coefficients of mass block and beam can be, respectively, expressed as

$$k_1 = \frac{ma}{\omega} = \frac{E b_0 h^3}{6L^3} \qquad (2.31)$$

$$k_2 = \frac{E b_0}{4} \left(\frac{h}{L}\right)^3 \qquad (2.32)$$

According to Eqs. (2.25), (2.30), and (2.31), we have the resonant frequency:

$$f_0 = \frac{1}{2\pi} \sqrt{\frac{k_1 + k_2}{M}} = 480 \text{ Hz} \qquad (2.33)$$

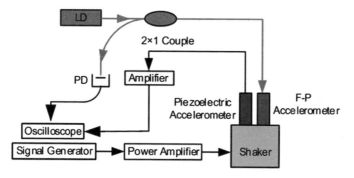

Fig. 2.32 Schematic diagram of test setup system

2.2.4 Test Results and Discussions

The schematic diagram of the F-P accelerometer test setup system is as shown in Fig. 2.32.

The time history and the corresponding frequency spectrum of acceleration measured by 45° F-P accelerometer under 250 Hz are as shown in Fig. 2.33a. For comparison, the time history of acceleration measured by the piezoelectric sensor is also as shown in Fig. 2.33b. The acceleration measured by the F-P sensor and the frequency spectrum are, respectively, in good agreement with those measured by the piezoelectric. Besides, the test result indicates that the feasibility of F-P sensor with sensitivity is about 150 mV/g and the calculated phase sensitivity is about 0.042 rad/g.

When the vibration frequency is swept manually from 80 to 1,250 Hz, it can be seen from Fig. 2.34. According to the figure, the sensitivity of F-P sensor is closely related to the vibration frequency, the on-axis sensitivity is almost flat from 80 to 250 Hz, and the natural frequency is about 400 Hz.

The frequency response of 45° F-P accelerometer is relatively flat, but the sensitivity is relatively low and has a big difference with the calculation result. The main reason is the airborne loss and the coupling loss cannot be accurately calculated. By improving the manufacture craft and improving the material size, we can further improve the sensitivity, such as coating process, the bonding process, and the diaphragm size.

Optical fiber sensors are widely used in structural health monitoring, offshore oil exploration, seismic wave detection, electricity, and other fields. These sensors have advantages of small size, lightweight, electromagnetic immunity, high sensitive, high resolution, and easy integration (Baldwin 2014; Gillooly et al. 2014; Ko et al. 2014; Moslehi et al. 2014; Skinner and Maida 2014). The F-P interference accelerometer is one of the most important members in the optical fiber sensor because of its simple structure. It just uses two reflective films to make up an F-P cavity and one signal optical fiber to generate an interference pattern (Pocha et al. 2007; Yang et al. 2013). The test signal is only embodied in the F-P cavity, and thus it can reduce the external

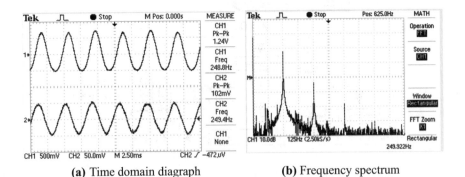

(a) Time domain diagraph   (b) Frequency spectrum

Fig. 2.33 Acceleration measured at 250 Hz

2.2 Optical Accelerometers Based on F-P Cavity

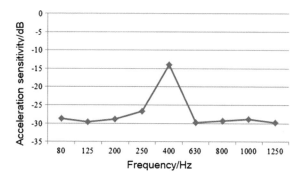

Fig. 2.34 Frequency response of accelerometer sensor

factors influencing the result of interference. Test signal is solved by changing the cavity length. But there are two common problems existing in the conventional fiber-optic F-P accelerometers: one is the problem between the sensitivity and the detection bandwidth, and the other one is the closed F-P cavity.

This paper proposes a novel F-P accelerometer structure, a 45° mirror fixed between two reflective films. The light of the mirror has a 45° change without closed cavity and it increases the optical path difference. The increased optical path difference has no effects on the resonance frequency of the system, but it can improve the sensitivity when the resonance frequency is kept unchanged.

2.3 Summary

This chapter mainly introduces the white light interference-type F-P optical fiber sensor and the acceleration sensor based on F-P cavity, and deeply studies the working principle, structural design, manufacturing process, and performance of the sensor, as well as the software and hardware design of the signal acquisition system. The main contents are as follows:

(1) According to the principle of white light interference, the strain and temperature sensors of white light interferometric F-P fiber are manufactured by using the sensing principle of perceiving changes in external factors through the change of F-P cavity length. Besides, the wavelength modulation/demodulation or white light interferometer is used to demodulate the signal, so that the demodulation by using the optical intensity method will not be influenced by external interferences to the greatest extent.

(2) This chapter describes a novel F-P accelerometer, a 45° mirror F-P cavity structure. The mirror can change the propagating direction of the light and increase the optical path difference without closing. The increased optical path difference has no effects on the resonance frequency of the system but can improve the sensitivity while the resonance frequency remains unchanged. Besides, the unclosed 45° structure can meet the special requirements of some areas about

the light propagating direction and extend the scope of the F-P accelerometer application. The sensitivity is 0.042 rad/g and the resonant frequency of the accelerometer is 400 Hz.

References

Baldwin C (2014) Optical fiber sensing in the oil and gas industry: overcoming challenges. In: OFS2014 23rd international conference on optical fiber sensors, 2 June 2014. SPIE, p 4. https://doi.org/10.1117/12.2072360

Dai JY, Zhang WT, Sun BC (2008) Analysis on a novel reflective intensity modulated fiber optic temperature sensor. In: Proceedings of the world forum on smart materials and smart structures technology, SMSST 07, pp 176–177

Escober P, Gusimeroti V, Mantineli M (1992) Fiber optic interferometric sensor for concrete structure. In: European conference on smart structure and materials, Glasgow, pp 215–218

Gillooly A, Hankey J, Hill M et al (2014) Design and performance of a high temperature/high pressure, hydrogen tolerant, bend insensitive single-mode fiber for downhole seismic systems and applications. In: SPIE sensing technology and applications, 18 June 2014. SPIE, p 7. https://doi.org/10.1117/12.2050458

Jackson DA (1994) Recent progress in monomode fibre-optic sensors. Meas Sci Technol 5(6):621

Ko MO, Kim S, Kim J et al (2014) Fiber optic dynamic electric field sensor based on nematic liquid crystal Fabry-Perot etalon. In: OFS2014 23rd international conference on optical fiber sensors, 2 June 2014. SPIE, p 4. https://doi.org/10.1117/12.2059037

Li Z (1992) Practical adhesive technology. New Age Press, Shanghai

Luo L (2002) Bonding engineering foundation. Standards Press of China, Beijing

Maaskant R, Alavie T, Measures RM et al (1997) Fiber-optic Bragg grating sensors for bridge monitoring. Cem Concr Compos 19(1):21–33. https://doi.org/10.1016/S0958-9465(96)00040-6

Moslehi B, Black RJ, Costa JM et al (2014) Fast fiber Bragg grating interrogation system with scalability to support monitoring of large structures in harsh environments. In: SPIE smart structures and materials + nondestructive evaluation and health monitoring, 17 Apr 2014. SPIE, p 8. https://doi.org/10.1117/12.2058221

Pocha DM, Meyer AG, McConaghy FC et al (2007) Miniature accelerometer and multichannel signal processor for fiberoptic Fabry-Pérot sensing. IEEE Sens J 7(5):285–292. https://doi.org/10.1109/JSEN.2006.888617

Rao YL (1999) Enabling fiber-optic sensor techniques for health monitoring of the Three Gorges Dam. In: ICHMCIS'99, pp 10–14

Rao YJ, Cooper MR, Jackson D et al (2000) Absolute strain measurement using an in-fibre-Bragg-grating-based Fabry-Perot sensor. Electron Lett 36:708–709. https://doi.org/10.1049/el:20000359

Skinner NG, Maida JL (2014) Downhole fiber optic sensing: the oilfield service provider's perspective: from the cradle to the grave. In: SPIE sensing technology + applications, 18 June 2014. SPIE, p 17. https://doi.org/10.1117/12.2049846

Yang YE, Sun BC, Wang QM et al (2005a) Research on the strain transferring characteristics of coated optical fiber sensor embedded in measured structure. Chin J Sens Actuators 18(2):363–366. https://doi.org/10.3969/j.issn.1004-1699.2005.02.039

Yang YE, Wang QM, Sun BC et al (2005b) Study on temperature and strain sensing characters of F-P fiber-optic sensor. J Transdu Technol 24(4):24–28. https://doi.org/10.13873/j.1000-97872005.04.009

Yang YE, Liu MZ, Wang QM et al (2006) The effect of the inaccuracy in line of fiber optic F-P sensing head on SNR in Fabry-Perot sensor. J Xidian Univ 33(3):462–465

Yang Z, Luo H, Xiong S (2013) High sensitivity fiber optic accelerometer based on folding F-P cavity. In: ISPDI 2013-fifth international symposium on photoelectronic detection and imaging. SPIE, p 7. https://doi.org/10.1117/12.2034550

Zhang WT, Dai JY, Sun BC (2008) Analysis on strain transfer of fiber optic Fabry-Perot sensors-based on FEM method. In: Proceedings of the world forum on smart materials and smart structures technology, SMSST 07, p 503

Chapter 3
Fiber Bragg Grating Sensor

According to the optical structure and sensing principle, fiber gratings are divided into phase shift grating, chirped grating, blazed grating, long period grating, and Bragg grating. Based on the basic principle and theoretical analysis of fiber Bragg grating, this chapter systematically introduces and analyzes the sensing principle, structure design and strain sensing performance of fiber Bragg grating temperature self-compensating strain sensor, fiber grating soil-pressure sensor, fiber grating displacement sensor, and fiber grating tilt sensor based on material thermal stress.

3.1 Basic Principle of Fiber Bragg Grating

3.1.1 Coupled Mode Theory for Fiber Bragg Grating

The optical fiber Bragg grating (FBG) forms a spatial phase grating in the core by utilizing the photosensitivity of optical fiber materials (permanent change of refractive index caused by interaction of external incident photons and germanium ions in the core), which can substantially form a narrow band (transmission or reflection) filter or mirror in the core, as shown in Fig. 3.1 (Li 2010).

The guiding principle of fiber grating is that the change in refractive index of fiber core causes the corresponding mode coupling of the light wave with a certain wavelength in the condition of fiber waveguide (Ma et al. 2001). See Fig. 3.2 for the distribution model of the refractive index.

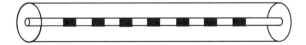

Fig. 3.1 Structure of fiber Bragg grating

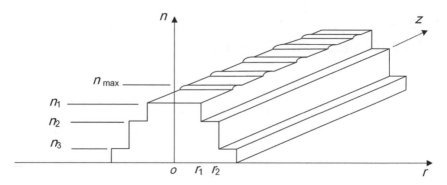

Fig. 3.2 Principle of refractive index distribution of FBG

In the entire optical fiber exposure area, the refractive index distribution of optical fiber grating can be generally expressed as

$$n(r, \phi, z) = \begin{cases} n_1[1+F(r, \phi, z)], & |r| < r_1 \\ n_2, & r_1 \leqslant |r| \leqslant r_2 \\ n_3, & |r| > r_2 \end{cases} \quad (3.1)$$

$F(r, \phi, z)$ is a function of the change in refractive index induced by the core light and has the following characteristics:

$$F(r, \phi, z) = \frac{\Delta n(r, \phi, z)}{n_1} \quad (3.2)$$

$$|F(r, \phi, z)|_{\max} = \frac{\Delta n_{\max}}{n_1}, \quad (0 < z \leqslant L) \quad (3.3)$$

$$F(r, \varphi, z) = 0, \quad (z > L) \quad (3.4)$$

where a_1 refers to the radius of the optical fiber core; a_2 refers to the radius of fiber cladding layer; n_1 refers to the initial refractive index of core; n_2 refers to the refractive index of cladding layer; $\Delta n(r, \phi, z)$ refers to the change in refractive index of optical signal; and $\Delta n_{\max}$ refers to the maximum change in refractive index.

Since the coating needs to be removed before manufacturing the optical fiber gratings, n_3 here is generally the air refractive index. r and ϕ coordinate term are the distribution of refractive index on the cross section. The general function used to describe the change in photo-induced refractive index is

$$F(r, \varphi, z) = \frac{\Delta n_{\max}}{n_1} F_0(r, \varphi, z) \sum_{q=-\infty}^{\infty} a_q \cos[k_g q + \varphi(z)z] \quad (3.5)$$

3.1 Basic Principle of Fiber Bragg Grating

$F(r, \phi, z)$ indicates that the nonuniformity of optical fiber cross section exposure caused due to the absorption of ultraviolet light by the fiber core, or the nonuniformity of grating axial refractive index modulation caused due to other factors, and because of $|F_0(r, \phi, z)|_{max} = 1$, these nonuniformities will affect the polarization and dispersion characteristics of the transmission light wave; $k_g = 2\pi/\Lambda$ is the propagation constant of grating; Λ refers to the period of grating; q refers to the order of spectral wave obtained by Fourier expansion in non-sinusoidal distribution, which leads to the inverse coupling of higher order Bragg wavelengths; a_q refers to the coefficient of expansion; and $\phi(z)$ refers to the grading function of periodic nonuniformity. For the gradualness of $\phi(z)$, it can be deemed as the quasi-periodic function. Combining with Eqs. (3.1) and (3.5), the actual refractive index distribution of grating region can be obtained as follows:

$$n(r, \varphi, z) = n_1 + \Delta n_{max} F_0(r, \varphi, z) \sum_{q=-\infty}^{\infty} a_q \cos[k_g q + \varphi(z) z] \quad (3.6)$$

Equation (3.6) regarded as the theoretical model of the fiber grating illustrates the refractive index distribution of the fiber grating, which is the basis for analyzing the grating characteristics. Typically, the refractive index change Δn_{eff} of the guided mode can be expressed as

$$\overline{\Delta n_{eff}(z)} = \Delta n_{eff}(z)\left[1 + s\cos\left(\frac{2\pi}{\Lambda}z + \phi(z)\right)\right] \quad (3.7)$$

where n_{eff} refers to the effective refractive index of FBG slightly smaller than the refractive index of core in a single-mode fiber actually used for substitute of effective refractive index of FBG; s refers to the fringe visibility related to the refractive index modulation, and s is in the range from 0.5 to 1 according to the grating reflectivity; Λ refers to the period of grating, which is about 0.2~0.5 μm; $\phi(z)$ refers to the chirp or phase shift of the grating period; $\Delta n_{eff}(z)$ refers to the average refractive index variation for the grating period by the order of 10^{-5}~10^{-3}, which may vary slightly along with the length direction z and corresponds to the apodizing (apodization) function of fiber grating. $\phi(z)$ and $\overline{\Delta n_{eff}(z)}$ can be used as the theoretical model of various fiber gratings.

At present, in most cases, optical fiber gratings are manufactured by using ultraviolet laser to form uniform interference fringes in the exposed region of optical fiber, causing the refractive index variation similar to the fringe structure on the fiber core (Meiarashi et al. 2002), as shown in Fig. 3.1. Figure 3.3 gives the refractive index distribution of a uniform periodic FBG without apodization. $\Delta n_{eff}(z)$ is same as Eq. (3.6), and $\phi(z) = 0$.

FBG with period of Λ is formed in the fiber segment with grating length L by UV exposure. Before UV exposure, the refractive index of core is n_1, the radius is a_1, and the refractive index of cladding is n_2, after UV exposure, the perturbation of refractive index for the core caused in the periodic structure can be expressed as

Fig. 3.3 Refractive index of uniform periodic fiber Bragg grating

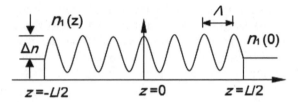

$$\Delta n_{\text{eff}}(z) = \Delta n_{\max} \cos\left(m\frac{2\pi}{\Lambda}z\right), \quad m = 1, 2, 3, \ldots \quad (3.8)$$

In most optical fibers, the change in refractive index $\Delta n(x, y, z)$ induced by ultraviolet light can be approximated as a change in the optical fiber core; however, it is negligible outside the core. The refractive index distribution of FBG can be expressed as

$$n(z) = \begin{cases} n_2 + \Delta n_{\max} \cos\left(\frac{2\pi}{\Lambda}mz\right), & -\frac{L}{2} \leqslant z \leqslant \frac{L}{2} \\ n_1, & z < -\frac{L}{2}, z > \frac{L}{2} \end{cases} \quad (3.9)$$

In an ideal fiber waveguide, the transverse mode field of optical field can be expressed as

$$E^{\text{T}}(x, y, z, t) = \sum_m \left[A_m(z)e^{i\beta n z} + B_m(z)e^{i\beta n z}\right]e_m^{\text{T}}(x, y)e^{-iwt} \quad (3.10)$$

where $A_m(z)$ and $B_m(z)$ refer to the slow-varying amplitudes of the mth mode propagating in the $+z$ and $-z$ directions, respectively, and the propagation constant is

$$\beta = \frac{2\pi}{\lambda}n_{\text{eff}} \quad (3.11)$$

The $e_m^{\text{T}}(x, y)$ may describe a core confinement mode, a radiating LP mode, or a cladding mode. In case of an ideal waveguide, these modes are orthogonal and do not exchange energy, but dielectric perturbations may cause them to couple, i.e., $A_m(z)$ and $B_m(z)$ can be expressed as

$$\begin{cases} \frac{dA_m}{dz} = i\sum_q A_q(C_{qm}^{\text{T}} + C_{qm}^{\text{L}})e^{i(\beta_q + \beta_m)z} + i\sum_q B_q(C_{qm}^{\text{T}} + C_{qm}^{\text{L}})e^{-i(\beta_q + \beta_m)z} \\ \frac{dB_m}{dz} = i\sum_q A_q(C_{qm}^{\text{T}} + C_{qm}^{\text{L}})e^{i(\beta_q + \beta_m)z} - i\sum_q B_q(C_{qm}^{\text{T}} + C_{qm}^{\text{L}})e^{-i(\beta_q + \beta_m)z} \end{cases} \quad (3.12)$$

The C_{qm}^{T} refers to the transverse coupling coefficient between the mth mode and the q-mode, which can be expressed by the following integral formula:

3.1 Basic Principle of Fiber Bragg Grating

$$C_{qm}^T(z) = \frac{w}{4} \iint_\infty \Delta\varepsilon(x,y,z) e_q^T(x,y) e_m^T(x,y) dx dy \tag{3.13}$$

The $\Delta\varepsilon(x,y,z)$ refers to the perturbation quantity of dielectric constant, which is approximately $2n\Delta n$ for $\Delta n \ll n$. In an optical fiber, the longitudinal coupling coefficient $C_{qm}^L(z) \ll C_{qm}^T(z)$ is negligible.

For FBGs, the coupling mainly occurs between two positive and reserve transmission modes with the same wavelength near the Bragg wavelength. Assuming that their amplitudes are $A(z)$ and $B(z)$, respectively, Formula (3.12) can be simplified to the following equation:

$$\begin{cases} \frac{dA^+}{dz} = i\zeta^+ A^+(z) + i\kappa B^+(z) \\ \frac{dB^+}{dz} = -i\zeta^+ B^+(z) - i\kappa A^+(z) \end{cases} \tag{3.14}$$

where $A^+(z) = A(z) e^{(i\delta_d z - \varphi/2)}$, $B^+(z) = B(z) e^{(-i\delta_d z + \varphi/2)}$ and ζ^+ refers to the coupling coefficient, which can be defined as

$$\zeta^+ = \delta_d + \zeta - \frac{1}{2} \times \frac{d\phi}{dz} \tag{3.15}$$

where δ_d refers to the detuning with respect to the Bragg wavelength independent of z, which can be defined as

$$\delta_d = \beta - \frac{\pi}{\Lambda} = 2\pi n_{\text{eff}} \left(\frac{1}{\lambda} - \frac{1}{\lambda_B} \right) \tag{3.16}$$

The grating is an ideal optical fiber Bragg grating, i.e., $\Delta n_{\text{eff}} \to 0$, and by solving the simultaneous Eq. (3.11), the center wavelength of FBG is known as

$$\lambda_B = 2 n_{\text{eff}} \Lambda \tag{3.17}$$

3.1.2 Principle and Sensitivity of Fiber Bragg Grating Temperature Sensor

The variation of the temperature around FBG has an influence on the grating constant Λ and the effective refractive index n_{eff}, which makes the Bragg wavelength λ_B shift. The effect of temperature variation on wavelength is as follows:

$$\frac{d\lambda_B}{dT} = 2\left(\frac{n_{\text{eff}} d\Lambda}{dT} + \frac{\Lambda dn_{\text{eff}}}{dT} \right) \tag{3.18}$$

If the temperature is changed, FBG is elongated or shortened due to the thermal expansion and contraction effect, thereby causing the grating period to change;

$$\frac{\Delta \Lambda}{\Lambda} = \alpha_f \Delta T \tag{3.19}$$

where α_f is the coefficient of thermal expansion of the optical fiber. For a typical germanium-doped silica fiber, α_f is $0.55 \times 10^{-6}/°C$. On the other hand, the thermo-optic effect changes the effective refractive index of the optical fiber:

$$\frac{\Delta n_{eff}}{n_{eff}} = \xi \Delta T \tag{3.20}$$

where ξ is called the thermo-optic coefficient of the optical fiber. The temperature sensitivity coefficient is

$$k_T = \frac{1}{\lambda_B} \frac{d\lambda_B}{dT} = \alpha_f + \xi \tag{3.21}$$

The temperature sensitivity coefficient reflects the relationship between the relative wavelength shift $\Delta\lambda/\lambda_B$ and ΔT. For a specific material, k_T is the constant related to the material coefficient from (3.21). If Eqs. (3.19) and (3.20) are substituted into Formula (3.18), the total effect of temperature on the FBG wavelength shift can be obtained:

$$\frac{\Delta \lambda_T}{\lambda_B} = (\xi + \alpha_t)\Delta T \tag{3.22}$$

where $\Delta \lambda_T$ is the drift amplitude of λ_B due to temperature variation; α_f is the coefficient of thermal expansion of the optical fiber material; ξ is the thermo-optical coefficient of the optical fiber material; ΔT is the variation of ambient temperature.

For the typical germanium-doped silica fiber, ξ is $6.5 \times 10^{-6}\,°C^{-1}$, and its thermal expansion coefficient $\alpha_f \approx 0.5 \times 10^{-6}\,°C^{-1}$. Therefore, for FBG with λ_B of 1,550 nm, the temperature sensitivity of grating $k_T \approx 10.85 \times 10^{-6}\,°C^{-1}$ can be calculated from Formula (3.21). It can be seen that the wavelength shift of FBG caused by temperature variation depends mainly on the thermo-optic effect accounting for about 92.85% of the wavelength shift.

According to the studies of Zhang et al. (2005a, b) and Jia et al. (2003a, b), the thermo-optic coefficient is dependent on the temperature, and the effective temperature sensitivity coefficient of the grating is

$$k_T = k_{T1} + k_{T2} \times \Delta T \tag{3.23}$$

$$k_{T1} = \frac{1}{\lambda_B} \frac{d\lambda_B}{dT} \tag{3.24}$$

$$k_{T2} = \frac{1}{\lambda_B} \frac{d^2\lambda_B}{dT^2} \tag{3.25}$$

k_{T1} and k_{T2}, respectively, represent the first-order temperature sensitivity coefficient and the second-order temperature sensitivity coefficient of the grating, and k_T represents the effective temperature sensitivity coefficient of the grating. According to the experimental results, $k_{T1} = 6.045 \times 10^{-6}/°C$, $k_{T2} = 10^{-8}/°C$, $k_T = 6.045 \times 10^{-6} + 10^{-8} \times \Delta T$.

3.1.3 Principle and Sensitivity of FBG Strain Sensor

According to the optical fiber mode coupling theory, the light with the wavelength only satisfying the Bragg condition can be reflected from the broadband light, while the other wavelength can be transmitted. If the Λ and n_{eff} in Formula (3.17) are changed due to the influence of external environment (temperature, stress, etc.), the reflected Bragg wavelength will be shifted, and Formula (3.17) can be derived:

$$\frac{d\lambda_B}{d\varepsilon} = \frac{2\Lambda dn_{eff}}{d\varepsilon} + \frac{2n_{eff}d\Lambda}{d\varepsilon} \qquad (3.26)$$

From Formula (3.17), it can be seen that the central reflection wavelength will change to $d\lambda_B$ with the Λ or n_{eff} being changed to $d\Lambda$ or dn_{eff}. Therefore, the FBG reflection wavelength is a good indicator of measured physical quantity.

When the external stress changes, the elastic strain induced will cause the variation of grating period Λ. At the same time, the effective refractive index n_{eff} of the grating will change due to the elastic-optical effect of the fiber, followed by the shift of the central reflection wavelength.

Where $dn_{eff}/d\varepsilon$ in Formula (3.26) represents the strain elasto-optical effect, and $d\Lambda/d\varepsilon$ represents the longitudinal elastic strain effect caused by stretching; the variation of grating period caused by axial strain is

$$d\Lambda = \Lambda d\varepsilon \qquad (3.27)$$

For isotropic core materials, the change in refractive index caused by strain is

$$\frac{dn_{eff}}{d\varepsilon} = -P_e n_{eff} \qquad (3.28)$$

where P_e is the valid elastic-optic constant. Therefore, the strain sensitivity coefficient is

$$k_\varepsilon = \frac{1}{\lambda_B}\frac{d\lambda_B}{d\varepsilon} = \frac{1}{n}\frac{dn}{d\varepsilon} + \frac{1}{\Lambda}\frac{d\Lambda}{d\varepsilon} = 1 - P_e \qquad (3.29)$$

The strain sensitivity coefficient reflects the relationship between the relative wavelength shift $\Delta\lambda/\lambda_B$ and $\Delta\varepsilon$. k_ε is the constant related to the material coefficient for a specific fiber material. For the germanium-doped silica fiber, $P_e \approx 0.22$, so $k_\varepsilon \approx 0.78$. The variation in reflection wavelength due to strain can be expressed as

$$\frac{\Delta \lambda_\varepsilon}{\lambda_B} = (1 - P_e)\varepsilon \qquad (3.30)$$

For a typical silica fiber, $P_e = 0.22$, so:

$$\Delta \lambda_\varepsilon = 0.78\varepsilon \times \lambda_B \qquad (3.31)$$

From the above formula, the value of the external strain ε can be obtained through $\Delta \lambda_B$. For a grating with λ_B of 1,550 nm, the change in wavelength of 1.2 pm/$\mu\varepsilon$ is calculated from Formula (3.31).

In fact, according to the study by Zhang et al. (2005a, b), the strain sensitivity coefficient is not a constant but varies with temperature. The results of this study show that

$$k_\varepsilon = 0.76318 + 0.03793 e^{-\frac{T}{124.3}} \qquad (3.32)$$

Therefore, for the ambient temperature of 25 °C (room temperature), the grating strain sensitivity coefficient is 0.7941995.

3.1.4 Theoretical Analysis of FBG Temperature–Strain Cross-Sensitivity

As can be seen from Formula (3.17), any physical quantity that changes n_{eff} and Λ will cause a shift in the reflection wavelength of the FBG. For temperature–strain sensing measurements, the reflection wavelength is a function of both. Taylor expansion of Formula (3.17) is performed, obtaining

$$\begin{aligned}
\lambda_B = {} & n(\varepsilon_0, T_0)\Lambda(\varepsilon_0, T_0) \\
& + \left[\Lambda \frac{\partial n}{\partial \varepsilon} + n \frac{\partial \Lambda}{\partial \varepsilon}\right]_{T=T_0, \varepsilon=\varepsilon_0} \Delta \varepsilon + \left[\Lambda \frac{\partial n}{\partial T} + n \frac{\partial \Lambda}{\partial T}\right]_{T=T_0, \varepsilon=\varepsilon_0} \Delta T \\
& + \left[\Lambda \frac{\partial^2 n}{\partial \varepsilon \partial T} + n \frac{\partial^2 \Lambda}{\partial \varepsilon \partial T} + \frac{\partial \Lambda}{\partial T}\frac{\partial n}{\partial \varepsilon} + \frac{\partial \Lambda}{\partial \varepsilon}\frac{\partial n}{\partial T}\right]_{T=T_0, \varepsilon=\varepsilon_0} \Delta \varepsilon \Delta T \\
& + \left[\Lambda \frac{\partial^2 n}{\partial \varepsilon^2} + n \frac{\partial^2 \Lambda}{\partial \varepsilon^2}\right]_{T=T_0, \varepsilon=\varepsilon_0} (\Delta \varepsilon)^2 + \left[\Lambda \frac{\partial^2 n}{\partial T^2} + n \frac{\partial^2 \Lambda}{\partial T^2}\right]_{T=T_0, \varepsilon=\varepsilon_0} (\Delta T)^2 + \cdots
\end{aligned} \qquad (3.33)$$

From Formula (3.33), it can be seen that $\Delta\varepsilon$, ΔT, their cross terms and high-order terms all cause wavelength shift. The contribution of higher order term to wavelength shift increases with the increase of $\Delta\varepsilon$ and ΔT. Upon $\Delta\varepsilon$ and ΔT exceeding a certain value, the variation of wavelength with $\Delta\varepsilon$ and ΔT is nonlinear; if $\Delta\varepsilon$ and ΔT variation is less than a certain value, the formula can be simplified as

$$\Delta\lambda_B(\varepsilon, T) = k_\varepsilon \Delta\varepsilon + k_T \Delta T + k_{\varepsilon,T} \Delta T \Delta\varepsilon \tag{3.34}$$

In the above formula, k_ε is strain sensitivity; k_T is the sensitivity of temperature; $k_{\varepsilon,T}$ is cross-sensitivity. $k_{\varepsilon,T}$ is related to both temperature and strain and indicates that the temperature sensitivity (or strain sensitivity) is not a constant but a parameter dependent of the strain (or temperature) at different strains, and its magnitude describes how the temperature sensitivity (or strain sensitivity) deviates from the constant. The relationship between sensitivity and sensitivity coefficient is

$$\begin{cases} k_T = K_T \times \lambda_B \\ k_\varepsilon = K_\varepsilon \times \lambda_B \\ k_{\varepsilon,T} = K_{\varepsilon,T} \times \lambda_B \end{cases} \tag{3.35}$$

In fact, according to the study by Zhang et al. (2005a, b), the temperature–strain cross-sensitivity coefficient is not a constant, while it varies with temperature. The results of this study show that

$$k_{\varepsilon,T} = -3.05 \times 10^{-4} \exp(-T/124.3) \tag{3.36}$$

Therefore, both temperature and strain variations cause an FBG wavelength shift. Thus, it is impossible to distinguish FBG wavelength shift induced by temperature from that induced by stain. Therefore, it is necessary to eliminate the shift of the Bragg wavelength caused by the change of the temperature.

3.2 Temperature Self-Compensated FBG Sensor Based on Thermal Stress

One of the most important obstacles to the practical application of FBG is the cross-sensitivity of temperature and strain of FBG. Because FBG is sensitive to both strain and temperature, it is difficult to distinguish the wavelength change caused by strain and temperature when the grating wavelength changes. When FBG is used as the optical communication device, the high stability of FBG wavelength is necessary, and the smaller the influence of temperature on the wavelength, the better. The FBG wavelength shift will seriously affect its application in laser frequency stabilization and wavelength division multiplexing. However, when FBG is used as a sensing element, the temperature–strain cross-sensitivity of the grating sensor makes it difficult to distinguish the change of temperature and strain. Therefore, without solving the problem of cross-sensitivity of temperature and strain of FBG, many unique advantages of FBG cannot be brought into play, which further restricts the application of FBG sensor in health monitoring of engineering structures.

3.2.1 Principle of Temperature Self-Compensation

At present, the most popular temperature compensation scheme of FBG is the pre-stressed encapsulation method (Li et al. 2014). The defect of such encapsulation method is that the single-point bonding process is easy to cause many problems arising from weak bonding. Although the negative expansion material encapsulation method can be used to fix the grating on the negative expansion material to solve the problem of weak bonding to some extent, the negative expansion material encapsulation method cannot measure the strain, while it can only be suitable for communication. Based on the advantages and disadvantages of the above two methods, the thermal stress encapsulation method with bimetallic encapsulation is proposed. Such method is hopeful to improve the long-term stability of the sensor and can effectively overcome the disadvantage that the negative expansion material encapsulation method cannot measure the strain. In addition, the temperature compensation strain sensor can also achieve the effect of strain enhancement.

The basic structure of the temperature compensation sensor consists of three components (Li 2010), as shown in Fig. 3.4. Component 1 is a temperature compensation element; component 2 functions as strain transmission and temperature compensation; component 3 is intended to withstand structural strain and transfer it to components 1 and 2.

In that design, it should be assumed that component 3 is not deformed by the stress of the internal component and is only affected by the external stress. This is because the sensor is usually stuck on the surface of large structure or embedded in the structure to be deformed synchronously with the large structure in practical engineering application. Therefore, the influence of stress of internal components of the sensor on the deformation of the large structure can be neglected.

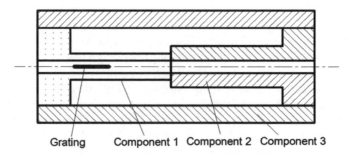

Fig. 3.4 Schematic diagram of designed sensor

The temperature compensation principle of the sensor is that when the temperature increases, the component made of metal material expands and generates thermal stress σ. Since the stiffness of component 2 is higher than that of component 1, the thermal stress of component 2 is more great than component 1, that is $\sigma_2 > \sigma_1$, which resulted in the fact that component 2 presses component 1, synchronously shortening component 1, followed by the drop of the grating wavelength, and offsetting the increased wavelength shift due to temperature rise. Thus, the temperature effect of the grating is removed. And so does the temperature drop. This is the so-called temperature self-compensation mechanism of FBG based on thermal stress of materials (Du et al. 2009).

When component 3 is subjected to external strain, both component 1 and component 2 will deform, and the strain polarity is the same. There is a certain proportional relationship between the deformation of component 1 and component 3. When the grating is deformed synchronously with component 1, component 3, and the structure, the strain of the structure can be attained according to the strain of the grating. In addition, through the reasonable structure design, the strain ε_1 generated on component 1 is greater than the external strain ε_3 received by component 3, and the additional effect of strain sensitivity enhancement can also be realized.

3.2.2 Structural Design

It is assumed that the temperature sensitivity of the encapsulated FBG is k_T pm/°C and the strain sensitivity is k_ε pm/µε. That is, the variation of FBG wavelength is k_T pm under temperature variation of 1 °C. Likewise, the variation of FBG wavelength is k_ε pm under strain of 1 µε. To achieve complete compensation of FBG temperature, there are

$$k_T \Delta T = -k_\varepsilon \Delta \varepsilon \tag{3.37}$$

where ΔT is the temperature variation; $\Delta \varepsilon$ is the strain applied to the FBG.

Therefore, to fully compensate the temperature of the grating, it is necessary to apply the compressive strain of k_T/k_ε µε to the grating as temperature increases by 1 °C. Assuming that there is no external stress, the sensor is only affected by temperature and the length L_3 of component 3 remains unchanged. Thereby the length of component 1 being shortened is equal to the elongate length of component 2.

Special attention should be paid here: When the sensor is free from external force and the ambient temperature changes, the length of component 3 changes due to thermal expansion; and when the sensor is stuck on a large structure, the measured structure usually made of large concrete or steel would also undergo thermal expansion (contraction) due to temperature rise (fall). The strain measured by the sensor should also include thermal expansion (or thermal contraction) of the structure due to temperature. Therefore, in the analysis of the temperature compensation, any change in the length of component 3 (whether deformation due to temperature change of the structure or deformation due to external stress of the structure) should be faithfully reflected by the strain sensor.

As can be seen from Formula (3.37), the strain of component 1 is

$$\varepsilon_1 = -\frac{k_T}{k_\varepsilon}\Delta T \qquad (3.38)$$

The strain of component 2 is

$$\varepsilon_2 = \frac{k_T}{k_\varepsilon}\frac{L_1}{L_2}\Delta T \qquad (3.39)$$

The following thermal stress equation is introduced:

$$\varepsilon = \frac{\sigma_T}{E} + \alpha_T \Delta T \qquad (3.40)$$

where σ_T is the thermal stress, α_T is the coefficient of thermal expansion of the material, and ΔT is the change in temperature.

The above formula can be rewritten as

$$\varepsilon = \frac{F_T}{ES} + \alpha_T \Delta T \qquad (3.41)$$

where F_T is the thermal binding force and S is the cross-sectional area of the material.

When the temperature increases, the thermal expansion coefficients of component 1 and component 2 are positive, and there is a tendency of expansion and elongation. However, component 1 is extruded by component 2 with bigger stiffness, resulting in the compressive strain of component 1.

The thermal stress equation of component 1 is

$$\varepsilon_1 = \frac{F_T}{E_1 S_1} + \alpha_{T1} \Delta T \qquad (3.42)$$

The thermal stress equation of component 2 is

$$\varepsilon_2 = \frac{F_T}{E_2 S_2} + \alpha_{T2} \Delta T \qquad (3.43)$$

Therefore, the structural design equation of the sensor is obtained as follows:

$$\frac{\alpha_{T1} + \frac{k_T}{k_\varepsilon}}{\alpha_{T2} - \frac{k_T}{k_\varepsilon}\frac{L_1}{L_2}} = \frac{E_2 S_2}{E_1 S_1} \qquad (3.44)$$

If component 1 and component 2 are made of the same material, the above formula can be simplified as

3.2 Temperature Self-Compensated FBG Sensor Based on Thermal Stress

$$\frac{\alpha_T + \frac{k_T}{k_\varepsilon}}{\alpha_T - \frac{k_T}{k_\varepsilon}\frac{L_1}{L_2}} = \frac{S_2}{S_1} \quad (3.45)$$

The above Formula (3.45) is the structural design formula of the sensor when complete temperature compensation is realized. As can be seen from the formula, the temperature effect of FBG can be avoided by selecting appropriate materials and structural dimensions. The elastic modulus, thermal expansion coefficient of the materials used for components 1 and 2, and the length ratio and cross-sectional area ratio of components 1 and 2 were main structural parameters affecting the temperature compensation.

In the following, the influence factors of temperature compensation are specifically analyzed in combination with the sensor made of hard aluminum material (components 1 and 2 are made of the same material) (Li 2010). From Figs. 3.5 and 3.6, it can be seen that the length of component 2 drops by 2.85 cm with the thermal expansion coefficient increasing by 1×10^{-6} °C^{-1}, and while k_T/k_ε varies by 1 unit, the length of component 2 is changed to 5.21 cm. Therefore, k_T/k_ε has a greater influence on the sensor structure than the coefficient of thermal expansion. Thus, the values of k_T/k_ε and thermal expansion coefficient must be more accurate.

Figure 3.7 shows the effect of the ratio of the length of component 1 to the length of component 2 on the temperature compensation effect. It can be seen from Eq. (3.48) that theoretical strain k_T/k_ε must be applied to the grating in order to fully compensate for the temperature-induced wavelength shift, but the strain applied to the grating is likely to drop due to structural machining errors and the like. It has been found in the literatures that the k_T/k_ε value is about 10. It can be seen in Fig. 3.7 the k_T/k_ε gradually approaches to the value of 10, and at that time the ratio of the length of component 1 to the length of component 2 has almost no effect on k_T/k_ε. Therefore, the design machining dimension of the sensor should be in the saturation state of the curve, that is, L_1/L_2 should be desired when k_T/k_ε approaches 10.

Figure 3.8b shows the effect of S_2/S_1 on temperature compensation. As can be seen from Fig. 3.8a, k_T/k_ε also increases with S_2/S_1 and gradually saturates. The

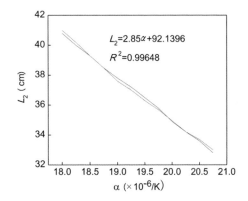

Fig. 3.5 Relation between L_2 and thermal expansion coefficient of compensated part

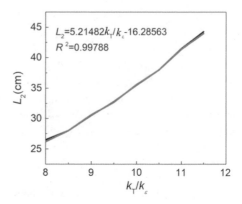

Fig. 3.6 Relation between L_2 and k_T/k_ε of compensated part

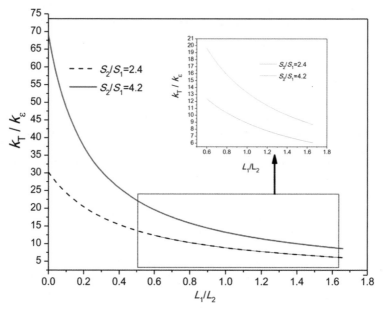

Fig. 3.7 Relation between k_T/k_ε and L_1/L_2 of compensated part

desired strain must be applied on the grating in order to fully offsetting for the FBG temperature-induced wavelength shift. However, the desired strain applied on component 1 may be decreased due to the structural machining errors. For example, when $L_1/L_2 = 0.75$ and 0.93, the slope and intercept of the linear fitting equations are, respectively, $k_T/k_\varepsilon = 4.561$, $S_2/S_1 = 0.14386$ and $k_T/k_\varepsilon = 3.759$, $S_2/S_1 = 0.33471$ in Fig. 3.8b. k_T/k_ε varies correspondingly from 9.02 to 11.06 with S_2/S_1 from 2.4 to 2.61, and thus the overcompensation stain is approximately 2.04 $\mu\varepsilon$. The variation in outer radius of component 2 from 7.9 to 8.59 mm is 0.69 mm. Therefore, it is necessary to improve the machining accuracy of the cross section, and it is suggested that the

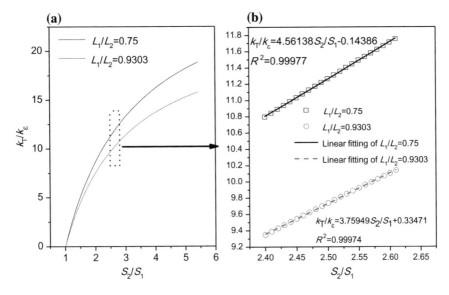

Fig. 3.8 Relation between k_T/k_ε and S_2/S_1 of compensated part

strain sensibility coefficient should not be enhanced by changing the cross-sectional area.

In summary, that design principle of the sensor can be concluded as follows:

(1) k_T/k_ε has a greater influence on the structure than the coefficient of thermal expansion.
(2) Under the premise that the temperature compensation effect is not greatly influenced, the strain sensitivity coefficient of the sensor can be increased by changing the length of component 2 while the length of component 1 remains unchanged.
(3) The strain sensibility coefficient should not be enhanced by changing the cross-sectional area of components 1 and 2.

3.2.3 Theoretical Analysis of Strain Sensing Characteristics

In order to analyze the strain sensing characteristics of the sensor, it is assumed that the sensor is not affected by temperature (Li and Sun 2015); the sensor is stuck on the structure under test; component 3 is deformed synchronously with the strain of the measured structure; the resulting strain is ε_3; the length is correspondingly changed to ΔL_3; and the following requirements are satisfied:

$$\Delta L_3 = \Delta L_1 + \Delta L_2 \tag{3.46}$$

The applied forces between components 1 and 2 are equal, i.e.,

$$F_1 = F_2 = F \tag{3.47}$$

Then the ratio of strain of component 1 to component 2 is

$$\frac{\varepsilon_1}{\varepsilon_2} = \frac{\frac{F}{S_1 E_1}}{\frac{F}{S_2 E_2}} = \frac{S_2 E_2}{S_1 E_1} \tag{3.48}$$

Also known

$$\frac{\varepsilon_1}{\varepsilon_2} = \frac{\frac{\Delta L_1}{L_1}}{\frac{\Delta L_2}{L_2}} = \frac{\Delta L_1}{\Delta L_2} \frac{L_2}{L_1} \tag{3.49}$$

Integrate Formulas (3.48) and (3.49) to obtain

$$\Delta L_2 = \frac{L_2}{L_1} \frac{S_1}{S_2} \frac{E_1}{E_2} \Delta L_1 \tag{3.50}$$

Therefore,

$$\varepsilon_3 = \frac{\Delta L_1 + \Delta L_2}{L_1 + L_2} = \frac{1 + \frac{L_2}{L_1} \frac{S_1}{S_2} \frac{E_1}{E_2}}{1 + \frac{L_2}{L_1}} \frac{\Delta L_1}{L_1} \tag{3.51}$$

The above formula can be rearranged as

$$\varepsilon_1 = 1 + \frac{\frac{L_2}{L_1}}{1 + \frac{L_2}{L_1} \frac{S_1}{S_2} \frac{E_1}{E_2}} \varepsilon_3 \tag{3.52}$$

If component 1 and component 2 are made of the same material, the above formula can be simplified as

$$\varepsilon_1 = \frac{1 + \frac{L_2}{L_1}}{1 + \frac{L_2}{L_1} \frac{S_1}{S_2}} \varepsilon_3 \tag{3.53}$$

Formulas (3.52) and (3.53) are the strain sensing formulas of the sensor. A is defined as the strain sensitivity coefficient of the sensor.

$$A = \frac{1 + \frac{L_2}{L_1}}{1 + \frac{L_2}{L_1} \frac{S_1}{S_2} \frac{E_1}{E_2}} \tag{3.54}$$

Since the strain sensitivity of the encapsulated FBG is k_ε pm/$\mu\varepsilon$, the strain sensitivity of the sensor is

3.2 Temperature Self-Compensated FBG Sensor Based on Thermal Stress

$$k_s = A \times k_\varepsilon \text{ pm}/\mu\varepsilon \tag{3.55}$$

To achieve strain sensitization $\varepsilon_1 > \varepsilon_3$, the following relationship must be satisfied:

$$\frac{1 + \frac{L_2}{L_1}}{1 + \frac{L_2}{L_1}\frac{S_1}{S_2}\frac{E_1}{E_2}} > 1 \tag{3.56}$$

i.e.,

$$\frac{S_1}{S_2}\frac{E_1}{E_2} < 1 \tag{3.57}$$

Therefore, it can be seen from Formula (3.57) that the sensor strain sensitivity is dependent on the cross-sectional area ratio of the component and the elastic modulus of the material, while it is independent of the length of the component (Du et al. 2008). However, the magnitude of strain sensitivity coefficient is related to the length of the component. In the structural design of the sensor, the cross-sectional area of component 2 should be maximized and the cross-sectional area of component 1 should be minimized; the material with high elastic modulus should be selected as the material of component 2 to increase its strain sensitization coefficient as much as possible.

Figure 3.9 is the plot of the area ratio of component 1 to component 2 and Fig. 3.10 is the curve of the influence of the length ratio of component 1 to component 2 on the strain sensitivity coefficient of the sensor. As can be seen from the figure, as the ratio S_1/S_2 of the cross-sectional area of component 1 to that of component 2 is gradually increased, the strain sensitivity coefficient is gradually decreased to 1. That is, there will be no strain sensitivity effect. Therefore, in order to increase the strain sensitivity coefficient of the sensor, S_1/S_2 must be minimized. As can be seen from the figure, when S_1/S_2 is 0.5 or more, the strain sensitivity coefficient does not change greatly. Therefore, when S_1/S_2 is 0.5 or less, it is favorable to changing the length for increasing the strain sensitivity coefficient.

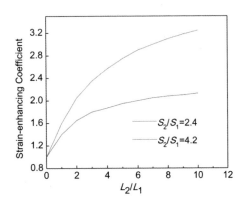

Fig. 3.9 Relation between sensor strain sensitivity coefficient and area ratio of component 1 to component 2

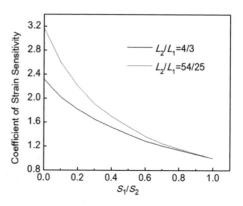

Fig. 3.10 Relation between sensor strain sensitivity coefficient and length ratio of component 1 to component 2

In the case of the same cross-sectional area ratio, the greater the L_2/L_1 is, the greater the strain sensitivity coefficient is. As L_2/L_1 increases gradually, the strain sensitivity coefficient increases gradually and finally reaches saturation. It can be seen that in the case where the length of component 1 is determined, the longer the length of component 2, the more sensitive it is to strain, rather than that there is a saturation point. The strain sensitivity coefficient can also be improved by increasing the cross-sectional area ratio of component 2 to component 1.

In any case, the ratio of the cross-sectional area of component 1 to that of component 2 should not be more than 0.5. The strain sensitivity coefficient of the sensor should be depended on the ratio of the length of component 2 to the length of component 1 with a saturation value.

3.2.4 Parameter Analysis of Temperature Compensation Structure Design

3.2.4.1 Test and Analysis of Temperature Sensitivity Coefficient of FBG

In order to ensure the accuracy of temperature sensitivity and that of sensor design size, the thermal–optical coefficient of grating is studied (Li 2010). Two ordinary germanium-doped silica FBGs with wavelengths of 1,530.013 and 1,534.146 nm are selected. The FBG's wavelength demodulation system composed of SLED, FBG and OSA measures FBG wavelength, as shown in Fig. 3.11; the temperature is measured accurately using digital Beckmann thermometer. The broadband light emitted by the light source is coupled into FBG by an optical splitter, and the light reflected from the grating is transmitted to the AQ-6140 spectrometer through optical splitter. The experiment was carried out in a high-precision high-low temperature tank. The temperature range was from room temperature to 60 °C for 4 cycles. As shown in Fig. 3.12, the temperature sensitivities of the two bare gratings are 10.77 and 10.92 pm/°C, respectively. When temperature changes

3.2 Temperature Self-Compensated FBG Sensor Based on Thermal Stress

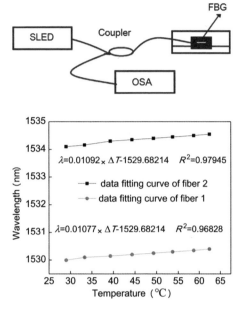

Fig. 3.11 Schematic diagram of optical fiber grating demodulation system

Fig. 3.12 Response curve of bare grating temperature

by 1 °C, $\Delta\lambda_{B1}(\varepsilon,T) = 10.77$ pm and $\Delta\lambda_{B2}(\varepsilon,T) = 10.92$ pm. The thermal expansion coefficient of quartz glass is 5.5×10^{-6} °C^{-1}.

According to Formula (3.35), the temperature sensitivity coefficient of the grating is $k_{T1} = 6.63 \times 10^{-6}$/°C, $k_{T2} = 6.73 \times 10^{-6}$/°C by substituting the above data into the formula.

In the design formula for the temperature compensation structure of the grating sensor, the temperature sensitivity of the grating is determined as the average of the measured values of the test results, i.e.,

$$k_T = 6.68 \times 10^{-6} \times \lambda_B \tag{3.58}$$

3.2.4.2 Test and Analysis of Thermal Expansion Coefficient of Packaging Materials

Considering thermal expansion coefficient and cost, hard aluminum alloy LY-12 and brass H59 are selected as the packaging materials of the sensor. In order to ensure the accuracy of thermal expansion coefficient of structural component and the accuracy of sensor design size, the thermal expansion coefficients of LY-12 hard aluminum alloy and H59 brass should be measured.

Six FBGs (wavelengths are, respectively, 1,527.9, 1,529.1, 1,536.2, 1,537.2, 1,543.2, and 1,546.0 nm) were bonded to the LY12 hard aluminum alloy hollow tube and H59 brass hollow tube with outer diameter of 6 mm and inner diameter of 4 mm. The adhesive is 353ND bicomponent epoxy resin adhesive developed by American

EPOXY Technology Company and used as special optical fiber adhesive. Then, it is cured in a blast drying box under the curing condition of 80 °C for 1 h. The operating temperature range of the epoxy adhesive is from −50 to 200 °C. Temperature is measured accurately using digital Beckmann thermometer. The test is carried out in a high-precision high-low temperature tank from range 10~60 °C for three cycles.

Figure 3.13 is the temperature response curve of the grating on hard aluminum alloy and brass. It can be seen that the FBG wavelength shift bonded to LY12 hard aluminum alloy is 36.3, 36.96 and 35.8 pm/°C, respectively. The wavelength shift of grating stuck in H59 brass pipe is 32.91, 32.61 and 32.1 pm/°C, respectively. The temperature sensitivity according to Formula (3.35) is altered into this following formula.

$$\Delta\lambda_B(\varepsilon, T) = k_\varepsilon \Delta\varepsilon + k_T \Delta T + k_{\varepsilon,T} \Delta T \Delta\varepsilon \qquad (3.59)$$

where $k_T = 6.68 \times 10^{-6} \times \lambda_B$, $k_\varepsilon = K_\varepsilon \times \lambda_B = 0.76318 + 0.03793 \exp(-T/124.3) \times \lambda_B$, $k_{\varepsilon,T} = k_{\varepsilon,T} \times \lambda_B - 3.05 \times 10^{-4} \exp(-T/124.3) \times \lambda_B$.

The temperature is taken as room temperature of 25 °C, and the above data are substituted and calculated as follows:

$$\alpha_{A1} = 21.49482791 \times 10^{-6}/°C, \; \alpha_{A2} = 21.68403 \times 10^{-6}/°C,$$
$$\alpha_{A3} = 21.067507 \times 10^{-6}/°C;$$

$$\alpha_{C1} = 18.56101841 \times 10^{-6}/°C, \; \alpha_{C2} = 18.1939374 \times 10^{-6}/°C,$$
$$\alpha_{C3} = 18.02112094 \times 10^{-6}/°C.$$

Therefore, the thermal expansion coefficient of LY12 duralumin and brass H59 is determined as the average value of the above data, namely: $\alpha_A = 21.41546 \times 10^{-6}/°C$, $\alpha_C = 18.25869 \times 10^{-6}/°C$.

3.2.4.3 Comparison of Strain Sensitivity of FBG Before and After Encapsulation

Before designing the specific structure of the temperature compensation packaging, the strain and temperature sensing characteristics of FBG must be tested, and the

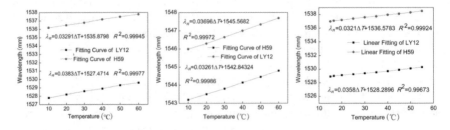

Fig. 3.13 Response curve of grating temperature on hard aluminum alloy and brass

3.2 Temperature Self-Compensated FBG Sensor Based on Thermal Stress

strain sensitivity of the grating must be measured to ensure the accuracy and precision of the parameters in the structure design. Because, compared with bare gratings, the strain sensitivity of the packaged gratings generally decreases to some extent (Jia et al. 2003a; Li et al. 2006). This is because the two-component epoxy resin adhesive used in the package has a creep effect and can absorb a part of strain energy, which leads to the FBG sensitivity decrease. In addition, the shrinkage of epoxy resin adhesive in the curing process will also change the grating pitch to a certain extent. Therefore, it is necessary to test the strain sensitivity of FBG before and after packaging to ensure the accuracy of each parameter in the theoretical formula of sensor structure design that is related to the accuracy of structure design.

FBG used in the test is produced by Beijing Geokon Instrument Co., Ltd. The packaging material is the processed hard aluminum alloy sensor tube 1. The adhesive is the 353ND bicomponent epoxy resin adhesive produced by EPOXY Technology Inc. As a special optical fiber adhesive, it is equipped with excellent adhesive property.

The main test equipment consists of the optical fiber interference signal three-dimensional adjustment instrument, AQ4315A broadband light source generator manufactured by ANDO, AQ6140 FBG multiwavelength analyzer (wavelength dispersive spectrometer) produced by ANDO to demodulate the wavelength of the grating, fiber-optic coupler, WDW-100 microcomputer controlled electronic universal testing machine with the maximum load of 100 kN, and resistance strain gauge. The FBG wavelength demodulation system is composed of a light source generator, a grating multiwavelength meter, and an optical fiber coupler, and is used to record the wavelength variation of the grating. The demodulation system transmits the signal light from the broadband light source to the Bragg grating through the coupler. Due to the wavelength selectivity of FBG, the reflected light wave is transmitted to FBG spectrometer through the coupler, and the reflected wavelength of FBG is analyzed by the spectrometer.

The bare fiber grating strain test: The both ends of the bare fiber grating are pasted on the working platform of the optical fiber interference signal adjuster with 502 ultrafast dry glue, and the tensile stress is applied to the bare grating fiber by three-dimensional adjusting knob to make it generate a certain degree of tensile strength. According to the adhesive length of 155.32 mm, 1% strain corresponds to 1.5532 mm, and the rotation of one lattice is 0.01 mm. The tensile strain can be read directly by the optical fiber interference signal adjuster. The fiber is loaded to a maximum strain of 8,369 $\mu\varepsilon$ and then is fully unloaded. A total of three cycles are performed.

Encapsulated grating strain test: The two gratings with the abovementioned bare grating strain characteristics are integrally stuck in two LY12 hard aluminum alloy hollow tubes with outer diameter of 6 mm and inner diameter of 4 mm, respectively, and then a resistance strain gauge parallel to the bonded grating direction is stuck on the hard aluminum alloy specimen. It is noted that the sticking direction of hard aluminum alloy pipe and strain gauge must be consistent strictly to ensure the same strain. In that tensile load test on the electronic universal test machine, the load cannot exceed the yield strength of hard aluminum alloy specimen, causing the specimen to deform unevenly. Also, the load should not be too small, resulting in relative amplification of the error. The strain of hard aluminum alloy specimen and the center wavelength of grating are recorded synchronously by resistance strain gauge and FBG demodulation system.

Figure 3.14 shows the relationship between wavelength and strain of bare grating. As can be seen from the figure, the strain sensitivity of the two bare gratings is, respectively, 1.19 and 1.16 pm/με, which is in good agreement with the theoretical calculation. In addition, the linearity of fitting curve is excellent, $R_2 = 0.99993$ and $R_2 = 0.99975$, and the strain sensitivity of bare FBG has better reliability.

Figure 3.15 shows the relationship between wavelength and strain of the encapsulated grating. It can be seen from this figure that the strain sensitivity of the grating after encapsulation is reduced to 1.09 and 1.06 pm/με, which is about 91% of that before encapsulation. This indicates that a certain degree of strain loss occurs during the transfer of strain, which reduces the strain sensitivity of the grating. This is mainly due to the thickness of the adhesive layer, the shear modulus of the adhesive, and the nature of the package substrate. The possible reasons for the strain loss of the grating are as follows:

(1) It is difficult to bond the grating to the inner wall of component 1 made of hard aluminum and brass, which may cause the grating to deviate from the axial

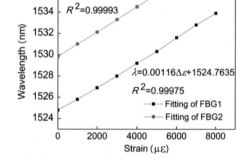

Fig. 3.14 Relation between wavelength of bare grating and applied strain

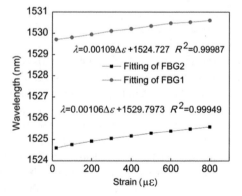

Fig. 3.15 Relation between wavelength and strain after encapsulation

3.2 Temperature Self-Compensated FBG Sensor Based on Thermal Stress

direction of the pipe, or be not closely attached to the inner wall, resulting in slight bending and incomplete strain transfer. It is desired to embed the grating into the capillary tube and the strain loss is smaller.

(2) Due to the inherent defects of slight creep of the adhesive itself, it absorbs part of the strain energy. Therefore, in the design formula of sensor temperature compensation structure, it is necessary to correct k_ε to make it conform to the actual situation.

In the theoretical formula for temperature compensation, it is assumed that the strain transfer efficiency is 100%, i.e., the deformation of component 1 is completely transferred to the grating. The experimental results show that the strain transfer efficiency is only 91%, so when the strain generated by component 1 is $\varepsilon_1' = k_T/0.91k_\varepsilon$, the complete temperature compensation of the grating can be realized.

In summary, the design formula of the temperature compensation structure of the sensor should be modified as follows:

$$\frac{\alpha_1 + \frac{k_T}{0.91 \times k_\varepsilon}}{\alpha_2 - \frac{k_T}{0.91 \times k_\varepsilon} \frac{L_1}{L_2}} = \frac{E_2}{E_1} \frac{S_2}{S_1} \tag{3.60}$$

The theoretical strain sensing formula must also be modified. The modified strain sensing formula is

$$\varepsilon_1 = 0.91 \frac{1 + \frac{L_2}{L_1}}{1 + \frac{L_2}{L_1} \frac{S_1}{S_2} \frac{E_1}{E_2}} \varepsilon_3 \tag{3.61}$$

Since the theoretical strain sensitivity of bare FBG is about 1.2 pm/με, the strain sensitivity formula of encapsulated FBG is as follows:

$$k_\varepsilon = 1.2 \times 0.91 \frac{1 + \frac{L_2}{L_1}}{1 + \frac{L_2}{L_1} \frac{S_1}{S_2}} \text{ pm/με} \tag{3.62}$$

3.2.5 FBG Strain Sensor with Integral Temperature Compensation Structure

In this section, sensor with integral temperature compensation structure is developed first, and its temperature self-compensation function is investigated. Then several sensors of various materials and structures are made and the performance tests are carried out.

3.2.5.1 Structural Design

In order to realize the temperature self-compensation function of grating, based on the idea of temperature compensation of internal thermal stress mechanism of material, the encapsulated structure with double metal package is proposed. In addition, the integral connection of the temperature compensating component 1 and component 2 can greatly improve the connection strength, ensure the efficiency of stress and strain transfer between component 1 and component 2, and solve the problems of bond creep and aging caused by adhesive and thread connection between component 1 and component 2. At the same time, the grating is fixed on the component, which can improve its bonding reliability and antiaging ability (Li 2010).

In order to verify whether the function of temperature compensation can be realized, three kinds of temperature compensation structures of sensor are designed. According to the structural design formula of the sensor, the metal material with greater thermal expansion coefficient is used for structural design. Therefore, the packaging materials of sensors are LY12 hard aluminum alloy and H59 brass which have good processing performance. In order to facilitate that machining, components 1, 2, and 3 are designed in the form of tubes, and the tubes 1, 2, and 3 correspond to components 1, 2, and 3, respectively. The connection between component 1 and component 2 adopts the integral connection method to improve the connection strength and solve the problem of strain loss caused by bonding and screwing between component 1 and component 2. At the same time, the grating is fixed on component 1, which can improve its bonding reliability and antiaging ability. Component 3 and components 1, 2 are made of the hard aluminum alloy LY12.

The dimensions are designed according to the temperature compensation Formula (3.60). Since the materials of components 1, 2 are the same, the formula can be simplified as

$$\frac{\alpha + \frac{k_T}{0.91 \times k_\varepsilon}}{\alpha - \frac{k_T}{0.91 \times k_\varepsilon} \frac{L_1}{L_2}} = \frac{S_2}{S_1} \tag{3.63}$$

where $k_T = K_T \times \lambda_B; k_\varepsilon = K_\varepsilon \times \lambda_B$. $K_T = 6.68 \times 10^{-6}/°C, K_\varepsilon = 0.7941995/°C$. Substituting the data, the design formula of the temperature compensation structure is

$$\frac{21.41546 + 9.24}{21.41546 - 9.24\frac{L_1}{L_2}} = \frac{S_2}{S_1} \tag{3.64}$$

The strain sensitivity formula for the sensor is

$$\varepsilon_1 = 91\% \times \frac{1 + \frac{L_2}{L_1}}{1 + \frac{L_2}{L_1}\frac{S_1}{S_2}} \varepsilon_3 \tag{3.65}$$

In that first type of aluminum sensor (1#), the outside diameter of the pipe 1 is 6 mm, and the inside diameter is 4 mm; Pipe 2 has an outer diameter of 8 mm and an

3.2 Temperature Self-Compensated FBG Sensor Based on Thermal Stress

inner diameter of 4 mm. Then the ratio $S_1:S_2$ of the area of the pipe 1 and the pipe 2 is 5:12, less than 0.5, conforming to the structural design principle of the sensor. The thermal expansion coefficient of hard aluminum alloy LY-12 is 21.41546×10^{-6} °C^{-1}. The ratio $L_1:L_2 = 0.9353194$ of the length of the pipe 1 to the pipe 2 can be calculated from Formula (3.64). Considering the grating length (16 mm) and packaging requirements, taking the length of the pipe 1 as 30 mm, the length of the pipe 2 shall be 32.07 mm. The length of the pipe 2 is increased to improve the strain sensitivity. Since the welding between pipe fitting 3 and the pipe fitting 2 and the pipe fitting 1 may cause the reduced strain transfer efficiency, it is necessary to increase the length of the pipe fitting 2 (40 cm is selected). Finally, it can be seen from Formula (3.65) that the strain sensitivity of the sensor is $1.365\varepsilon_3$, i.e., 1.638 pm/με, which is more than that of bare FBG.

The pipe fitting 3 has a length of 70 mm (the sum of the lengths of the pipe fitting 1 and the pipe fitting 2), an outer diameter of 18 mm, and an inner diameter of 10 mm. To facilitate the clamping of the electronic universal testing machine to the sensor, the length of both ends of the sensor shall be 20 mm each. The cross-sectional area of the pipe fitting 3 is much more than the cross-sectional areas of the pipe fitting 1 and the pipe fitting 2. This ensures that the internal stress of the pipe fitting 1 and the pipe fitting 2 does not affect the pipe fitting 3. The design structure of 1# hard aluminum alloy sensor is as shown in Figs. 3.16 and 3.17.

Meanwhile, to investigate the influence of the same material and different structures on the temperature compensation effect, the following structures are designed:

In that second type of aluminum sensor (2#), the outside diameter of the pipe fitting 1 is 6 mm, and the inside diameter is 4 mm; Pipe fitting 2 has an outer diameter of 10 mm and an inner diameter of 4 mm. Therefore, the ratio of the area of pipe fitting 1 to pipe fitting 2 $S_1:S_2 = 5:21$, and the thermal expansion coefficient of the hard aluminum alloy LY-12 is 21.41546×10^{-6} °C^{-1}. The ratio $L_1:L_2 = 1.5277642$ of the length of the pipe fitting 1 to the pipe fitting 2 can be calculated from Formula (3.64). If the length of the pipe fitting 1 is 30 mm, the length of the pipe fitting 2 shall be 19.6365 mm. To increase the strain sensitivity coefficient, the length of the pipe

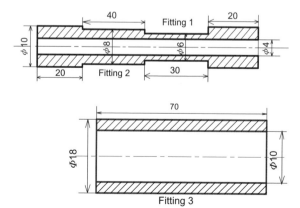

Fig. 3.16 Diagram of internal structure of duralumin sensor 1#

Fig. 3.17 Diagram of external packaged part of duralumin sensor 1#

fitting 2 needs to be increased. At the same time, considering that the pipe fitting 3 and the welding between the pipe fitting 2 and the pipe fitting 1 may reduce the strain transfer efficiency, it is necessary to increase the length of the pipe fitting 2 by 40 mm to ensure the temperature compensation effect. At the same time, to investigate the influence of the same length and different cross-sectional area on the temperature compensation effect, the length of the pipe fitting 2 is specially selected to be 40 mm. The design structure of 2# hard aluminum alloy sensor is shown in Fig. 3.18.

Finally, it can be seen from Formula (3.65) that the strain sensitivity of the sensor is $1.612\varepsilon_3$, i.e., 1.93 pm/με, which is more than that of bare FBG and indicates that the designed can improve the strain sensitivity of FBG strain sensor.

The length of the pipe fitting 3 is 70 mm (the sum of the lengths of pipe fitting 1 and pipe fitting 2); the outer diameter is 18 mm, and the inner diameter is 10 mm. To facilitate the clamping of the sensor by the fixture of the electronic universal tester, the length of both ends is 20 mm. Figure 3.19 is a photograph of the structural parts of the hard aluminum sensor.

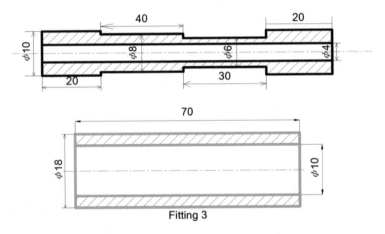

Fig. 3.18 Diagram of external packaged part of duralumin sensor 2#

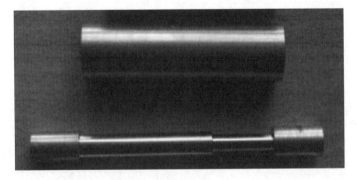

Fig. 3.19 Photo of structural parts for hard aluminum LY12 sensor processing

3.2 Temperature Self-Compensated FBG Sensor Based on Thermal Stress

To investigate the influence of different materials on the temperature compensation effect, a sensor structure with H59 brass as packaging material is designed as follows:

In that third type of aluminum sensor, the outside diameter of the pipe fitting 1 is 6 mm, and the inside diameter is 4 mm; Pipe fitting 2 has an outer diameter of 8 mm and an inner diameter of 4 mm. Therefore, the ratio of the area of pipe fitting 1 to pipe fitting 2 $S_1:S_2 = 5:12$, and the thermal expansion coefficient of the brass H59 is 18.25869×10^{-6} °C^{-1}. The ratio $L_1:L_2 = 1.5277642$ of the length of the pipe fitting 1 to the pipe fitting 2 can be calculated from Formula (3.64). If the length of the pipe fitting 1 is 25 mm, the length of the pipe fitting 2 shall be 33.966 mm. To increase the strain sensitivity coefficient, the length of the pipe fitting 2 needs to be elongated. At the same time, considering that the pipe fitting 3 and the welding between the pipe fitting 2 and the pipe fitting 1 may cause the problem of reducing the strain transfer efficiency, it is necessary to increase the length of the pipe fitting 2 by 54 cm to ensure the temperature compensation effect (as shown in Fig. 3.20). Finally, it can be seen from Formula (3.61) that the strain sensing characteristic of the sensor is $1.497\varepsilon_3$, i.e., 1.796 pm/µε, which is more than that of bare FBG and indicates that the designed can improve the strain sensitivity of FBG strain sensor.

The length of the pipe fitting 3 is 70 mm (the sum of the lengths of pipe fitting 1 and pipe fitting 2); the outer diameter is 18 mm, and the inner diameter is 10 mm. To facilitate the clamping of the sensor by the fixture of the electronic universal tester, the length of both ends is 20 mm. All pipe fittings are completed on precision mechanical lathes and drills.

The pipe fitting 1 and the pipe fitting 2 are integrally connected to avoid the large strain loss caused by adhesive and screw connection between the pipe fitting 1 and the pipe fitting 2, thereby improving the connection strength. At the same time, the whole grating is fixed on the material, which can improve its bonding reliability and antiaging ability.

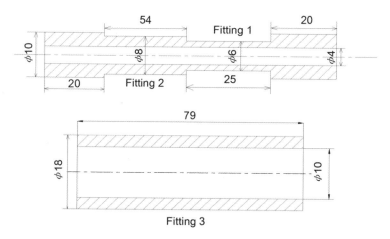

Fig. 3.20 Structure diagram of brass sensor

The integral structure of pipe fitting 1 and pipe fitting 2 is welded together with pipe fitting 3 by argon arc welding. To reduce welding residual stress and deformation of welding structure, welding process which can reduce welding heat input should be selected as much as possible. This welding method solves the problem of large strain loss caused by adhesive and screw connection by improving the connection strength and ensures that the structural strain of the outer pipe 3 can be faithfully transferred to the pipe fitting 1.

The structural parameters of three sensors are shown in Table 3.1.

3.2.5.2 Manufacturing Process

The adhesive is the 353ND bicomponent epoxy resin adhesive produced by EPOXY Technology Inc, ideal optical fiber adhesive. The main characteristics are: elevated temperature resistance, maximum design operating temperature of 200 °C, high strength, tensile strength up to 70 MPa, shear strength up to 20 MPa, high antiaging performance and long-term stability, high corrosion resistance and excellent resistance to a variety of solvent and chemicals, and good fluidity. The resin was approved for use in the space flight program, according to tests by NASA. The steps of the sensor packaging are as follows.

1. Preparations before Packaging

To improve the bonding strength and bonding reliability of the grating, the inner wall of the pipe fitting 1 needs to be carefully polished with a small file and sandpaper, then the inner wall is cleaned with anhydrous ethanol and anhydrous acetone, and the factors affecting the bonding strength like oil stain are removed. The pipe fittings 2 and 3 are also carefully scrubbed with absolute ethanol and acetone to remove oil stain.

2. Packaging of the Grating

The bonding position of grating in the process of pipe connection is easily moved due to the long curing time of 353ND. Therefore, the position of grating should be fixed first. And then pull the grating straight, make it close to the inner wall of the pipe fitting 1 as far as possible. Fix both ends of the grating with 502 quick-drying glue. The glue should not be applied too much to avoid affecting the contact between 353ND glue and the grating. The 502 glue can be used only for temporary fixing due to the poor long-term stability, so the bonding effect of the grating and the pipe fitting 1 depends on the bonding of the 353ND adhesive.

Carefully pour 353ND glue into the pipe fitting 1. Glue should be instilled in moderation. Too little glue cannot guarantee the firm bonding between the grating and the pipe fitting 1, while too much glue will increase its creep effect due to too thick glue layer, and may also lead to changes in the elastic modulus of the pipe fitting 1, affecting the temperature compensation effect.

3.2 Temperature Self-Compensated FBG Sensor Based on Thermal Stress

Table 3.1 Structural parameters for three kinds of sensors

Packaging materials	L_1	L_2	L_3	Φ_1		Φ_2		Φ_3		Strain sensitivity (pm/με)	Temperature compensation strain (με)
				Inner diameter	Outer diameter	Inner diameter	Outer diameter	Inner diameter	Outer diameter		
Duralumin 1#	30	40	70	4	6	4	8	10	18	1.638	10.71
Duralumin 2#	30	40	71	4	6	4	10	10	18	1.934	16.51
Brass	25	54	80	4	6	4	8	10	18	1.796	12.11

Note Subscripts 1, 2, and 3, respectively, refer to structural parameters of the tubes 1, 2, and 3. The unit of structural parameter is mm

3. Connection and Packaging of the Sensor

The sensor shown in Fig. 3.21 was placed in a constant temperature blast drying box and cured at 80 °C for 1 h. The optical fiber cleaver and fusion splicer are used to connect both ends of the optical fiber with single-mode jumper, so that the grating can be easily connected to the wavelength demodulation system and the optical fiber connection can be protected by shrinkable tube. For the test convenience, the bare optical fiber can be directly connected with the demodulation system by using the optical fiber adapter temporarily.

3.2.5.3 Sensing Characteristics

1. Optical Fiber Sensor Based on Hard Aluminum Alloy Shell
(1) Test Results of 1# Hard Aluminum Alloy Sensor

Figure 3.22 is the initial spectrum of the grating without loading, while Fig. 3.23 is the spectrum of the grating loaded to 12 kN. The grating wavelength is shifted from 1,544.535 to 1,545.683 nm, and the strain correspondingly generated on the sensor is about 790 $\mu\varepsilon$, so the strain sensitivity is approximately 1.47 pm/$\mu\varepsilon$.

Figure 3.24 is the strain sensing characteristic curve of the sensor. The fitting curve of grating wavelength and strain is linear, the equation is: $\lambda = 0.00151\Delta\varepsilon + 1544.5174$, and the fitting degree is higher (0.99444). Therefore, the strain sensitivity of the sensor is 0.00151 nm/$\mu\varepsilon$, i.e., 1.51 pm/$\mu\varepsilon$, which is in good agreement with the theoretical calculated value 1.638 pm/$\mu\varepsilon$. The strain sensitivity decreases to some extent, and the strain transmission efficiency is 92.2%. Since the welding of the sensor also reduces the strain transmission efficiency, it is necessary to revise the structural design formula of the temperature compensation theory, that is,

$$\frac{\alpha_A + 9.24/0.922}{\alpha_A - 9.24/0.922\frac{L_1}{L_2}} = \frac{\alpha_A + 10.02}{\alpha_A - 10.02\frac{L_1}{L_2}} = \frac{S_2}{S_1} \quad (3.66)$$

Fig. 3.21 Photo of hard aluminum LY12 sensor

3.2 Temperature Self-Compensated FBG Sensor Based on Thermal Stress

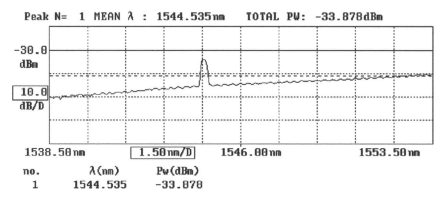

Fig. 3.22 Reflected spectrum of grating loaded by 0.425 kN

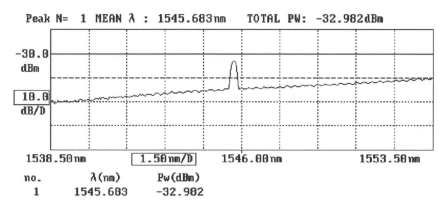

Fig. 3.23 Reflected spectrum of grating loaded by 12 kN

Fig. 3.24 Strain response of 1# FBG sensor made from LY12

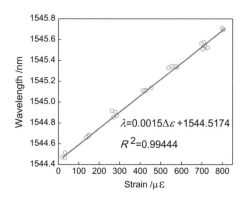

The meaning of the above Formula (3.66) is that the theoretical strain of the grating is 10.02 μɛ when the relationship between L and S conforms to the formula. If the structure (L, S) is changed, the theoretical strain to which the grating is subjected

will be changed, assuming that the theoretical strain is x, the following relationship is satisfied:

$$\frac{\alpha_A + x}{\alpha_A - x\frac{L_1}{L_2}} = \frac{S_2}{S_1} \quad (3.67)$$

Figure 3.25 is the temperature test data curve of the sensor. It can be seen that the curve fitting equation of external tube grating temperature characteristic is $\lambda = 0.03689\Delta T + 1,545.56346$, and the thermal expansion coefficient of hard aluminum alloy LY-12 is 21.41546×10^{-6} K^{-1}. The fitting equation of temperature sensitivity curve of inner tube grating is: $\lambda = 0.0324 \Delta T + 1,543.45964$. The temperature sensitivity of the outer tube grating is 36.89 pm/°C, and the temperature sensitivity of the inner tube is 32.4 pm/°C. Qualitatively, the temperature sensitivity of the sensor is less than the temperature sensitivity of the outer tube, so the sensor has the temperature compensation effect.

In this structure design, the sensor made of hard aluminum alloy has $L_1/L_2 = 3/4$, $S_1/S_2 = 5/12$, which are substituted into the above formula, obtaining:

$$x = -1/2\alpha_A \quad (3.68)$$

The thermal expansion coefficient of the alloy is about 21.41546×10^{-6} °C^{-1}, so $x = 10.71$ με. The grating can be fully temperature compensated as long as x reaches 10.02 με, so there is about 0.7 με overcompensation, that is, the overcompensation is about 0.83 pm/°C. The temperature compensation effect of this experiment is 32.4 pm. According to the calculating formula $-1.51 \times 21.41546 = -0.06$ pm/°C, there is an overcompensation of 0.06 pm/°C, which is basically in agreement with the theoretical calculation.

From another point of view, it is verified whether the structure requirements of the grating are consistent with the actual structure. According to the temperature test, the coefficient of thermal expansion of aluminum alloy is 21.41546×10^{-6} °C^{-1}. When the temperature compensation effect of grating is -0.06 pm/°C, the strain of grating

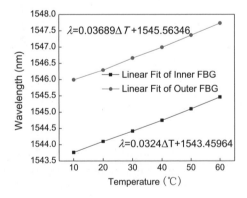

Fig. 3.25 Temperature response of 1# FBG sensor made from LY12

3.2 Temperature Self-Compensated FBG Sensor Based on Thermal Stress

is $-0.06/1.1 + 10.02\,\mu\varepsilon = 10.07\,\mu\varepsilon$. The ratio of the length of the pipe fitting 1 to the length of the pipe fitting 2 obtained by substituting them into the above formula is 0.82 while keeping the cross-sectional area of the two inner pipes constant, which is slightly different from the actual size ratio $3:4 = 0.75$.

Figure 3.26 is the temperature compensation effect curve of the sensor. As can be seen from the figure, the fitting equation of bare grating temperature sensitivity is $\lambda = 0.01156\Delta T + 1543.7566$, and temperature sensitivity is 11.6 pm. The fitting equation of temperature sensitivity is $\lambda = 0.00049178\Delta T + 1543.8067$. After data processing, the temperature sensitivity is 0.49 pm/°C, which is reduced to 5.8% of bare grating. The temperature compensation effect is much better. From the test data of three cycles, the repeatability and reliability of the sensor are also better.

(2) Analysis Results of 2# Hard Aluminum Alloy Sensor

Figure 3.27 is the strain sensing characteristic curve of 2# hard aluminum alloy sensor. In that figure, the fit curve of wavelength and strain is linear, and the equation is $\lambda = 0.00175\Delta\varepsilon + 1,528.45271$ with high fitting degree (0.99994). The strain sensitivity of the sensor is 0.00175 nm/$\mu\varepsilon$, i.e., 1.75 pm/$\mu\varepsilon$, which is consistent with the theoretical calculating value 1.934 pm/$\mu\varepsilon$, and the strain sensitivity decreases to some extent. Therefore, the deformation of the structure can be transmitted to the grating with a strain transfer efficiency of 90.4%, and the strain loss rate is less.

Since the welding of the sensor also reduces the strain transfer efficiency, it is necessary to revise the structural design formula of the temperature compensation theory (see Formula 3.69).

$$\frac{\alpha_A + 9.24/0.904}{\alpha_A - 9.24/0.904\frac{L_1}{L_2}} = \frac{\alpha_A + 10.22}{\alpha_A - 10.22\frac{L_1}{L_2}} = \frac{S_2}{S_1} \quad (3.69)$$

Figure 3.28 is the temperature response curve of the sensor. The fitting equation of the temperature characteristic of the inner tube grating is: $\lambda = 0.02926\Delta T + 1,527.7419$, while the temperature sensitivity of the outer tube grating is 36.89 pm/°C and the temperature sensitivity of the inner tube is 32.4 pm/°C.

Fig. 3.26 Processed temperature response of 1# FBG sensor made from LY12

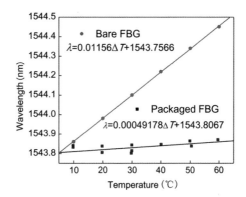

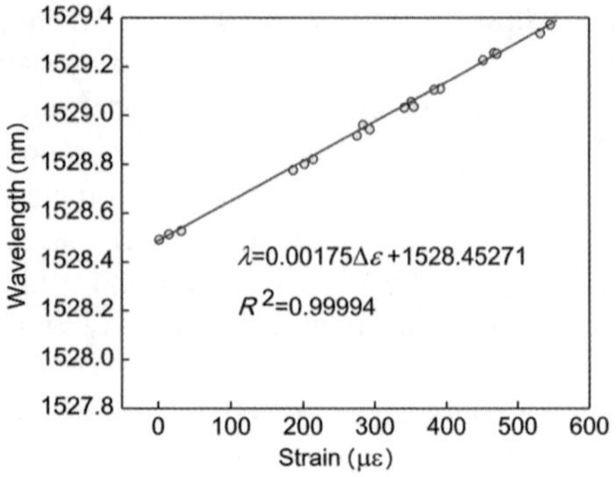

Fig. 3.27 Characteristic curve of strain sensing of hard aluminum 2# sensor

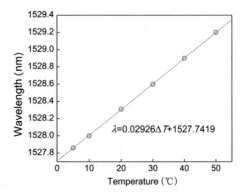

Fig. 3.28 Curve of temperature compensation test for hard aluminum 2# sensor

The coefficient of thermal expansion of hard aluminum alloy LY-12 is known to be $21.41546 \times 10^{-6}\,°C^{-1}$. From the quantitative analysis, the meaning of Formula (3.69) is: When the relationship between L and S conforms to the formula, the theoretical strain of the grating is 10.22 µε. If the structure (L, S) is changed, the theoretical strain of the grating will be changed, and the theoretical strain of x is satisfied as follows:

$$\frac{\alpha_A + x}{\alpha_A - x\frac{L_1}{L_2}} = \frac{S_2}{S_1} \tag{3.70}$$

The sensor made of hard aluminum alloy has $L_1/L_2 = 3/4$, $S_1/S_2 = 5/21$, which are substituted into the above formula, obtaining:

$$x = -0.771\alpha_T \tag{3.71}$$

From the temperature characteristics, the thermal expansion coefficient of the alloy is about $21.41546 \times 10^{-6}\,°C^{-1}$, so $x = 16.51$ µε. The grating can be fully temperature compensated as long as x reaches 10.22 µε, so there has about 6.3 µε overcompensation, that is, the overcompensation is about 7.6 pm/°C. The temper-

3.2 Temperature Self-Compensated FBG Sensor Based on Thermal Stress

ature compensation effect of this experiment is 29.26 pm $-1.75 \times 21.41546 = -8.21$ pm/°C, with 8.21 pm/°C overcompensation, which is in good agreement with the theoretical calculation value of 7.6 pm/°C.

From another point of view, it is verified whether the sensor dimensions calculated by the temperature-compensated data are consistent with the actual structure. For example, according to the temperature test, the coefficient of thermal expansion of aluminum alloy is 21.4146×10^{-6} °C^{-1}. When the temperature compensation effect of grating is -8.21 pm/°C, the strain of grating is $-8.2/1.1 + 10.22$ με $= 17.67$ με. The ratio of the length of the pipe fitting 1 to the length of the pipe fitting 2 obtained by substituting the above Formula (3.70) is 0.69 while keeping the cross-sectional area of the two inner pipes constant, which is slightly different from the size ratio $3:4 = 0.75$ (30:40 mm) of the actual structure.

Figure 3.29 is the temperature compensation effect curve of the sensor, the temperature compensation effect is -8.21 pm/°C, and the curve fitting degree is 0.98908. From the test data of three cycles, the repeatability of the sensor is better, and the reliability of the sensor and the reliability are higher. Therefore, the proposed temperature compensation mechanism and structural design formula are correct and feasible from the analysis of the test results of sensors with the same material and different structures.

2. Optical Fiber Sensor Based on Brass Outer Shell

Figure 3.30 is the strain sensing characteristic curve of the brass sensor. The fitting curve of FBG wavelength and strain test data is linear. The equation is: $\lambda = 0.00164\Delta\varepsilon + 1{,}532.9304$, and the fitting degree is high (0.99308). The strain sensitivity of the sensor is 0.00164 nm/με, i.e., 1.64 pm/με, which is in good agreement with the theoretical calculated value 1.796 pm/με. The strain sensitivity decreases to some extent. The deformation of the structure can be transmitted to the grating with a strain transfer efficiency of 91.3% and the strain loss rate is lower.

Since the strain transmission efficiency of the sensor is reduced to some extent, it is necessary to revise the structural design formula of the temperature compensation theory, that is,

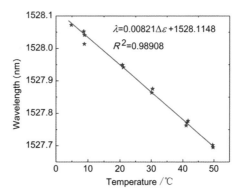

Fig. 3.29 Characteristic curve of strain sensing of hard aluminum 2# sensor

Fig. 3.30 Curve of temperature compensation test for hard aluminum 2# sensor

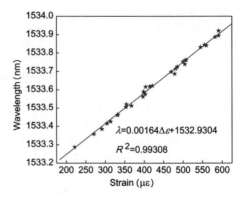

$$\frac{\alpha_A + 9.24/0.913}{\alpha_A - 9.24/0.913\frac{L_1}{L_2}} = \frac{\alpha_A + 10.12}{\alpha_A - 10.12\frac{L_1}{L_2}} = \frac{S_2}{S_1} \qquad (3.72)$$

Figure 3.31 is the temperature test data curve of the sensor. The curve fitting equation of temperature characteristic of sensor grating is: $\lambda = 0.002679\Delta T + 1{,}532.60547$. The coefficient of thermal expansion of brass H59 is known to be $18.25869 \times 10^{-6}\ °C^{-1}$. The temperature sensitivity of the external tube grating is 26.79 pm/°C. Quantitatively, the meaning of Formula (3.72) is: When the relationship between L and S conforms to the formula, the theoretical strain of the grating is 10.12 με. If the structure (L, S) is changed, the theoretical strain of the grating will change, and the theoretical strain of x is satisfied as follows:

$$\frac{\alpha_A + x}{\alpha_A - x\frac{L_1}{L_2}} = \frac{S_2}{S_1} \qquad (3.73)$$

In this design, the sensor made of brass has $L_1/L_2 = 3/4$, $S_1/S_2 = 5/12$, which are substituted into the above formula, obtaining:

Fig. 3.31 Curve of temperature compensation test for brass 59# sensor

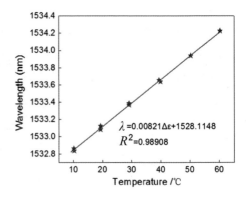

$$x = -0.663\alpha_A \qquad (3.74)$$

The thermal expansion coefficient of the alloy is about 18.25869×10^{-6} °C^{-1}, so $x = 12.11$ με. The grating can be fully temperature compensated as long as x reaches 10.12 με, so there is about 2 με overcompensation, that is, the overcompensation is about 2.4 pm/°C. The temperature compensation effect of this experiment is $26.79 - 1.64 \times 18.25869 = -3.15$ pm/°C, with 3.15 pm/°C overcompensation, which is in good agreement with the theoretical calculation value of 2.4 pm/°C.

From another point of view, it is verified whether the structure requirements of the grating are consistent with the actual structure. According to the temperature test, the coefficient of thermal expansion of brass is 18.25869×10^{-6} °C^{-1}. When the temperature compensation effect of grating is -3.15 pm/°C, the strain of grating is $3.15/1.2 + 10.12 = 12.745$ με. The ratio of the length of the pipe fitting 1 to the length of the pipe fitting 2 obtained by substituting them into Formula (3.73) is 0.42 while keeping the cross-sectional area of the two inner pipes constant, which is slightly different from the actual size ratio $25:54 = 0.46$.

Figure 3.32 shows the temperature compensation effect curve of the sensor. It can be seen from the figure that the temperature sensitivity fitting equation of the sensor is $\lambda = 0.00315\Delta T + 1{,}532.91629$. The temperature sensitivity after data processing is -3.15 pm/°C, that is, there is 3.15 pm/°C overcompensation, which is in agreement with the theoretical calculation value 2.4 pm/°C. The repeatability and reliability of the sensor are better for three cycles.

In a word, the test results of sensors with varied materials and different structures also show that the proposed temperature compensation mechanism and structural design formula are correct and feasible.

3. Secondary Revision of Structural Design Formula for Temperature Compensation

According to the strain sensing characteristic test of the FBG, the strain transfer efficiency of the grating before and after packaging is about 92%, and the strain transfer efficiency of the sensor will decrease further due to welding and machining

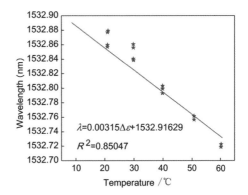

Fig. 3.32 Curve of temperature compensation effect data of brass sensor after processing

error, and the transfer efficiency will decrease to about 91%. Therefore, it is necessary to revise the theoretical formula of temperature compensation and strain sensing of the sensor to make it conform to the actual situation. In the theoretical formula for temperature compensation, it is assumed that the strain transfer efficiency is 100%, and the deformation of the pipe fitting 1 is completely transferred to the grating. According to the experiment, the strain transfer efficiency is only 92% × 91%, so to realize the complete temperature compensation to the grating, the structural design formula of temperature compensation should be revised as follows:

$$\frac{\alpha_{\gamma 1} + \frac{k_\gamma}{0.92 \times 0.91 k_\varepsilon}}{\alpha_{\gamma 2} - \frac{k_\gamma}{0.92 \times 0.91 k_\varepsilon} \frac{L_1}{L_2}} = \frac{E_2}{E_1} \frac{S_2}{S_1} \tag{3.75}$$

Meanwhile, the theoretical strain sensing Formula (3.37) must also be revised. The modified strain sensing formula is

$$\varepsilon_1 = 0.92 \times 0.91 \times \frac{1 + \frac{L_2}{L_1}}{1 + \frac{L_2}{L_1} \frac{S_1}{S_2} \frac{E_1}{E_2}} \varepsilon_3 \tag{3.76}$$

The theoretical strain sensitivity of bare fiber grating is about 1.2 pm/με, so the strain sensitivity formula of packaged FBG is as follows:

$$k_\varepsilon = 0.92 \times 0.91 \times 1.2 \times \frac{1 + \frac{L_2}{L_1}}{1 + \frac{L_2}{L_1} \frac{S_1}{S_2} \frac{E_1}{E_2}} \text{ pm/με} \tag{3.77}$$

3.2.6 Small FBG Strain Sensor

To meet the specific requirements, it is necessary to continue to study the miniaturization technology of fiber grating strain sensor, such as the selection of package material, the design of package structure, the package technology, and so on.

3.2.6.1 Structural Design

According to the theoretical analysis and the test results of the integrated strain sensor with temperature compensation structure, FBG strain sensor can not only realize temperature compensation but also achieve the effect of enhanced strain sensitivity (Li 2010). However, since the size of the sensor is large, when the sensor is embedded in a structure, it may adversely affect the strength of the structure and the like. Especially for cable force monitoring of large cable-stayed bridge, the sensor should be embedded in stay cable for long-term monitoring. The diameter of stay cable is generally 5~15 mm, and the diameter of strain sensor embedded in stay cable cannot exceed a certain limit. According to the above numerical analysis, the maximum

3.2 Temperature Self-Compensated FBG Sensor Based on Thermal Stress

dimension of the sensor cannot exceed 2 mm; otherwise, the tensile strength of the stay cable will be reduced. The capillary tube packaging of FBG can realize the miniaturization of the sensor ensuring that the monitored structure is not affected.

The sensor consists of three capillary tubes with different lengths and sizes shown in Fig. 3.33. The fiber grating is adhered to the inner wall of the capillary tube 1, and the capillary tubes are bonded together using cold welding glue.

Considering the service environment of the sensor, the packaging material is required to have high strength, chemical corrosion resistance, and fatigue resistance. The materials of tubes 1, 2, and 3 are made of austenitic stainless steel 304 (Cr18Ni9), which is widely used in engineering, and have a good fatigue strength (Gang 1992).

According to the current production technology, capillary tubes with a minimum outside diameter of 0.1 mm can be made. However, since the outer diameter of FBG with the coating layer is 0.25 mm, the minimum inner diameter of the capillary tube is 0.31 mm.

The basic principle of temperature compensation is still that the temperature change causes the thermal stress which can press the capillary tube 1 to produce strain. And then a synchronous strain is occurred in fiber grating bonded to the capillary tube 1 to counteract the wavelength drift of the fiber grating due to temperature change. Figure 3.34 is a photograph of a capillary tube for packaging. According to the temperature compensation formula, the specific specifications of the capillary tube are obtained, and the dimensions are as shown in Table 3.2.

Because the diameter of bare optical fiber is 0.125 mm and the diameter of coated optical fiber is 0.25 mm, the inner diameter of capillary 1 is 0.31 mm. If the inner

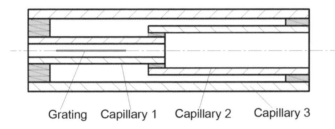

Grating   Capillary 1   Capillary 2   Capillary 3

Fig. 3.33 Structure diagram of temperature self-compensation for miniature grating

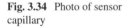

Fig. 3.34 Photo of sensor capillary

Table 3.2 Dimensional parameters of mini-sensor

Structure	Diameter (mm)	Outer diameter (mm)	Wall thickness (mm)	Length (mm)
Capillary 1	0.31	0.55	0.12	30
Capillary 2	0.7	1.05	0.175	60
Capillary 3	1.6	2	0.2	90

diameter is too small, it is difficult for the optical fiber to be inserted into the tube, and it is also difficult to make sufficient contact between the adhesive and the tube wall. If there is weak bonding, the premature failure of the sensor may easily occur. The capillary tube 2 has the wall thickness of 0.175 mm, which is thicker than the capillary tube 1. This is to increase the cross-sectional area of the capillary tube 2 as much as possible to generate a greater compressive thermal stress on the capillary tube 1, so as to achieve temperature compensation and also to help improve the strain sensitivity effect. A certain gap is left between the capillary tube 1 and the capillary tube 2, and between the capillary tube 2 and the capillary tube 3, so as to facilitate sufficient contact between the adhesive agent and the pipe component and improve the bonding strength. The grating length is 8 mm. To prevent stress concentration at both ends of the capillary tube 1 from affecting the strain sensitivity of the grating, the length of the tube 1 should be a little longer, and the length of the capillary tube 1 should be 30 mm. The length of the capillary tube 2 can be calculated from the structural design formula of temperature compensation (3.48). To improve the testing accuracy of the grating and consider the influence of the capillary encapsulating adhesive on the length, the length of the capillary 2 is 60 mm. The capillary tube 3 is required to just have enough length to be packaged into the tube 1 and the tube 2. The ratio of the cross-sectional areas of the capillary tube 2 and the capillary tube 1 is 2.9675:1. According to the revised grating strain sensitivity Formula (3.77), the theoretical strain sensitivity of the sensor is 1.89 pm/$\mu\varepsilon$.

3.2.6.2 Determination of Thermal Expansion Coefficient of Packaging Materials

In general, there are much errors in theory parameters for specific packaging materials. To ensure the accuracy of thermal expansion coefficient and sensor design dimension in the temperature compensation structural design formula, it is necessary to test the thermal expansion coefficient of stainless steel 304.

Four FBGs (wavelengths are 1,537.36, 1,532.94, and 1,528.8 nm, respectively) are pasted into stainless steel tubes with outer diameter of 0.55 mm and inner diameter of 0.31 mm, and then placed in a blast oven for curing under conditions of curing at 80 °C for 1 h. The adhesive used is 353ND bicomponent epoxy resin adhesive. The test is carried out in a high-precision temperature constant temperature tank. The temperature range is 10~60 °C for three cycles.

3.2 Temperature Self-Compensated FBG Sensor Based on Thermal Stress

From Fig. 3.35, the wavelength shift of the grating bonded to stainless steel 304 is 25.41, 25.62, and 25.93 pm °C^{-1}, respectively. The research result of temperature sensitivity coefficient is $\Delta\lambda_B(\varepsilon,T) = k_\varepsilon \Delta\varepsilon + k_T \Delta T + k_{\varepsilon,T} \Delta T \Delta\varepsilon$, where $k_T = 6.68 \times 10^{-6} \times \lambda_B$, $k_\varepsilon = K_\varepsilon \times \lambda_B = 0.76318 + 0.03793 \exp(-T/124.3) \times \lambda_B$, $k_{\varepsilon,T} = K_{\varepsilon,T} \times \lambda_B = -3.05 \times 10^{-4} \exp(-T/124.3) \times \lambda_B$.

The temperature is taken as room temperature of 25 °C, and the above data are substituted into the calculation to obtain: $\alpha_{S1} = 12.4564 \times 10^{-6}$ °C^{-1}, $\alpha_{S2} = 12.6858 \times 10^{-6}$ °C^{-1}, and $\alpha_{S3} = 12.8222 \times 10^{-6}$ °C^{-1}. Take the average of the above data as the coefficient of thermal expansion of stainless steel 304, i.e., $\alpha_S = 12.6548 \times 10^{-6}$ °C^{-1}.

3.2.6.3 Packaging Process

The 353ND bicomponent epoxy resin adhesive with thickness of 0.3~2 mm is used for the bonding of gratings. The adhesive can achieve sufficient bonding between gratings and capillary tubes and improve the bonding strength and reliability. Due to the large gap between tube 1, tube 2 and tube 3, cold welding glue is selected as the ideal bonding material after many tests. The adhesive is a two-component adhesive filler, which is the repairing agent for filling, and is generally used for repairing and maintaining metal materials, ceramic materials, and equipment. Curing time was 5 min at room temperature. The operating temperature of the adhesive is from −50 to 250 °C, and the anti-shearing force is up to 180 kg/mm^2.

The packaging steps of the strain sensor are as follows.

1. Preparations before Packaging

The length of the capillary tube 1 is 30 mm, so the length of FBG packaged in the capillary tube 1 is also 30 mm. The optical grating is marked with an oil pen at both ends with the length of 30 mm, so that it is easy to package. Because the capillary tube is extremely thin, the dispensing equipment must also be miniaturized. After the bicomponent 353ND glue is thoroughly mixed, it is poured into a medical syringe, and the glue is dispensed with a needle. The capillary tube and grating are cleaned with anhydrous ethanol and anhydrous acetone, and oil stains and dirty marks are erased.

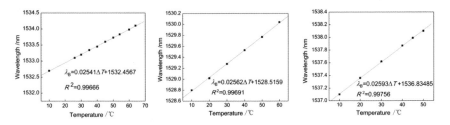

Fig. 3.35 Temperature response curve of grating of stainless steel tube

2. Packaging of the FBG

Lightly glue the grating with a needle within 30 mm of the length marked with an oil pen. The coated grating segment is inserted into the capillary tube 1 and the glue outside the overflow tube is erased. Place it for 1 min until the adhesive is in sufficient contact with the inner wall of the capillary tube, then pull out the grating and continue to apply the adhesive within 30 mm. Repeat the above steps several times to ensure that the glue is fully filled into the capillary tube and the excessive glue on the optical fiber is wiped off. Then, the packaged capillary tube 1 and grating are placed in a temperature constant temperature blast drying box, cured at 80 °C for 1 h, and then removed. The reason why the grating is cured in advance is to prevent the grating from being touched carelessly and change the position of the grating package.

3. Packaging of the Capillary Tube

Use the needle of the syringe to carefully apply a small amount of the mixed cold welding glue to both ends of the capillary tube which need to be bonded. The application length is about 2 mm. Since the curing time of the cold welding glue is short, the packaging must be fast. Finally, at room temperature, the packaged sensor can be cured for 24 h without external interference before performing other operations.

4. Completion of the Sensor

The solidified sensor is adhered to the stainless steel block (200 mm × 80 mm × 10 mm) with cold welding glue in the longitudinal direction for temperature compensation and strain sensing test. Also, the packaging must be fast. The packaged sensor cured for 24 h at room temperature without external interference before performing other operations. Figure 3.36 is a photograph of the packaged micrograting strain sensor with temperature self-compensation function.

3.2.6.4 Performance Evaluation of Strain Sensor

In that specific process of the experiment, the resistance strain gauge is stuck on the surface of the stainless steel specimen along the direction of FBG arrangement to

Fig. 3.36 Photo of encapsulated miniature sensor

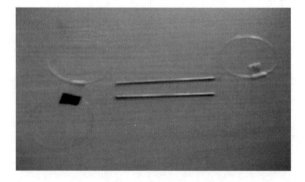

3.2 Temperature Self-Compensated FBG Sensor Based on Thermal Stress 119

sense the strain of the stainless steel plate. The optical fiber cleaver and fusion splicer are used to connect both ends of the optical fiber with single-mode jumper, so that the grating can be easily connected to the wavelength demodulation system and the optical fiber connection can be protected by shrinkable tube. For the convenience of the test, the bare optical fiber can be directly connected with the demodulation system by using the optical fiber adapter temporarily.

Before pasting the sensor on the stainless steel plate, first place it in a high-precision thermostatic bath, set the temperature range of 5~60 °C, and carry out three cycles in total to ensure the accuracy of the data. Clamp both ends of the stainless steel test piece on the anchor of the universal testing machine, and gradually load stress with a force of 3 kN, with a maximum load of 30 kN, for two cycles in total. The strain measured by the resistance strain gauge stuck on the stainless steel plate is compared with the wavelength shift of FBG stuck in the capillary tube, and the actual strain sensitivity of the sensor is measured.

By analyzing the strain characteristic and temperature characteristic of the sensor, the grating wavelength of the three sensors has a good linear relationship with the strain of the structure, as shown in Fig. 3.37. The strain sensitivity is 1.45, 1.49, and 1.39 pm/με, respectively, which is different from the theoretically calculated value of 1.89 pm/με. However, compared with the strain sensitivity of bare grating of 1.2 pm/με, the strain sensitivity is increased to a certain extent, and the strain measurement accuracy of the sensor is improved. At the same time, from the test data of three cycles, the stability and reliability of the sensor are better.

Since the strain transmission efficiency of the sensor is reduced, it is necessary to revise the structural design formula of the temperature compensation theory, that is,

$$\frac{\alpha_A + 11.014}{\alpha_A - 11.014\frac{L_1}{L_2}} = \frac{S_2}{S_1} \tag{3.78}$$

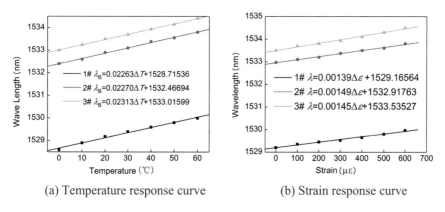

(a) Temperature response curve   (b) Strain response curve

Fig. 3.37 Temperature and strain response curves of grating sensors for stainless steel tube 1#, 2#,

Since the grating cannot fully sense the strain experienced on the pipe fitting 1, the pressure of the pipe fitting 2 on the pipe fitting 1 is not entirely transferred to the grating, so the temperature compensation effect also fails to reach the theoretical effect. Therefore, it is necessary to find a more efficient packaging process and modify the sensor structure to achieve the full compensation effect. In the specific application, it can be improved from the following aspects:

(1) Selection of adhesive agents. The adhesive should have good fluidity and wettability. However, when that capillary tubes are connected to each other, the adhesive needs to be thicker and the curing time to be as short as possible. Under different bonding conditions, different adhesives can be selected to meet different performance requirements of the sensor.
(2) Connection of the capillary tube. In this test, the connection between the capillary tubes can be bonded by resin adhesive. The resin adhesive has some disadvantages such as poor creep resistance, poor aging resistance, and low connection strength to a certain extent. Therefore, the capillary tube can be connected by welding, and it is noted that the conventional arc welding cannot be used, but brazing or laser welding should be adopted to prevent excessive thermal deformation of the capillary tube.
(3) Minimize the outer diameter of the sensor. In line with the current trend of lightweight FRP bars, the outer diameter of the sensor should be designed to be smaller so as not to affect the strength of FRP bars. The capillary tube having small diameter may be used for packaging. In addition, that diameter of the conventional optical fiber at present is 250 µm (including coat layer), which restricts the miniaturization of the capillary tube, and the optical fiber grate with smaller diameter can be used for packaging, so that the outer diameter of the sensor can be greatly reduced.
(4) Protection of tail fiber. Because the sensor will be buried in FRP bars, it is very important to protect the tail fiber. Generally, heat-shrinkable tube plus single-mode jumper is used to deal with it to avoid bare fiber.

3.3 FBG Soil-Pressure Sensor Based on Dual L-Shaped Levers

The variation of stress field is considered as a key parameter to estimate the stable state of surrounding soil and rock. It is crucial in monitoring the soil pressure that can represent the stress field.

Optical fiber sensors are extensively studied for pressure monitoring, for they have a series of obvious advantages compared with conventional sensing techniques, such as immunity to electromagnetic interference, compact size, long distance transmission, multiplexing capabilities, and so on. Among various applications, FBGs are widely used in soil pressure monitoring. Three different forms of FBG-based soil

3.3 FBG Soil-Pressure Sensor Based on Dual L-Shaped Levers

2009), in which FBG was bonded directly, making it easy to cause stress arisen. An FBG soil-pressure sensor based on diamond transmission mechanism was reported (Wang et al. 2013); and Hu et al. (2010) developed an optical fiber grating soil-pressure sensor with temperature compensation, which can transfer soil pressure to the wavelength drifting of FBG through two flanges fixed to the flat diaphragm. The above two gauges were improved in sensitivity and packaging process. However, due to the introduction of special transferring schemes, the thickness of reported sensors increased accordingly. According to reference (Louise 2012), the accuracy of monitoring was affected largely by thickness-to-diameter ratio of the sensor. One way around the problem for Geokon Inc., USA is to add a secondary sensitive part that is separated from the first sensitive part embedded in geotechnical structure. However, the transition part between the first sensitive part and the secondary sensitive part tends to be long.

In this paper, the authors present a new type of FBG soil-pressure sensor based on dual L-shaped levers, making it possible to reduce thickness-to-diameter ratio of the sensor largely. The design, calibration, and field test of the new type soil-pressure sensor are presented in detail. The proposed sensor has recorded a clear change of soil pressure successfully after a rainstorm in Beijing.

3.3.1 Structure and Principle of the Soil-Pressure Sensor

The proposed soil-pressure sensor, made of stainless steel, consists of a circular diaphragm with clamped edges, a flange, and L-shaped levers (Li et al. 2013). The flange, fixed in the center of the circular diaphragm, would transfer the displacement of circular diaphragm. The L-shaped lever is used to transfer longitudinal displacement to horizontal displacement, driving wavelength shift of FBG fixed to the end of L-shaped lever, as shown in Fig. 3.38.

Figure 3.39 is the schematic diagram of proposed soil-pressure sensor. With uniform soil pressure applied, it is easy to figure out the central displacement of clamped circular diaphragm using the small deflection theory, given by

$$\omega = \frac{qR^4}{64D} - \frac{FR^2}{16\pi D} \tag{3.79}$$

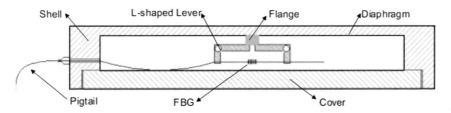

Fig. 3.38 Structure of FBG soil-pressure sensor based on dual L-shaped levers

where F is the force of flange, R is the radius of circular diaphragm, q is the pressure acting on the diaphragm of soil-pressure sensor, and D is the deformation coefficient of circular diaphragm and defined as (Mi and Li 2001)

$$D = \frac{Et^3}{12(1-\mu^2)} \qquad (3.80)$$

where E, t, and μ are the Young's modulus, the thickness, and the Poisson's ratio of circular diaphragm, respectively.

When L-shaped levers rotate, the moment gets balance, and the following relationship can be achieved:

$$\frac{1}{2} F L_2 = T L_1 \qquad (3.81)$$

where T is the axial force of the optical fiber, and L_1 and L_2 are the length of horizontal part and vertical part of the L-shaped lever, respectively.

For optical fiber, T can be described as

$$T = E_f \cdot A_f \cdot \varepsilon \qquad (3.82)$$

and

$$\varepsilon = \frac{\Delta L}{L} \qquad (3.83)$$

where E_f and A_f are the Young's modulus and the cross section area of the optical fiber, respectively, ε is the strain on the FBG, and L and ΔL are the fixed length and its variation of FBG, respectively.

Supposing the L-shape lever rotation, the relationship between ω and ε can be determined by

$$\varepsilon = \frac{\omega L_1}{L L_2} \qquad (3.84)$$

So, combining Eqs. (3.79)~(3.83), we can get

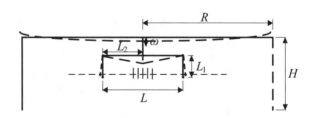

Fig. 3.39 Schematic diagram of proposed FBG soil-pressure sensor

3.3 FBG Soil-Pressure Sensor Based on Dual L-Shaped Levers

$$\varepsilon = \frac{qR^4 \pi L_1 L_2}{64\pi DLL_2^2 + 8E_f A_f R^2 L_1^2} \tag{3.85}$$

The sensitivity of the proposed soil-pressure sensor can be given by

$$S = \frac{\Delta \lambda}{q} = (1 - P_e)\lambda \frac{\pi R^4 L_1 L_2}{64\pi DLL_2^2 + 8E_f A_f R^2 L_1^2} \tag{3.86}$$

where λ and $\Delta \lambda$ are the central wavelength and its shift of FBG, and P_e is the effective photoelastic constant of the optical fiber.

It is easy to find out that the sensitivity of soil-pressure sensor can be adjusted by changing thickness t and radius R of the diaphragm or the proportional relations between L_1 and L_2. Furthermore, another FBG in a free state is installed to realize temperature compensation.

3.3.2 Design and Strength Check of Soil-Pressure Sensor

Considering the practical application, two major factors should be taken into account. One is the variation of stress field caused by installation, and the other one is the variation of stress field caused by the deformation of diaphragm.

Generally speaking, in order to decrease the effect of installation, the thickness-to-diameter ratio should be less than 0.2. Moreover, the matching between the stiffness of the pressure sensor and surrounding soil must be taken into consideration. Here, we introduced a parameter of flexibility factor α, which is given by Clayton and Bica (1995)

$$\alpha = \frac{E_s d^3}{E t^3} \tag{3.87}$$

where E_s and E are the Young's modulus of soil and diaphragm, respectively, and d and t are diameter and thickness of the diaphragm, respectively. According to the accuracy of practical application, it is better to decrease the flexibility factor α.

After comprehensive consideration and dynamic adjustment, the parameters of proposed sensor are determined and as shown in Table 3.3 to make the proposed sensor having a good comprehensive index. And the FBG soil-pressure sensor based on dual L-shaped levers is designed as Fig. 3.40.

The yield strength of stainless steel diaphragm is $\sigma_{0.2} = 440$ MPa, and the permissible stress is $[\sigma] = \sigma_{0.2}/3 = 147$ MPa. For the diaphragm center, tangential stress σ_t is equal to the radial stress σ_r. If it is made that

$$\sigma_r = \sigma_t = \frac{3pR^2}{8t^2}(1+\mu) = 145 \text{ MPa} \tag{3.88}$$

Table 3.3 Parameters of FBG soil-pressure sensor

Parameter	Value
Sensor thickness H (mm)	20
Sensor diameter D (mm)	120
Diaphragm thickness t (mm)	5
Diaphragm radium R (mm)	100
Length of horizontal part L_1 (mm)	5
Length of vertical part L_2 (mm)	10
Bonding length of FBG L (mm)	21
Young's modulus of fiber E_f (GPa)	72
Cross section area A_f (mm^2)	0.0123
Young's modulus of diaphragm E (GPa)	203
Central wavelength of FBG λ (nm)	1,560
Effective photoelastic constant P_e	0.22
Poisson's ratio of diaphragm μ	0.30

Fig. 3.40 Photo of FBG soil-pressure sensor based on dual L-shaped levers

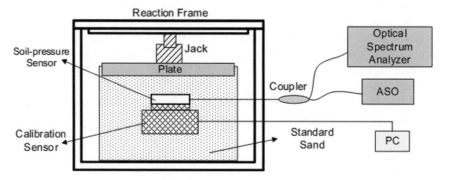

Fig. 3.41 Schematic diagram of laboratory calibration test

We can obtain $p = 2.97$ MPa. According to this value, we choose 2 MPa as the measurement range first. During deformation, the edge position of diaphragm has a maximum shear stress and maximum radial stress, given by

$$\sigma_r = -\frac{3pR^2}{4t^2} = -150 \text{ MPa} \tag{3.89}$$

$$\sigma_t = -\frac{3pR^2}{4t^2}(1-\mu^2) = -136.5 \text{ MPa} \tag{3.90}$$

According to the fourth strength theory, the equivalent stress of diaphragm can be achieved

$$\sigma_{eg} = \sqrt{\sigma_r^2 + \sigma_t^2 - \sigma_r \sigma_t} = -143.7 \text{ MPa} \tag{3.91}$$

which meets the strength requirement. So, we considered 2 MPa as the measurement range finally.

Substituting the corresponding parameters into Eq. (3.86), a sensitivity of 1,217 nm/MPa can be obtained.

3.3.3 Laboratory Calibration Tests

At present, there are three kinds of laboratory calibration methods for soil-pressure sensor, gas pressure calibration, oil pressure calibration, and soil medium calibration, respectively. Among them, the soil medium calibration is the most ideal one, which is closer to the real service environment. Finally, we chose soil medium calibration as the laboratory calibration method, depicted as Fig. 3.32, which consists of a tin used for storage of standard sand, a loading system made up of a jack and a reaction frame, a bearing plate used for transferring the applied load to the soil medium uniformly, and a standard spoke force transducer used as the calibration sensor with an accuracy level of 0.05, and a measurement range of 10~30,000 N. And the FBG interrogator is AQ 6317B produced by ANDO with a wavelength resolution of 1 pm. The setup of laboratory calibration test is shown in Figs. 3.41 and 3.42.

First, preloading is implemented twice. After two cycles of loading and unloading, the measurements are given as Fig. 3.43.

According to the calibration data, we can get a repeatability error of 1.8%, and the formal calibration equation is

$$\lambda = 1.131p + 1558.75 \tag{3.92}$$

with the least-square linearity of 0.7%. The measured sensitivity is about 1.131 nm/MPa, with an error factor of 2.5%. Considering the resolution of 1 pm, we can obtain the pressure resolution of 0.884 kPa. A comparison of comprehensive parameters is as shown in Table 3.4.

Fig. 3.42 Setup for laboratory calibration test

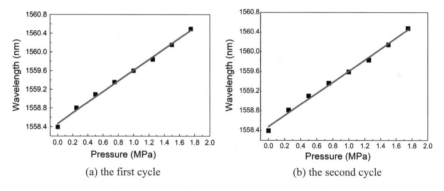

Fig. 3.43 Wavelength shift of soil-pressure sensor

It can be seen that the proposed sensor has an excellent comprehensive index, especially in the aspect of thickness-to-diameter ratio; meanwhile, the range and sensitivity of the sensor, with a good match, are also competitive.

3.3.4 Field Tests

3.3.4.1 Installation of Optical Soil-Pressure Sensor at Earth Dam Site

After laboratory calibration, some sealing measures were taken to ensure that the sensor can serve in a long-term. An artificial earth dam was selected due to its

Table 3.4 Comparison of comprehensive parameters for different technologies for FBG soil pressure sensor

Technology	Sensor thickness (mm)	Sensor diameter (mm)	Sensitivity (nm/MPa)	Range (MPa)
Flat diaphragm 1	35	78	2.02	0.8
Flat diaphragm 2	25	160	0.25	0.6
Planar circular sheet + cantilever 3	25	90	1.819	2
Dual L-shaped levers	20	120	1.131	2

settlement for more than one year. The sensing diaphragm of soil-pressure sensor was perpendicular to the surface of earth dam. And concrete was grouted in the bottom part in order to provide a rigid supporting. This work was completed on May 13, 2012 before the rainy season arrived.

The wavelengths were recorded after field installation of the soil-pressure sensor. The stress of the surrounding soil medium can be calculated by Eq. (3.88).

3.3.4.2 Monitoring Results

The monitoring data were collected using SM-125 interrogator with a resolution of 1 pm produced by MOI Company (Micron Optics International) on May 13, May 20, June 3, June 17, July 1, July 15 and July 29 of 2012. In consideration of these data, the stress variation of the surrounding soil medium can be computed by Eq. (3.88).

The changes of soil pressure with time obtained by the proposed FBG soil-pressure sensor are as shown in Fig. 3.44.

It is found that the variation of soil pressure is small in most time. However, the monitoring results on 29 July changed obviously. This phenomenon indicates that the residual pushing force of the earth dam has increased, and that maybe the earth dam has occurred internal slip after an experience of severe rainstorm on July 22 in Beijing.

In summary, the authors developed an FBG soil-pressure sensor based on dual L-shaped levers. By introducing L-shaped lever, the thickness of sensor is reduced largely, minimizing the stress variation caused by the effect of installation. After laboratory calibration, a field test has been carried out. Monitoring data have been collected for eight times, and it has been recorded for the rainstorm process on July 22 in Beijing successfully. The observation and results imply that the proposed FBG pressure sensor is a reliable technique for soil-pressure monitoring in a harsh environment.

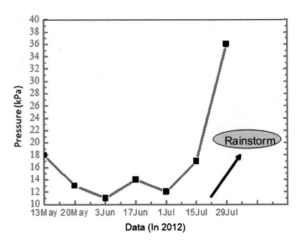

Fig. 3.44 Soil pressure results against time obtained from soil-pressure sensor

3.4 Fiber Bragg Grating Displacement Sensor

With the development of high-speed railway and heavy haul railway, the railway security requirement is higher and higher. Turnout, as an important part of railway infrastructure, with a number of and complex structure, is short in service life and limit the train speed, and because of the low security, it has become a weak link of the track (García Márquez et al. 2003). Eventually, mostly the disease of turnout comes out in the form of incomplete closing of tongue rail. Therefore, it is vitally important to monitor whether the switch is close to the track. At present, monitoring the closing of tongue rail is a heavy labor. The existing complex circuit can only achieve switch signal monitoring (Zhang et al. 2008) (e.g., turnout gap detection).

In addition to the same functions as traditional electric sensors, the fiber grating sensor also has some special characteristics such as distributed sensing, resistant to electromagnetic interference, high precision, and long-term stability. Wei et al. (2002) designed FBG (FBG) sensor. However, its range is small. Therefore, a wide range of FBG sensors are designed in this paper.

3.4.1 Sensor Design

The proposed FBG sensor is designed as shown in Fig. 3.45. The shell of sensor is fixed at the bottom of stock rail, and one end of the inner sleeve is fixed at the bottom of switch rail. When the switch rail moves, it drives the moving of inner sleeve, and the amount of spring compression changed . The equal strength beam is bent due to the restoring force of spring, then the central wavelength of FBG glued on the equal strength beam is changed (Qiu et al. 2006). Through measuring

3.4 Fiber Bragg Grating Displacement Sensor

the central wavelength of the FBG sensor, the displacement of switch point can be detected.

The proposed FBG is designed as shown in Fig. 3.46. FBG1 and FBG2 were axially affixed along the upper and lower surface of the equal strength beam. When bending deformation occurs to the equal strength beam, the FBG of affixed along the upper surface of equal strength beam is pressed. Because the temperature coefficient of two FBGs is the same, this approach for testing and demodulation of FBG sensors is proposed. It not only can serve the purpose of temperature compensation but also doubles the sensitivity of FBG sensor (Zhou and Fan 2005).

The proposed equal strength beam structure is as shown in (3.93). The strain of FBG is the same as that on the surface of equal strength beam:

$$\varepsilon = \frac{\sigma}{E} = \frac{6Fl}{b_0 h^2 E} \tag{3.93}$$

where σ, E, l, b_0, and h are the stress, the Young's modulus, the length, the width of fixed end, and the thickness of equal strength beam, respectively; F is the restoring force of spring.

By Hooke's law

$$F = k(x - w_B) \tag{3.94}$$

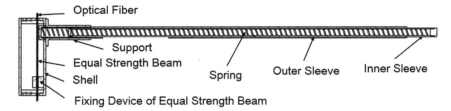

Fig. 3.45 Structure of FBG displacement sensor

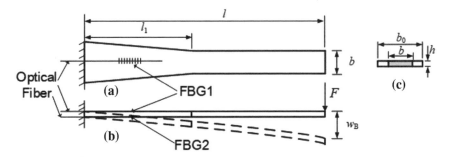

Fig. 3.46 Equal strength beam affixed with FBG: **a** top view, **b** front view, and **c** side view

where k is the elastic coefficient, x is the displacement of the switch rail and w_B is the deflection of the equal strength cantilever beam. The deflection of the equal strength cantilever beam is given by

$$w_B = \frac{6Fll_1^2}{Eb_0h^3} + \frac{4Fl^3}{Ebh^3} - \frac{2Fl_1^2}{Ebh^3}(3l - l_1) \tag{3.95}$$

where l_1 and b are the length of equal strength part and the width of free end of the equal strength cantilever beam, respectively.

The relationship between the shift of the Bragg central wavelength and strain is given by

$$\frac{\Delta\lambda}{\lambda} = (1 - P_e)\varepsilon \tag{3.96}$$

where λ and $\Delta\lambda$ are the central wavelength and its shift of FBG, respectively, and P_e is the effective photoelastic constant of the optical fiber. The sensitivity of the proposed soil-pressure sensor can be given by

$$S = \frac{\Delta\lambda_1 - \Delta\lambda_2}{x} = \frac{12(1 - P_e)\lambda bklh}{Eb_0bh^3 + 6bkll_1^2 + b_0k(4l^3 - 6ll_1^2 + 2l_1^3)} \tag{3.97}$$

where $\Delta\lambda_1$ and $\Delta\lambda_2$ are the shift of the Bragg central wavelength affixed along the upper surface of the equal strength beam and the shift of the Bragg central wavelength which is affixed along the down surface of the equal strength beam.

3.4.2 Tests and Results

The FBG sensor was calibrated on the mobile platform with displacement measurement ranging from 0 to 1,000 mm, the accuracy is 0.1 mm. Figure 3.47 shows the setup of laboratory calibration test, which consists of a mobile platform, an amplified spontaneous emission (ASE) light source and an FBG interrogator that is 86140B and produced by Agilent (America) with a wavelength resolution of 1 pm. The twice measurements are given in Fig. 3.48 (Xu et al. 2013).

According to the calibration data, we can get a repeatability error of 0.3%, and the formal calibration equation is obtained by fitting at least squares principle, given by

$$\Delta\lambda = 60.262\,x + 1967.2 \tag{3.98}$$

The test results show that the formal calibration equation has a good linear degree ($R^2 = 0.999$), and the sensitivity is 60.262 pm/mm. Considering the resolution of 1 pm, we can obtain the displacement resolution of 0.017 mm.

3.4 Fiber Bragg Grating Displacement Sensor

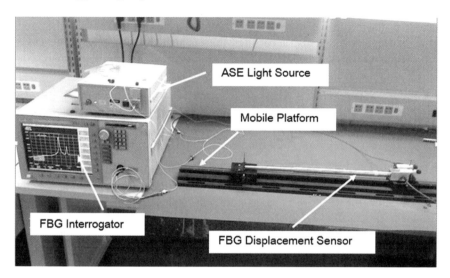

Fig. 3.47 Setup of laboratory calibration test

Fig. 3.48 Wavelength shift of FBG displacement sensor

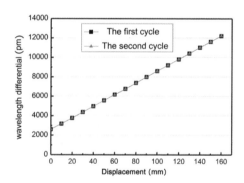

In this section, an FBG displacement sensor is developed. By introducing the spring and equal strength beam, both the sensitivity and the measurement range of the sensor can be improved, and within the range of 160 mm, the sensor accuracy is 0.017 mm. The test results and theoretical calculation agree well, indicating that the proposed sensor can meet the requirement of railway turnout contact monitoring system.

3.5 Fiber Bragg Grating Tilt Sensor

The variation of deep horizontal displacement is important information reflecting the stable state of high-steep slope and is a key index for predicting the occurrence

focused on the monitoring of its deep horizontal displacement. At present, the monitoring of deep horizontal displacement of slope is indirectly measured by tilt sensor. This section will discuss the feasibility of optical fiber tilt sensor in slope monitoring in the view of structure design and sensing characteristics.

3.5.1 Structure Design of Fiber Bragg Grating Tilt Sensor

The optical fiber tilt sensor is the core component of optical fiber inclination measurement device. According to the principle of fiber Bragg grating strain sensing, an optical fiber tilt sensor with temperature self-compensation is designed based on lead hammer structure. By analyzing and comparing the performance parameters of materials and selecting appropriate components, FBG tilt sensor is developed. According to the technical specifications, the measurement accuracy of deep horizontal displacement of slope is 0.7 mm (within 1 m gauge length), and the sensitivity of sensor shall be $\geq 10,000$ pm/rad. The tilt sensor is shown in Fig. 3.49, with an internal structure shown in Fig. 3.50. The optical fiber tilt sensor consists of the upper cover plate, housing, upper bracket, bracket, pressing plate, lower bracket, lead hammer, lower cover plate, etc. For this sensor, the upper cover plate is connected with the house through screws, and a ball bearing is installed between the low bracket and the bracket. The ball bearing can effectively reduce the friction force between the lead hammer part and the bracket. One end of the fiber Bragg grating is fixed on the upper bracket by a mechanical packaging method through a pressing piece, and the other end is fixed on the lower bracket portion of the weight by the same structure and packaging method. If the sensing probe is tilted, the gravity of hammer still keeps the trend of plumb line, and the gravity component of hammer along the optical fiber direction causes the center wavelength of the optical fib grating to drift. By establishing the relationship between the wavelength shift of FBG and the inclination of the sensor, the inclination can be monitored by measuring the wavelength of the FBG. The hammer structure is the core element of the whole sensor probe. In order to improve the sensitivity of the sensor probe, lead is selected as its material. Considering that the lead hammer has low hardness and the connection with stainless steel bearing will produce large friction, the hammer can be divided into two parts: stainless steel lower support and lead hammer, as shown in Figs. 3.51 and 3.52. The lower support of the hammer structure is connected with the bearing through large washer, both of which are of stainless steel structure, so that the friction force can be reduced as much as possible and the hysteresis of the sensor probe can be reduced. The lower support and the lead hammer are fixed by screws to complete the design of the whole hammer structure. In addition, the housing mainly functions as protection of the sensor probe and installation of the connecting components. The upper support plays a role of fixing the FBG and the support has protective effect on the upper support and the hammer structure. Each part is made of 304 stainless steel material with strong corrosion resistance.

3.5 Fiber Bragg Grating Tilt Sensor

Fig. 3.49 FBG tilt sensor

1. Temperature Compensation Design

Since the grating has high temperature sensitivity, the influence of ambient temperature on it cannot be neglected. Therefore, the temperature compensation design of optical fiber inclination sensor is needed. A differential temperature compensation structure (Fig. 3.53) is designed, in which two FBGs are packaged on the sensor probe. When the sensor probe is tilted, the hammer structure still keeps the trend of plumb line. One FBG is pulled and the center wavelength increases. The prestress of another FBG is gradually released, and the center wavelength is decreased. Because the temperature has the same influence on them, the difference between them can eliminate the influence of temperature. This structure not only realizes the self-compensation of temperature but also can double the sensitivity of the sensor.

2. Theoretical Analysis

For optical fibers,

$$\varepsilon = \frac{\sigma}{E} = \frac{F}{\pi R^2 E} = \frac{\Delta l}{l} \quad (3.99)$$

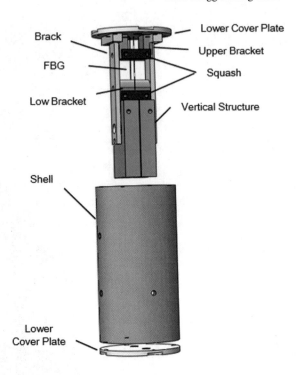

where σ is the strain of the optical fiber; E is the elastic modulus of the optical fiber; F refers to the tensile force of optical fiber; R is the radius of the optical fiber; Δl is the elongation of the optical fiber; l is the length of the stressed optical fiber.

The tensile force F of the optical fiber is provided by the component of the plumb gravity G. When the angle between the plumb weight and the vertical direction is α, the tensile force of the optical fiber is

$$F = nG \sin \alpha = nG \alpha \qquad (3.100)$$

where G is the gravity of the plumb; α is the angle between lead hammer and vertical direction; n is the ratio of the acting radial vectors.

The stretched length of the optical fiber under force is equal to the arc length of the relative rotation angle, as shown in Fig. 3.54. The geometric arc length formula shows that

$$\Delta l = r(\theta - \alpha) \qquad (3.101)$$

where θ is the deflection angle of the angle sensing probe (inclinometer); α is the angle between the plumb hammer and the perpendicular direction when the deflection angle of the sensor probe is θ; r is the effective radius of rotation of the plumb.

As can be seen from the above formula,

3.5 Fiber Bragg Grating Tilt Sensor

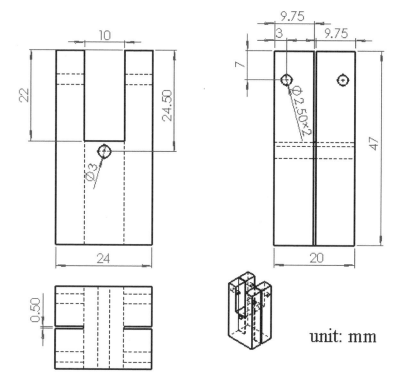

Fig. 3.51 Schematic diagram of lower stent

$$\alpha = \frac{\pi r E R^2}{\pi r E R^2 + \ln G} \theta \quad (3.102)$$

The relationship between the center wavelength and strain of FBG is as follows:

$$\frac{\Delta \lambda}{\lambda} = (1 - P_e)\varepsilon \quad (3.103)$$

In that formula, $\Delta \lambda$ is the variation of the central wavelength of the FBG; λ is the center wavelength of FBG; P_e is the elastic-optical coefficient of optical fiber material

The final sensitivity of the sensor probe is

$$S = \frac{\Delta \lambda}{\theta} = (1 - P_e)\lambda \frac{nGr}{\pi r E R^2 + \ln G} \quad (3.104)$$

It can be arranged as

$$S = \frac{\Delta \lambda}{\theta} = (1 - P_e)\lambda \frac{r}{\frac{\pi r E R^2}{nG} + l} \quad (3.105)$$

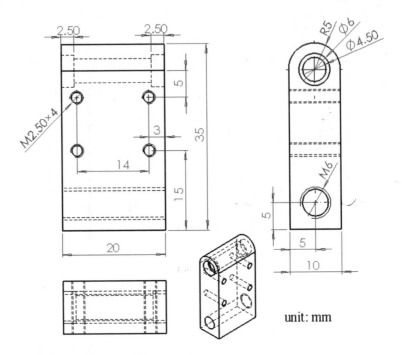

Fig. 3.52 Schematic diagram of plumb part

Table 3.5 Materials selection of sensing probe materials

Name	Parameter	Remarks
Effective length of fiber	20 mm	Quartz
Fiber radius	0.0625 mm	
Fiber elastic modulus	70 GPa	
Plumb density	11.34 g/cm³	Lead
Plumb volume	20 × 20 × 45 mm	
Plumb gravity	2 N	
Sensor probe length	90 mm	Stainless steel

As can be seen from the above formula, the sensitivity of the optical fiber inclination sensor probe is proportional to the effective rotation radius r of the lead hammer structure, and the r value can be changed by changing the design of the lead hammer structure, thereby changing the sensitivity of the optical fiber inclination sensor probe.

The material selection of each element in the fiber tilt sensor is shown in Table 3.5.

3. Packaging Process of Sensor Probe

Optical fiber grating is the main sensing element of optical fiber inclination sensing probe. How to collect the inclination signal safely and accurately is a very important

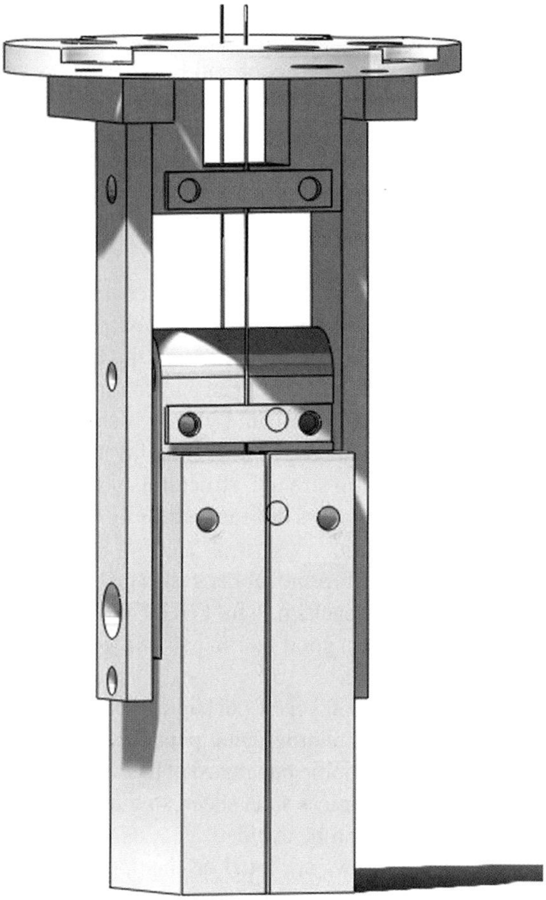

3.5 Fiber Bragg Grating Tilt Sensor

Fig. 3.53 Design of temperature compensation for probe

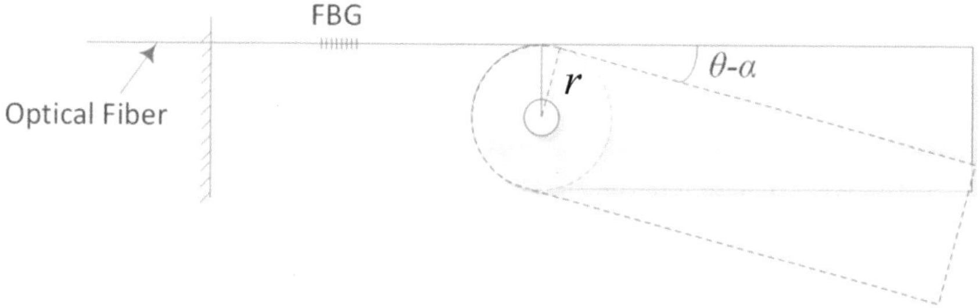

Fig. 3.54 Schematic diagram of optical fiber stretching

problem. In order to ensure the safety and accuracy, the fiber grating needs to be packaged in a protective manner, and the packaging process should also meet the technical requirements that it cannot affect the transmission of the optical signal in the fiber grating.

As the tablet and the upper and lower brackets are made of stainless steel, their fixation will certainly result in gap, which is not conducive to the fixation of fiber grating. At the same time, the rigid fixation of two metal materials can damage the fragile fiber grating and affect the transmission effect of optical signal. In order to protect the fiber grating and ensure that the fiber grating is tightly fixed with the two, use the three-layer packaging process with TS811 structural adhesive, fluorine rubber gasket, and photo paper gaskets.

First, it is necessary to remove the coating layer near the fiber grating and the naked gate part of the fiber grating should be fixed. The fiber inclination sensor probe is subjected to different stress in the test state. The effect of stress will result in the relative sliding of bare gate and coating layer, thus causing test error.

Second, the TS811 structure glue, which is mixed by the quality ratio of 4:1, is coated in the fiber grating and the first layer of FBG is packaged. TS811 structural glue is a kind of high strength structural adhesive, which has the advantages of high strength, good toughness, uniform stress distribution, no thermal effect on parts, and thermal deformation.

Third, paste the fluorine rubber gasket on the coated TS811 structure glue to realize the second layer of packaging for FBG. Fluorine rubber is a kind of synthetic polymer elastic rubber. It has good thermal stability, oxidation resistance, oil resistance, and corrosion resistance.

Fourth, the third layer of packaging for FBG is realized by covering the fluoro rubber gasket with photographic paper gasket. Compared with the fluorine rubber gasket, the photographic paper gasket has a higher hardness, and it is smaller than the hardness of the stainless steel sheet, so that the fiber grating can be better protected and the light loss can be avoided.

Finally, the screws are fixed on the upper bracket to complete the packaging of fiber grating.

4. Packaging Process of Inclinometer

Cable will go through the inclinometer so the shell size of optical fiber inclination sensing probe should be less than the inner diameter of inclinometer joint. However, it is not convenient for the connection of sensing probe and inclinometer joint. Therefore, the external support is designed. The two external supports can be fixed on both sides of the sensor probe shell. The outer diameter is exactly the same as the inner diameter of inclinometer joint, thus improving the convenience of installation. Since it is only encapsulated on both sides of the sensor probe, the cable can go through the other two sides of the sensor probe. This design not only guarantees the packaging effect but also does not affect the cable layout. It has a strong scientific nature.

3.5.2 Sensing Performance of Fiber Bragg Grating Tilt Sensor

According to JJF 1352-2012 *Angular Displacement Sensor Calibration Specification*, the small angle static characteristic parameters (sensitivity, hysteresis, repeatability, linearity, etc.) of the sensing probe were tested by a precise angular displacement stand (see Fig. 3.55).

Zolix KSMG15-65 angular displacement stand and SM125 optical fiber grating demodulator manufactured by MOI Company in the United States are adopted for static property tests at the step length of 0.5° and the full scale of 3° and −3°. A round trip of positive and negative stroke is a circle. According to the requirements of JJF 1352-2012 *Angular Displacement Sensor Calibration Specification*, three cycles are tested totally, and then the data are processed in accordance with the specifications. The basic error of sensor is required to be ≤1%.

Tables 3.6, 3.7, and 3.8 show the test result of FBG tilt sensor. Figure 3.56 is the sensitivity curve of the FBG tilt sensor probe.

1. Range

The maximum value that can be detected by the FBG tilt sensor probe is the upper limit, and the minimum value is the lower limit. The difference between the upper limit and the lower limit is the range.

This FBG tilt sensor probe has a measuring range of −3° to 3° and a range of 6°.

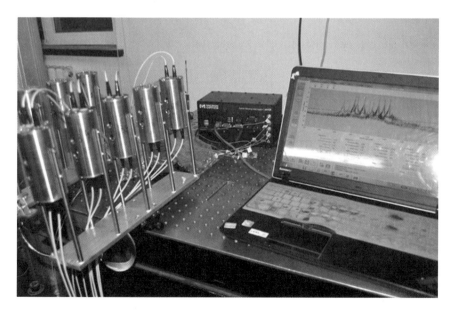

Fig. 3.55 Laboratory test of FBG tilt sensor

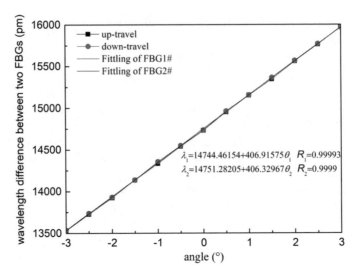

Fig. 3.56 FBG sensor wavelength versus angle

2 Sensitivity

Sensitivity is the output generated from unit input of FBG tilt sensor probe in the working conditions, namely, the variable quantity of grating center wavelength caused by the unit angle input. Sensitivity can be defined as.

Table 3.6 FBG tilt sensor probe of $-3°$ to $3°$ during first cycle

Deflection angle	Forward stroke		Backward stroke	
	λ_1 (nm)	λ_2 (nm)	λ_1 (nm)	λ_2 (nm)
−3	1,536.679	1,550.211	1,536.685	1,550.221
−2.5	1,536.583	1,550.315	1,536.588	1,550.322
−2	1,536.4	1,550.324	1,536.498	1,550.41
−1.5	1,536.297	1,550.437	1,536.382	1,550.524
−1	1,536.202	1,550.536	1,536.216	1,550.574
−0.5	1,536.101	1,550.654	1,536.113	1,550.68
0	1,536.02	1,550.741	1,536.03	1,550.767
0.5	1,535.907	1,550.864	1,535.915	1,550.876
1	1,535.802	1,550.964	1,535.82	1,550.977
1.5	1,535.717	1,551.06	1,535.717	1,551.082
2	1,535.614	1,551.162	1,535.617	1,551.182
2.5	1,535.509	1,551.273	1,535.511	1,551.287
3	1,535.417	1,551.385	1,535.414	1,551.381

3.5 Fiber Bragg Grating Tilt Sensor

Table 3.7 FBG tilt sensor probe of $-3°$ to $3°$ during second cycle

Deflection angle	Forward stroke		Backward stroke	
	λ_1 (nm)	λ_2 (nm)	λ_1 (nm)	λ_2 (nm)
−3	1,536.689	1,550.226	1,536.687	1,550.216
−2.5	1,536.599	1,550.329	1,536.588	1,550.327
−2	1,536.505	1,550.419	1,536.484	1,550.423
−1.5	1,536.395	1,550.527	1,536.409	1,550.536
−1	1,536.289	1,550.635	1,536.302	1,550.647
−0.5	1,536.189	1,550.738	1,536.207	1,550.741
0	1,536.09	1,550.839	1,536.103	1,550.843
0.5	1,535.99	1,550.94	1,535.986	1,550.95
1	1,535.89	1,551.05	1,535.9	1,551.041
1.5	1,535.798	1,551.142	1,535.793	1,551.149
2	1,535.683	1,551.26	1,535.681	1,551.266
2.5	1,535.591	1,551.365	1,535.589	1,551.371
3	1,535.491	1,551.458	1,535.494	1,551.472

Table 3.8 FBG tilt sensor probe of $-3°$ to $3°$ during third cycle

Deflection angle	Forward stroke		Backward stroke	
	λ_1 (nm)	λ_2 (nm)	λ_1 (nm)	λ_2 (nm)
−3	1,536.674	1,550.199	1,536.671	1,550.216
−2.5	1,536.578	1,550.298	1,536.572	1,550.312
−2	1,536.467	1,550.404	1,536.472	1,550.409
−1.5	1,536.361	1,550.504	1,536.379	1,550.523
−1	1,536.276	1,550.601	1,536.263	1,550.628
−0.5	1,536.178	1,550.702	1,536.179	1,550.725
0	1,536.082	1,550.805	1,536.086	1,550.813
0.5	1,535.969	1,550.923	1,535.968	1,550.936
1	1,535.877	1,551.017	1,535.874	1,551.042
1.5	1,535.772	1,551.123	1,535.766	1,551.137
2	1,535.671	1,551.229	1,535.675	1,551.226
2.5	1,535.577	1,551.327	1,535.583	1,551.333
3	1,535.465	1,551.442	1,535.49	1,551.442

$$\lim_{x \to 0} \left(\frac{Y}{X} \right) = \frac{dY}{dX} \qquad (3.106)$$

According to the fitting result of test curve, the average sensitivity of the designed FBG tilt sensor probe in the range of $-3°$ to $3°$ is about 406.6 pm/deg with a linear fitting of 0.9999.

3 Resolution

The minimum variable quantity in the measured value that can be detected by the FBG tilt sensor probe is the resolution.

According to test results, the FBG demodulator commonly used in the current market is adopted with a demodulation accuracy of 1 pm. The designed FBG tilt sensor has a resolution of 8.9″ at an angle range of $-3°$ to $3°$.

4 Linearity

Linearity is an index to measure the performance of linear sensor probe and the least squares linearity is generally adopted:

$$\xi_L = \frac{|\overline{y_i} - y_i|_{max}}{y_{max} - y_{min}} \times 100\% \qquad (3.107)$$

where $\overline{y_i}$ is the arithmetic mean value of the test point; $\overline{y_i}$ is the calculated value of the fitting line, $y_i = a + bx_i$.

The test results are calculated according to the calibration specification of JJF 1352-2012 *Angular Displacement Sensor Calibration Specification*. The linearity of the designed FBG tilt sensor probe in the range of $-3°$ to $3°$ is 0.62%.

5. Hysteresis

For a certain input quantity, the output quantity of the sensor probe in the positive stroke is obviously and regularly different from the output in the same input quantity in the reverse stroke, which is called hysteresis. The hysteresis can be calculated by the maximum difference between the forward stroke and the reverse stroke of the sensor probe and is expressed as the percentage of the full range output:

$$\xi_H = \frac{y_{max}}{y_{max} - y_{min}} \times 100\% \qquad (3.108)$$

where y_{max} is the average value of three cyclic positive and reverse stroke outputs when the displacement is supreme. y_{min} is the average of three cyclic positive and reverse stroke outputs for the displacement to the lower limit.

The test results are calculated according to the requirements of JJF 1352-2012 *Angle Displacement Sensor Calibration Specification*. The designed hysteresis of the FBG tilt sensor probe is 0.86% in the range of $-3°$ to $3°$.

3.5 Fiber Bragg Grating Tilt Sensor

6. Repeatability

According to the test data of three cycles, the output value is measured three times at the ith calibration point from the positive and negative same-direction travels to find the maximum mutual difference among the same-direction travels. Take the point with the largest difference in the same trip for the Δi to represent:

$$\xi_R = \frac{0.61\Delta i}{y_{\max} - y_{\min}} \times 100\% \qquad (3.109)$$

The test results are calculated in accordance with JJF 1352-2012 *Angular Displacement Sensor Calibration Specification*. The repeatability of the designed optical fiber grating inclination sensing probe within $-3°$ to $3°$ is 0.85%.

3.5.3 Indoor Simulation Experiment of Fiber Bragg Grating Tilt Sensor

In order to verify the reliability of the sensor, an indoor simulation experiment was carried out on the installation and test of the sensor and it was compared with the traditional electrical sensor.

Fiber-optic inclinometer mainly consists of an FBG tilt sensor probe, a connecting part, and an inclinometer. The optical FBG tilt sensor probe is the core part of the device. It mainly reflects and characterizes the sliding condition of deep slope through the drift of the central wavelength of the optical fiber grating encapsulated in the fiber-optic tilt sensor probe; the connecting part shall ensure the effective connection of the sensor probe and the inclinometer connecting head. Through reasonable design, the connecting part can quickly realize the connection between the sensor probe and the inclinometer, and ensure the internal seal, so as to facilitate the monitoring of the deep displacement of the slope; the inclinometer is made of ABS materials, which can better couple the inclinometer with the loess layer of the slope and facilitate the displacement transmission. The optical fiber inclinometer is shown in Fig. 3.57.

When the slope slides, the displacement generated in the deep part can cause the inclination (deflection change) of inclinometer, so that the inclinometer generates a deflection angle. The optical fiber inclination sensor packaged in the inclination measuring device can sense the deflection angle, the central wavelength of the optical fiber grating will drift, the demodulation system on the ground can demodulate the drift amount of the central wavelength of the optical fiber inclination sensor and then the inclination angle of the position can be calculated. Through the drift of a series of sensor probe, the displacement–depth curve can be drawn and the judgment of deep sliding of slope is realized.

The model experimental device is shown in Fig. 3.58. Use predesigned support to suspend the whole fiber-optic dip angle inclinometer. Apply a transverse load on it to produce a deflection Angle. The horizontal displacement of the inclined pipe

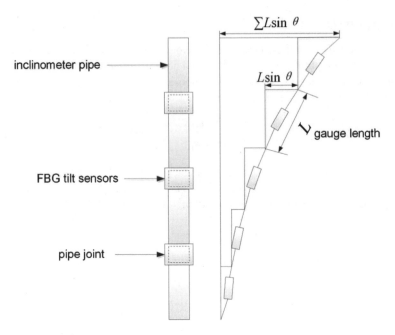

Fig. 3.57 Optical fiber inclinometer and its measuring principle

at the horizontal line is recorded with a dial indicator. The center wavelength drift of the optical fiber sensor probe is demodulated with the demodulator. The voltage signal of the electrical inclination sensor is collected by the optical fiber laser sensor network analyzer. Calibrate the two with dial indicator to obtain the error of the whole fiber dip angle inclinometer and the error of the electrical inclination sensor probe. Based on the comparison between the two errors, the performance of the optical inclinometer is measured and evaluated according to the facts and the indoor simulation of fiber-optic dip angle inclinometer is completed.

In the test as a contrast, the use of electrical tilt sensor probe is CX-7A fixed inclination probe manufactured by Wuhan Qinshen Inclinometer Co., Ltd., as shown in Fig. 3.59. The sensitivity of the fixed inclination probe is 0.075 V/deg, and the nominal total accuracy is 0.09%.

With the optical fiber sensor network analyzer developed by the Institute of Semiconductors of Chinese Academy of Sciences, the voltage signal of the probe is collected. As shown in Fig. 3.60, the collection resolution of the analyzer is 0.0001 V.

The testing process is also specified in accordance with JJF 1352-2012 *Angular Displacement Sensor Calibration Specification*. As shown in Table 3.9, it can be seen that the basic error of FBG tilt sensor is significantly smaller than that of traditional electrical inclination sensor.

3.6 Summary

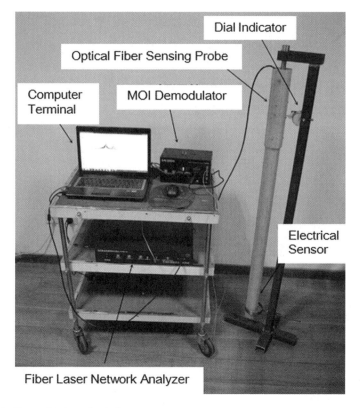

Fig. 3.58 Photo of indoor simulation test

Table 3.9 Comparison of FBG inclination sensors with traditional electrical inclination sensors

Sensor type	Linearity (%)	Delaying (%)	Repeatability (%)
FBG tilt sensor	0.62	0.86	0.85
Wuhan-based depth inclinometer	1.52	1.03	1.38

3.6 Summary

This chapter studies the fiber grating temperature self-compensation strain sensor, fiber grating soil-pressure sensor, fiber grating displacement sensor, and fiber grating tilt sensor. The system analysis and detailed discussion are carried out from sensing principle, structure design, and strain sensing characteristics. The main contents are as follows:

(1) The optical fiber grating temperature compensation mechanism is proposed based on thermal stress of material and the temperature compensation structure is designed. With the integral packaging method, the fiber grating strain sensor

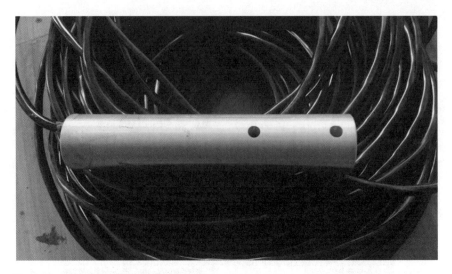

Fig. 3.59 Photo of CX-7A fixed tilt sensor

Fig. 3.60 Fiber sensor network analyzer

with temperature self-compensation is developed independently. A single FBG sensor is used to realizes automatic temperature compensation and increases the strain sensitivity making up the defects of temperature compensation sensor. It has the unique advantages of simple structure, easy realization, and obvious temperature compensation effect.

(2) FBG soil-pressure sensor based on dual L-shaped lever is proposed, which makes it possible to reduce thickness-to-diameter ratio of the sensor largely. The design, calibration, and field test of the new type soil-pressure sensor are presented in detail. The proposed sensor has recorded a clear change of soil pressure successfully after a rainstorm in Beijing.

(3) An FBG displacement sensor is developed. By introducing the spring and equal strength beam, both the sensitivity and the measurement range of the sensor can be improved. Within the range of 160 mm, the sensor accuracy is 0.017 mm. The test results and theoretical calculation agree well, indicating that the proposed sensor can meet the requirement of railway turnout contact monitoring system.

(4) An FBG tilt sensor is designed to measure the slippage of slope. The performance of FBG tilt sensor are verified from theoretical analysis, packaging process, and static sensing performance test and laboratory simulation test. The deep sliding condition of slope is reflected and characterized by measuring angle.

References

Clayton CRI, Bica AVD (1995) The design of diaphragm-type boundary total stress cells. Geotechnique 45(2):349–351. https://doi.org/10.1680/geot.1995.45.2.349

Correia R, Li J, Staines S et al (2009) Fibre Bragg grating based effective soil pressure sensor for geotechnical applications, vol 7503. https://doi.org/10.1117/12.835751

Du YL, Liu CX, Li JZ (2008) Investigation into new optical-fiber-grating strain sensors based on temperature compensation. Chin J Mech Eng 6(1):19–22. https://doi.org/10.15999j.cnki.311926.2008.01.029

Du YL, Sun BC, Li JZ (2009) Fiber Bragg grating temperature self-compensation strain sensor based on thermal stress mechanism. China patent CN2,008,100,796,383, 18 Mar 2009

Gang Y (1992) China stainless steel corrosion handbook. Metallurgical Industry Press, Beijing

García Márquez FP, Schmid F, Conde Collado J (2003) A reliability centered approach to remote condition monitoring. A railway points case study. Reliab Eng Syst Saf 80:33–40. https://doi.org/10.1016/S0951-8320(02)00166-7

Hu Z, Wang Z, Ma Y et al (2010) Soil pressure sensor based on temperature compensation FBG. J Appl Opt 31(1):110–113

Jia ZA, Qiao XG, Fu HW (2003a) Study on temperature sensitivity coefficient of fiber Bragg gratings. J Optoelectron Laser 14(05):454–456. https://doi.org/10.16136/j.joel.2003.05.004

Jia ZA, Qiao XG, Li M et al (2003b) Nonlinear phenomena of fiber Bragg grating temperature sensing. Acta Photo Sin 32(07):844–847

Li J (2010) Study on FBG sensing technology-based fiber reinforced polymer smart stayed cables. Dissertation, Beijing Jiaotong University

Li JZ, Sun BC (2015) Theory analysis of novel fiber Bragg grating temperature compensated method based on thermal stress. High Power Laser Part Beams 27(2):84–90. https://doi.org/10.11884/HPLPB201527.024115

Li B, Fu Y, Wei H et al (2006) Long-term stabilization of fiber gratings by high-temperature packaging. J Optoelectron Laser 17(07):798–802

Li F, Du Y, Zhang W et al (2013) Fiber Bragg grating soil-pressure sensor based on dual L-shaped levers. Opt Eng 1:6. SPIE. https://doi.org/10.1117/1.oe.52.1.014403

Li JZ, Sun BC, Du YL (2014) Union self-compensated packaging of FBG strain sensor. Optoelectron Lett 10(1):30–33. https://doi.org/10.1007/s11801-014-3179-7

Louise W (2012) The use of a novel miniature strain device to investigate the response of tendon cells to in vitro tensile strain. Dissertation, University of Sheffield

Ma W, Shi W, Fu H et al (2001) Progress in the study of fiber Bragg grating sensors. Stud Opt Commun 106(04):58–62. https://doi.org/10.13756/j.gtxyj.2001.04.016

Meiarashi S, Nishizaki I, Kishima T (2002) Life-cycle cost of all-composite suspension bridge. J compos constr 6(4):206–214. https://doi.org/10.1061/(ASCE)1090-0268(2002)6:4(206)

Mi H, Li C (2001) Elastic mechanics. Chongqing University Press, Chongqing

Qiu W, Jiang Q, Sui Q (2006) An experiment of simultaneous displacement and temperature sensing with FBG. Opt Fiber Electric Cable Appl ThE90. https://doi.org/10.1364/ofs.2006.the90

Wang J, Jiang D, Liang Y et al (2007) A differential optical fiber grating pressure cell and the temperature characteristic. J Optoelect Laser 18(04):390–391

Wang Z, Sui Q, Wang J et al (2013) A highly sensitive fiber grating soil pressure gauge

Wei ZB, Liu Y, Wang CT et al (2002) Application of fiber displacement sensor in railway system. Instr Tech Sens (04):4–5

Xu HB, Zhang WT, Du YL et al (2013) Fiber optic displacement sensor used in railway turnout contact monitoring system. In: Asia Pacific optical sensors conference 2013, 15 Oct 2013. SPIE, p 4. https://doi.org/10.1117/12.2031258

Zhang X, Wu Z, Zhang B et al (2005a) Experimental investigation on temperature sensitivity of fiber Bragg gratings. Opt Tech 31(04):497–499. https://doi.org/10.13741/j.cnki.11-1879/o4.2005.04.006

Zhang X, Wu Z, Zhang B et al (2005b) Experimental study on cross-sensitivity of temperature and strain of fiber optic Bragg gratings. J Optoelectron Laser 16(5):566–569

Zhang H, Sun Y, Hu Y (2008) Research on turnout condition monitoring system. Railw Sig Comm 44(11):7–9. https://doi.org/10.13879/j.issn1000-7458.2008.11.004

Zhou X, Fan B (2005) Dynamic demodulation technology of fiber grating with high resolution based on cantilever beam. J Trans Technol 24(04):54–56

Zhou Z, Wang H, Ou J (2006) A new kind of FBG-based soil-pressure sensor. In: Optical fiber sensors, Cancun, 2006/10/23. OSA technical digest (CD). Optical Society of America, p ThE90. https://doi.org/10.1364/ofs.2006.the90

Chapter 4
Fiber Laser Sensor

With the continuous development of photoelectron technology, based on the traditional passive FBG (FBG) sensor, a new generation of sensor with distributed feedback (DFB) fiber laser as sensing element has emerged and becomes a hot topic in the field of optical fiber sensing. Fiber laser has the advantages of common FBG sensor like simple structure, strong anti-electromagnetic interference, small size, and easy to set up sensor network by WDM multiplexing. It also has the unique advantages of single frequency, narrow line width, high power, and ultra-low noise. Combined with high-resolution wavelength demodulation technology, it can realize high sensitivity signal detection. It has an incomparable advantage in weak signal detection and has broad application prospects in oil exploration, earthquake prediction, and safety monitoring. This chapter will introduce the basic principle, structure, and sensing characteristics of acoustic emission receiver based on DFB and DFB fiber laser accelerometers.

4.1 Acoustic Emission Receiver Based on DFB

Nowadays, acoustic emission (AE) detection has been widely used for the nondestructive testing of concrete structures, such as bridges and viaducts or masonry historical buildings (Carpinteri et al. 2007; Tan et al. 2009). This method can predict the damage evolution and the time to structural collapse. Most traditional AE sensors utilized consist of piezoelectric elements undergoing transduction, which are commercially available (Nair and Cai 2010). But these conventional sensors are neither capable of nor flexible for making the appropriate measurements in the case of electromagnetic interference environment or remote monitoring applications (Kazuro et al. 2005). A number of fiber optic acoustic emission sensors suitable for these applications were developed in the past 30 years. The majority of these sensors such as Mach–Zahnder (Matsuo et al. 2006; Liang et al. 2009), Michelson (Tsuda et al. 1999), Sagnac (Qin et al. 2010), or Fabry–Perot (Finkel et al. 2001; Liang et al. 2008) detectors are based on optical fiber interferometry. These sensors have considerable advantages over conventional piezoelectric sensors such as immunity to electromagnetic interference, feasibility for long-term and long-distance structural condition monitoring, but they are pretty large in size and are difficult to multiplex.

In the past 10 years, FBGs have been successfully used for AE sensing applications (Ian et al. 2002; Baldwin and Vizzini 2003; Takeda et al. 2003; Lee and Tsuda 2005; Minardo et al. 2005; Wild and Hinckley 2007; Tsuda et al. 2010; Jiang et al. 2012). FBGs offer distinct advantages for remote sensing such as ease of multiplexing, and simultaneous measurement of several parameters such as temperature and strain. FBGs are small in size and light in weight and can be embedded into material structure. Above all, their directional responses make them suitable for determining principal strains and the direction of propagation of acoustic waves (Thursby et al. 2008). However, in some cases, due to the limited sensitivity, FBGs cannot detect the ultra slight waves of the structure, and higher strain sensitivity is necessary. Recently, the new generation of optical fiber sensor based on distributed feedback (DFB) fiber laser has received considerable research interests in this field (Hill et al. 2005). The DFB fiber laser sensor has the advantages of small dimensions, ultra-narrow line width, and low-noise properties. It can reach ultra-high sensitivity in the detection of weak signals with high-resolution wavelength demodulation technique. The acoustic emission detection using DFB fiber lasers offers not only the advantages associated with FBG sensors, but also better performance for ultra slight acoustic emission (Ye and Tatam 2005).

We reported the realization of DFB fiber laser and DFB fiber laser rosette for acoustic emission detection (Huang et al. 2012b, 2013). The acoustic emission directional characteristics of DFB fiber laser are demonstrated in experiments. According to our analysis on location method based on the directional sensitivity of DFB fiber laser, the location precision is mainly limited by the intensity of AE source with high noise level. And, the previous location method will fail if the source strength changes.

This paper focuses on the use of DFB fiber laser rosette as acoustic emission receiver. The principle of AE detection using DFB fiber laser is presented. A method of digital signal analysis, which can improve signal-to-noise ratio (SNR), is used to investigate the directional sensitivity of DFB fiber laser. And, the normalization method is processed to eliminate the influence of distance and intensity of AE source. Then, we configured three DFB fiber lasers into a rosette for acoustic emission source detection and location based on a new location algorithm. And the test results are given below.

4.1.1 Operation Principles

DFB Fiber laser consists of a length of Er^{3+}-doped or Yb^{3+}/Er^{3+}-doped fiber with Bragg gratings. By introducing a π phase shift, the grating resonance is moved to the center of the grating reflection band, which makes DFB fiber laser operating robustly in a single longitudinal mode. In our configuration, a phase-shifted grating is formed into a length of Er^{3+} fiber, whose ends are spliced to a matching passive fiber for reducing splice loss. The grating is pumped with a 980 nm or 1,480 nm semiconductor laser. The wavelength of DFB fiber laser is determined by the central wavelength in the reflective spectrum of phase-shifted grating, which is shown as

$$\lambda_B = 2n_{eff}\Lambda \tag{4.1}$$

where λ_B is the lasing wavelength, Λ is the period of grating, and n_{eff} is the effective index of fiber core. The period Λ and the effective reflective index n_{eff} are changed with the environmental conditions such as strain, temperature, and acoustic wave.

Figure 4.1 shows the schematic diagram of a digital phase-generated carrier (PGC) based on wavelength demodulation system for DFB fiber laser sensors. The PGC scheme is used in the system to recover the phase signal and overcome the bias drift-induced fading. One arm of the interferometer is wrapped onto a PZT tube and the phase modulation of PGC scheme is introduced to the interferometer by electrically modulating the PZT. The interferometric signals are then received by a photodetector and an amplifier. The electrical output of the amplifier is digitalized in accordance with the carrier signal using an A/D convertor. The PGC phase demodulation is accomplished on an FPGA board or in a computer.

The strain resolution of DFB fiber laser acoustic sensor is limited by system noise level, which is determined by frequency noise and relative intensity noise (RIN) of DFB fiber laser, electrical circuit noise, and environmental noise on the interferometer. The noise level is measured in a quiet laboratory with the isolation of vibrations and acoustic noise. Figure 4.2 shows the noise level in the frequency range from 20 to 2,000 Hz, and it can be observed that the noise level is below 1×10^{-6} pm/$\sqrt{Hz}$@1 kHz, which will result in a strain resolution of about 10^{-12}.

It has been proved that FBGs exhibit the maximum sensitivity to acoustic waves when the direction of maximum strain is parallel to the fiber axis and minimum when it is normal to the axis (Thursby et al. 2008). The directional responses of FBGs make them suitable for determining principal strains and the propagation direction of incident acoustic waves. In our research, DFB fiber lasers have shown a similar trend. Here, the directional responses of DFB fiber lasers, instead of FBGs, are analyzed when they are used as acoustic emission receivers.

4.1.2 Investigation of AE Directional Sensitivity of DFB Fiber Laser

The acoustic emission experiment is carried out in a large marble stone. The DFB fiber laser is glued to the marble plate and operate at different center wavelengths with 300 GHz (2.4 nm) spacing from 1,529 to 1,546 nm, and pumped through a single fiber by a 980-nm pump laser. The center wavelength of fiber laser will be modulated

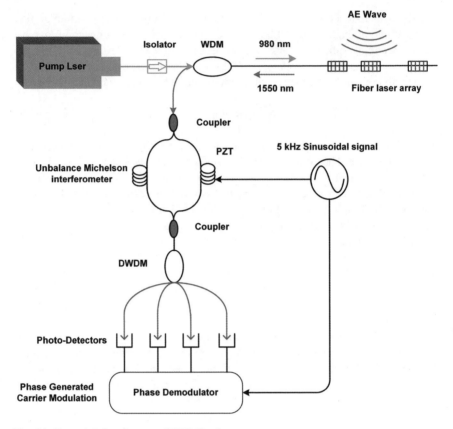

Fig. 4.1 Demodulation diagram of DFB fiber lasers

by acoustic waves produced by an acoustic emission source and was demodulated by the demodulation system. Generally, a PZT sensor is also glued to the marble plate close to fiber laser as a reference.

Primarily, an effective coupled mode should be selected to record AE signal with the maximum sensitivity in the experimental process, which will much more closely reflect the features of AE source. First, DFB fiber lasers are bonding on the surface of marble plate with three different coupled modes. A PZT AE sensor is used to generate continuous sine acoustic waves. Then, the DFB fiber laser of every coupled mode is used to detect the sine wave. We can find that the response of DFB fiber laser coupled with 353ND fiber adhesive is more prominent than the other two kinds of cases, showing that the coupled mode of 353ND fiber adhesive has the maximum AE sensitivity, which suits to AE detection experiment.

It has been proved that FBGs exhibit maximum sensitivity to acoustic waves when the direction of maximum strain is parallel to the fiber axis and minimum when it is normal to the axis (Thursby et al. 2008). The directional responses of FBGs make them suitable for determining principal strains and the propagation direction

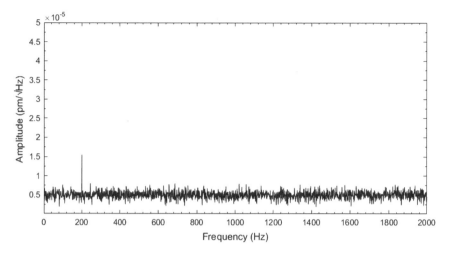

Fig. 4.2 Noise level of DFB fiber laser sensing system

of incident acoustic waves. In our research, DFB fiber lasers have shown a similar trend. The directional sensitivity of DFB fiber laser is investigated by calculating location coefficient using a method of digital signal analysis using a PZT sensor or steel ball as an exciting source, respectively, which is shown in Fig. 4.3. We use two kinds of analyzing method to calculate the location coefficient for periodic AE source location and burst AE source location.

For a periodic AE source, the relationship between the wavelength drift of DFB fiber laser and the angle of AE source is tested. The directivity of DFB fiber laser has been tested by mounting PZT source transducers on the marble plate in a circular array with an DFB fiber laser at the center. The amplitude of signal detected by the

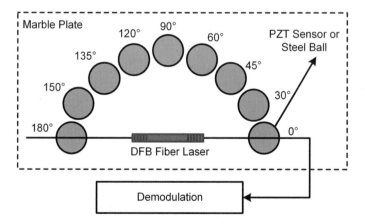

Fig. 4.3 Investigation of AE directivity of DFB fiber laser

DFB fiber laser was then measured as each PZT was excited in turn. Sine acoustic waves were generated by a PZT AE transducer (Soundwel Technology Co., Ltd., SR10-10112-B), which was driven directly by a synthesized function generator. An autocorrelation approach is used to extract the periodic AE signal. We can get a sine AE wave with high SNR by the proposed method, which is shown in Fig. 4.4. The autocorrelation is shown as

$$R_{xx}(\tau) = \lim_{T \to \infty} \frac{1}{T} \int_0^T x(t)x(t+\tau)dt \qquad (4.2)$$

where $R_{xx}(\tau)$ is the result of autocorrelation, $x(t)$ is the AE source, and T is the time of sample data of the AE source.

In order to research the directional characteristic based on amplitude response of DFB fiber laser, the amplitude normalization, which can eliminate the influence of distance and intensity of AE source, is processed. The relationship between the wavelength shift and the angle of an AE source are shown in Fig. 4.5. Curve fitting shows a sine-squared relationship between the signal amplitude and the angle

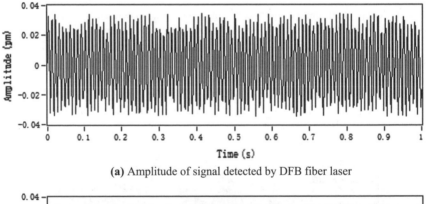

(a) Amplitude of signal detected by DFB fiber laser

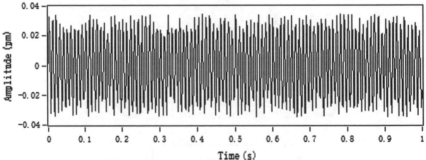

(b) Sine AE signal obtained by autocorrelation method

Fig. 4.4 Autocorrelation approach used to extract periodic AE signal

4.1 Acoustic Emission Receiver Based on DFB 155

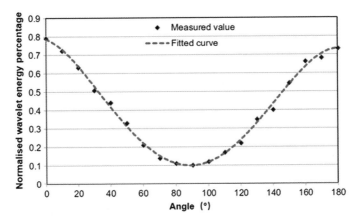

Fig. 4.5 Relationship between the wavelength drift and the angle of periodic AE source

between the wave propagation direction and the DFB fiber laser axis. So, we can get the direction of a periodic AE source according to the size of the amplitude response of DFB fiber laser. Linear normalization is shown as

$$\text{Nor} = \frac{\text{Act} - \text{Minvalue}}{\text{Maxvalue} - \text{Minvalue}} \quad (4.3)$$

where Nor is the value after normalization, Act is the measured value. Maxvalue is the maximum sample value, Minvalue is the minimum sample value.

For a burst AE source, the relationship between the percentage of wavelet packet energy to the response of DFB fiber laser and the angle of the impact response is established. Here, a steel ball with 9 mm diameter and 1.2 g in weight instead of PZT AE sensor is used to generate impact signal. The ball was dropped from a height of 20 cm above the horizontal marble plate. Then, the impact response is measured by DFB fiber laser, which is shown in Fig. 4.6. We can see from the frequency domain response of DFB fiber laser that the energy of dominant frequency around 200 Hz decrease and low-frequency component increase, indicating that we can get the direction of a burst AE source according to the size of dominant frequency component.

Then, we can calculate the wavelet packet energy percentage when the ball impact at different angles, which is shown in Fig. 4.7. We can see that the energy percentage of low-frequency component increases with the angle. This method can also eliminate the influence of distance and intensity of AE source. Here, the dominant frequency is about 200 Hz and the sampling frequency is 10 kHz, so the layer of wavelet packet decomposition is chosen as five. And, the frequency of 200 Hz will be in the third layer.

The relationship between the percentage of wavelet packet energy and the angle of a burst AE source are shown in Fig. 4.8. And, it can be seen from the test results that the curve fitting is in a tube shape.

Fig. 4.6 Relationship between the percentage of wavelet packet energy to response of DFB fiber laser and the angle of impact response

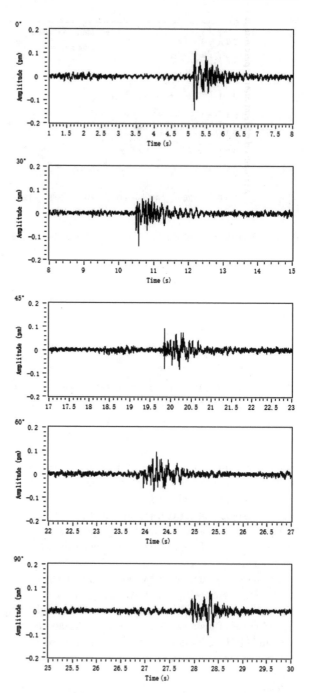

4.1 Acoustic Emission Receiver Based on DFB

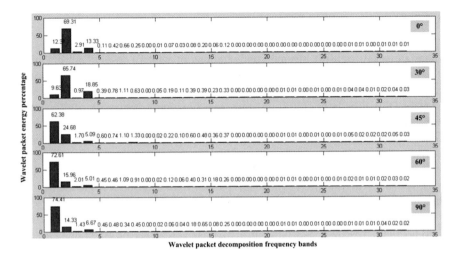

Fig. 4.7 Wavelet packet energy percentage at different angles

According to the directional sensitivity of DFB fiber laser, we can use three DFB fiber lasers as a rosette which can determine the source location of an acoustic wave. Here, we can get the location coefficient of a periodic AE source from Fig. 4.5 and get the location coefficient of a burst AE source from Fig. 4.8. Then, we can calculate the information for AE source location using an appropriate algorithm based on the location coefficient. This process is directly analogous to the use of ESG or FBG rosettes to determine principal strain magnitudes and directions.

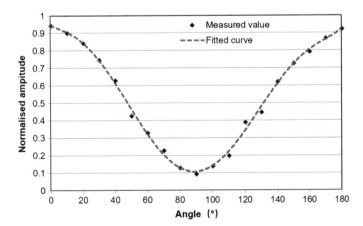

Fig. 4.8 Relationship between the percentage of wavelet packet energy and the angle of burst AE source

4.1.3 Location Algorithm

There are many well-established configurations in which conventional electrical strain gages (ESG) may be formed. Among them, one is the Delta (Δ) rosette configuration that is used to determine the direction and magnitude of the principal strain. Because DFB fiber lasers fulfill an equivalent function to their electrical counterparts, it seems appropriate to investigate the possibility of producing an DFB fiber laser rosette, which is shown in Fig. 4.9. There fiber lasers is the same piece of fiber, each of a different center wavelength, while a fourth may be added to provide either temperature measurement or compensation. The purpose of configuring DFB fiber lasers into a rosette is to detect and locate the source of acoustic wave based on their directional characteristics of strain response.

The principle diagram of triangulation approach for AE source location based on DFB fiber laser rosette is shown in Fig. 4.10. First, we can get the direction of the AE source through each fiber laser according to the directivity of DFB fiber laser which is proved in the above experiments. Then, three fiber lasers with different angles are used and we can get three lines which intersect at A (x_4, y_4), B (x_5, y_5), and C (x_6, y_6). These points make a triangle ABC whose center of gravity is the location of AE source.

Assume that the coordinates of the three fiber lasers are $F_{L1}(x_1, y_1)$, $F_{L2}(x_2, y_2)$, $F_{L3}(x_3, y_3)$, respectively. According to the directional characteristics of fiber laser, we can determine the included angles between the fiber lasers and AE source are α, β, γ, respectively. Then, the three lines $F_{L1}A$, $F_{L2}B$, and $F_{L3}C$ are expressed in the following way, respectively.

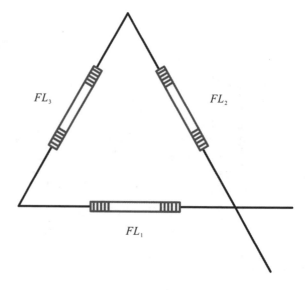

Fig. 4.9 DFB fiber laser rosette

4.1 Acoustic Emission Receiver Based on DFB

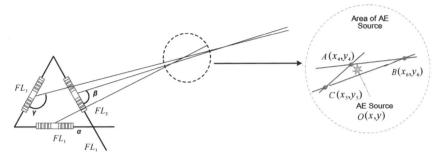

Fig. 4.10 Principle of triangulation approach for AE source location based on DFB fiber laser rosette

$$\begin{cases} y = x \tan \alpha + y_1 - x_1 \tan \alpha \\ y = x \tan \beta + y_2 - x_2 \tan(\beta - 120°) \\ y = x \tan \gamma + y_3 - x_3 \tan(\gamma - 60°) \end{cases} \quad (4.4)$$

Then, the coordinates of $A(x_4, y_4)$, $B(x_5, y_5)$, and $C(x_6, y_6)$ can be calculated, respectively. And, the coordinates of the center of gravity of the triangle ABC are shown as

$$\begin{cases} x = \frac{x_4+x_5+x_6}{3} = \frac{1}{3} \left\{ \frac{y_1-y_3-[x_1 \tan\alpha - x_3 \tan(\gamma-60°)]}{\tan\alpha - \tan(\gamma-60°)} \right. \\ \qquad + \frac{y_1-y_2-[x_1 \tan\alpha - x_2 \tan(\beta-120°)]}{\tan\alpha - \tan(\beta-120°)} \\ \qquad \left. + \frac{y_2-y_3-[x_2 \tan(\beta-120°) - x_3 \tan(\gamma-60°)]}{\tan(\beta-120°) - \tan(\gamma-60°)} \right\} \\ y = \frac{y_4+y_5+y_6}{3} = \frac{1}{3} \left[\frac{y_3 \tan\alpha - y_1 \tan(\gamma-60°) - (x_3-x_1) \tan\alpha \tan(\gamma-60°)}{\tan\alpha - \tan(\gamma-60°)} \right. \\ \qquad \frac{y_2 \tan\alpha - y_1 \tan(\beta-120°) - (x_2-x_1) \tan\alpha \tan(\beta-120°)}{\tan\alpha - \tan(\beta-120°)} \\ \qquad \left. + \frac{y_3 \tan(\beta-120°) - y_2 \tan(\gamma-60°) - (x_3-x_2) \tan(\beta-120°) \tan(\gamma-60°)}{\tan(\beta-120°) - \tan(\gamma-60°)} \right] \end{cases} \quad (4.5)$$

So, we can determine the location of AE source as we know the coordinate of $O(x, y)$. Then, a specific location algorithm based on the directional sensitivity of DFB fiber laser and the triangulation approach is shown in Fig. 4.11. The advantage of proposed algorithm over the traditional methods based on FBG includes the capability of having higher strain resolution for AE detection and taking into account of the two different types of AE source. The location calculates algorithm for two different types of AE source which consists of three steps: ① judge the type of AE source; ② calculate the location coefficient for AE source location; ③ use the triangulation approach to determine the location of AE source. Here, normalization is processed to eliminate the influence of distance and the intensity of AE source.

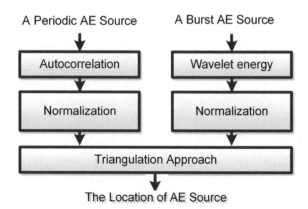

Fig. 4.11 Specific location algorithm based on the directional sensitivity of DFB fiber laser and the triangulation approach

4.1.4 Tests and Results

Here, a Delta (Δ) rosette configuration based on three DFB fiber lasers is established according to the directivity of DFB fiber laser. And, the novel acoustic emission source location algorithm is used and tested to determine the direction and the location of acoustic emission waves. In this experiment, the DFB fiber laser rosette is glued to the marble plate using 353ND fiber adhesive. A PZT sensor is used as an excitation of continuous AE source and a steel ball is used as an excitation of low velocity impact source, which is as shown in Fig. 4.12. The location experiment based on DFB fiber laser rosette is carried out by mounting PZT transducers or steel ball in turn on the marble plate in different places beyond the rosette. Then we can evaluate the location accuracy when the AE source is generated at different distance.

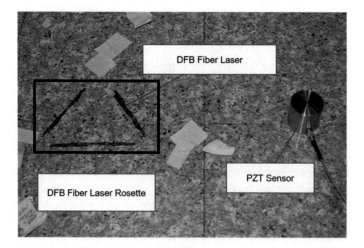

Fig. 4.12 DFB fiber laser rosette for acoustic emission location

4.1 Acoustic Emission Receiver Based on DFB

The AE source location result based on DFB fiber laser rosette is shown in Fig. 4.13. From the test result, we can see that the location error is less than 2 cm within the circumference of 50 cm where the fiber laser rosette is the center circle. And, the location accuracy and distance is limited by the distance because of the attenuation and interference of AE signals when the acoustic waves spread in the marble plate. DWA is the location algorithm based on the relationship between wavelength drift and the angle of a continuous acoustic emission source. PWPEA is the location algorithm based on the relationship between the percentage of wavelet packet energy and the angle of an impact response.

To improve the location accuracy and range, two or more rosettes based on DFB fiber lasers are necessary. An DFB fiber laser rosette array is tested for AE location in Fig. 4.14. The location of AE source obtained by one DFB fiber laser rosette is modified by another rosette. And the location result is shown in Fig. 4.15. From the test result, we can conclude that a higher location accuracy and a larger location range is achieved when using an DFB fiber laser rosette array in the experiment.

However, the experiment is conducted under the condition of quiet environment. In fact, the environmental vibration and temperature fluctuations may lead to a wrong result of AE source location. We can eliminate the influence of temperature fluctuations by digital filter processing. But the environmental vibration cannot be filtered simply by digital processing method. A specific packaging structure, which is immune to environmental strain, is required for engineering application in health monitoring of civil structures. And, a safety guard should be used to protect the DFB fiber laser because of the fragility of fiber.

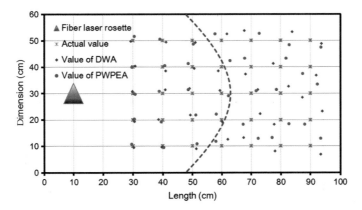

Fig. 4.13 Results of location using DFB fiber laser rosette

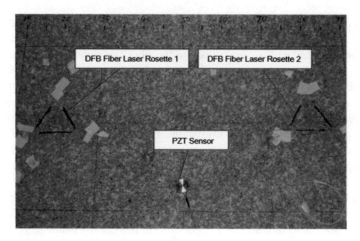

Fig. 4.14 DFB fiber laser rosettes array for acoustic emission location

4.2 DFB Fiber Laser Accelerometers

With the growth of high speed and heavy transportation on railway tracks, rail corrugation becomes an increasingly serious problem which causes deterioration in ride quality, restricts running speed, generates noise, and has the potential to result in serious accidents. Therefore, rail corrugation should be regularly monitored and repaired with appropriate methods to maintain good ride quality and stability. Rail corrugation is conventionally been measured using exclusive track inspection vehicles. However, this method is costly and is therefore not widely employed on local and submain lines (Kojima et al. 2006). Recently, accelerometers installed on bogie or axle box has been used to cope with this issue, which is proved to be simple and

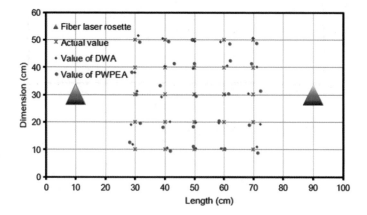

Fig. 4.15 Results of location using a pair of DFB fiber laser rosettes

4.2 DFB Fiber Laser Accelerometers

high efficiency (Suda et al. 2002; Takeo 2008; Molodova et al. 2011). This method can work better if the sensors are immune to electromagnetic interference (EMI) and have a larger dynamic range (Du and Zhang 2009; Wei et al. 2012). In the past 10 years, a number of optical fiber sensors, with many advantages such as high sensitivity, small size, large dynamic range, and immunity to EMI, have been developed to be used in railroad (Mandriota et al. 2004; Tam et al. 2005; Du and Zhang 2009; Filograno et al. 2010). Recently, a new generation optical fiber sensor based on distributed feedback (DFB) fiber laser has received considerable research interests in this field (Huang et al. 2012a). It can reach ultra-high sensitivity in the detection of weak signals for rail corrugation with high-resolution wavelength demodulation technique.

On the other hand, although the inertial method that the double integration of acceleration gives its displacement has already been achieved for the estimation of rail corrugation in the measured acceleration, the inertial algorithms meet some problems for rail corrugation measurement. One difficulty is that the previous analog inertial measuring system remained wave distortion because the system used analog high-pass filters (HPFs), which had a nonlinear phase property. To avoid wave distortion caused by analog HPFs, the digital filtering technique for multi-resolution analysis method was developed (Kojima et al. 2006; Zumpano and Meo 2006; Berggren et al. 2008; Kobayashi et al. 2008; Lee Jun et al. 2012). The wavelet analysis is herein used as a tool for the analysis of the vibration signals due to the rail corrugation, following a new trend in this research area. Wavelet analysis has been proved to be more adequate for the processing of nonstationary signals, such as these, for which the classical Fourier analysis presents limited results.

This paper presents a rail corrugation measurement system based on DFB fiber laser accelerometers (FLAs). The principle of rail corrugation detection and the theoretical model and test results of FLAs are given. An inertial algorithm based on double integration and wavelet denoising is proposed for accurate estimation of rail corrugation in the measured acceleration. And field test results are also given below.

4.2.1 Principles

4.2.1.1 Principle of Rail Corrugation Detection

In many cases, track geometry irregularities, such as rail corrugation which is presented rough waveform in rail top surface, can be approximately described using single or multiple Jane harmonic. A single harmonic irregularity, which is shown in Fig. 4.16, can be simplified to use cosine function to describe the rail surface appearance.

So, the rail corrugation can be expressed in the following way:

$$\eta(t) = \frac{1}{2}(1 - \cos \omega t) \quad \left(0 \leq t \leq \frac{\lambda}{V}\right) \tag{4.6}$$

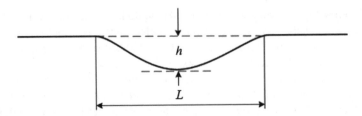

Fig. 4.16 Single harmonic irregularity

in which $\omega = \frac{2\pi V}{\lambda}$ where λ is the wavelength of rail corrugation, h is the deepness of rail corrugation, and V is the speed of the train.

The dynamical model of wheel-rail contains two parts of locomotive and track. The locomotive mainly involves unsprung mass and the part of body can be simplified. And, the track is simplified into a single degree of the equivalent system. Then, we can give the simple lumped parameter model of wheel-rail, which is shown in Fig. 4.17. Here, the harmonic type irregularity is introduced in the simple lumped parameter model of wheel-rail.

While m_1 is the unsprung mass, m_2 is the equivalent mass of rail, K_1 is the equivalent spring rate of wheel-rail contact, K_2 is the vertical stiffness of track, c_2 is the vertical damping of track, Z_1 and Z_2 are the displacement of unsprung mass and track relative to their static equilibrium position, respectively.

Then, according to the dynamic principle, we can get the relationship between rail corrugation and acceleration recorded by accelerometer installed on the bogie or axle box of the train, which is shown as

$$Z_1 = a = \sum A_i e^{-p_i t} + D \cos(\omega t + \psi) \tag{4.7}$$

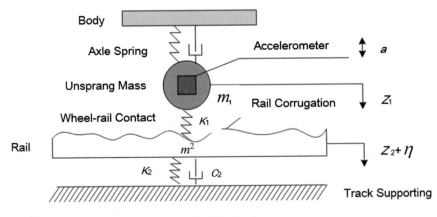

Fig. 4.17 Simple lumped parameter model of wheel-rail

4.2 DFB Fiber Laser Accelerometers

where a is axle box acceleration, A_i, p_i, and D are constants, and Ψ is the initial phase.

From the Formula (4.7), we can see that the frequency of acceleration is the same with the frequency of rail corrugation and is the quadratic differential of rail corrugation. So, we can get the rail corrugation by double integration if we have detected the vertical axle box acceleration of the train using accelerometer.

4.2.1.2 DFB Fiber Laser Accelerometer

The optical fiber sensor based on distributed feedback (DFB) fiber laser has attracted considerable research interests in the past 10 years. The fiber laser consists of a length of Er^{3+}-doped or Yb^{3+}/Er^{3+}-doped fiber written in Bragg gratings. By introducing a π phase shift, the grating resonance is moved to the center of grating reflection band, which makes DFB fiber laser operate robustly in a single longitudinal mode. The ultra-narrow line width (<3 kHz) of DFB fiber laser will result in a low equivalent noise level when using interferometric demodulation method. The DFB fiber laser sensor also has the advantages of small dimensions, ultra-narrow line width, and low-noise properties.

In this paper, the proposed ultra-thin fiber laser accelerometer (FLA), as shown in Fig. 4.18, has been improved to become a corrugation detector. The mass has been enlarged to make the nature frequency lower to detect the low-frequency rail corrugation signal. The fiber laser is fixed by a kind of double diaphragm structure as shown in Fig. 4.18a. The stiffness perpendicular to the axis of fiber laser, in this structure, is far greater than axial stiffness theoretically. And the transverse sensitivity can be reduced to about 40 dB.

The FLA is encapsulated in an inner sensor shell to be protected and there is a protection shell outside the sensor shell, as shown in Fig. 4.18b. Between the inner sensor shell and the protection shell, it's filled with polyurethane foam, which makes the FLA airtight and isolates the acoustic noise (see the inset of Fig. 4.18b).

The center wavelength of the DFB fiber laser is within 1,525~1,565 nm. The outer diameter of the FLA is 12 mm. The length of the FLA is about 85 mm.

And the acceleration sensitivity of FLA can be written as

$$M_a = \frac{\Delta \lambda_B}{a} = \frac{0.78 \lambda_B m \left[1 - \left(\frac{r}{R}\right) \frac{1-\left(\frac{r}{R}\right)^2 + 4\ln^2\left(\frac{r}{R}\right)}{1-\left(\frac{r}{R}\right)^2}\right]}{\frac{32\pi DL}{R^2} + E_f A \left[1 - \left(\frac{r}{R}\right)^2 \frac{1-\left(\frac{r}{R}\right)^2 + 4\ln^2\left(\frac{r}{R}\right)}{1-\left(\frac{r}{R}\right)^2}\right]} \tag{4.8}$$

in which $D = \frac{E_m t^3}{12(1-v^2)}$ where m is the mass, a is the acceleration, A is the cross area of the fiber, E_f is the Young's modulus of the fiber, t is the thickness of the diaphragm, R is the radius of the diaphragm, r is the contact radius of the mass and the diaphragm, E_m is the Young's modulus of the diaphragm, v is the Poisson's ratio of the diaphragm, and L is the length of the fixed fiber.

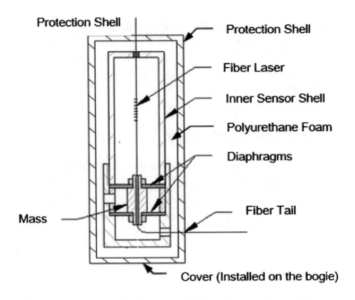

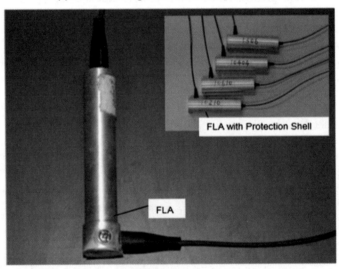

(b) Photo of FLA

Fig. 4.18 Schematic diagram and photo of FLA

The mechanical model of FLA can be equivalent to Fig. 4.19. Then, the resonant frequency of the mechanical model of FLA can be written as

4.2 DFB Fiber Laser Accelerometers

Fig. 4.19 Mechanical model of FLA

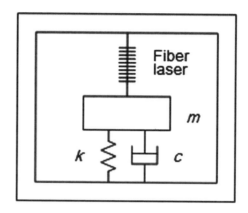

$$f_0 = \frac{1}{2\pi}\sqrt{\frac{kL + E_f A}{mL}} \qquad (4.9)$$

in which $k = \dfrac{32\pi D}{R^2\left[1-\left(\frac{r}{R}\right)^2 \frac{1-\left(\frac{r}{R}\right)^2 + 4\ln^2\left(\frac{r}{R}\right)}{1-\left(\frac{r}{R}\right)^2}\right]}$.

Formula (4.8) provides a guidance direction for the design and improvement of FLA. The material properties and geometric parameters have significant effects on the sensitivity of FLA. Here, both ends of fiber laser are fixed using UV light adhesive and make sure that the fiber laser retains small prestressed. The material properties and geometric parameters of FLA are shown in Table 4.1. The sensitivity of FLA is calculated to be 51.636 pm/g. According to Formula (4.1) and Table 4.1, we can also know that the resonant frequency of FLA is 880.4 Hz.

The sensitivity of FLA is tested in the laboratory before the field demonstration. The calibration of FLA is carried out according to Chinese Standard GB/T 13823 (*Calibration Method of Vibration and Impact Sensor*). Figure 4.20 shows the sensitivity test result of FLA. We can see that the longitudinal sensitivity of FLA is

Table 4.1 Parameters for mechanical and geometric properties of FLA

Parameter	Value
A	0.012 3 mm^2
E_f	72 GPa
t	0.08 mm
R	4.5 mm
r	1.5 mm
m	11.1 g
E_m	112 GPa
ν	0.37
L	45 mm
λ_B	1,550 nm

50 pm/g (34 dB, ref: 0 dB = 1 pm/g) which is consistent with the theoretical value and has a flat frequency response in the 10–300 Hz band range.

The interrogation of FLA is achieved by using phase-generated carrier (PGC) demodulation, which is shown in Fig. 4.21. The unbalanced interferometer can convert the FLAs wavelength shifts to the interferometer phase shifts, and the wavelength demodulation can be achieved with high-resolution PGC algorithm. This set-up can also use a commercially available dense wavelength division multiplexer (DWDM) as a wavelength filter at the output of the fiber interferometer to interrogate multiple sensors.

The equivalent noise level is tested to be 1×10^{-6} pm/(Hz)$^{1/2}$. With a sensitivity of 50 pm/g of the FLA, a minimum detectable signal of 20 ng can be achieved.

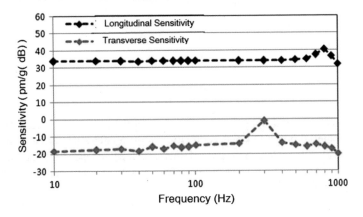

Fig. 4.20 Longitudinal sensitivity and transverse sensitivity of FLA

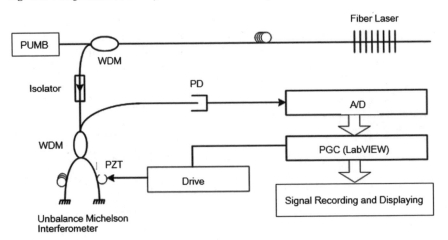

Fig. 4.21 Schematic diagram of wavelength demodulation for FLA

In the field of railway engineering, many technology reports and research papers show that the wavelength of rail corrugation is about 50~300 mm. Here, the running speed of the train is about 60 km/h. So, we can realize that the main frequency of rail corrugation is below 300 Hz. The resonance frequency of fiber laser accelerometer (FLA) is about 900 Hz. And FLA has a flat frequency response in the 10~300 Hz band range, which is important to obtain a correct response of rail corrugation. So this device which works below resonance is suitable for detection of rail corrugation.

4.2.2 Wavelet Denoising

Wavelet transform decomposes a signal into a series of base functions of dilated and translated versions of the mother wavelet function. The expression of the DWT is shown below

$$X_{\text{DWT}}(j, k) = \sum_{j,k} x_{j,k}(t) 2^{-j/2} \psi \left(\frac{t - 2^j k}{2^j} \right) \tag{4.10}$$

In this equation, j and k are dilation and translation factors, respectively. Mallat proposed a method using a bank of low-pass and high-pass quadrature mirror filters (QMFs) to compute the DWT (Mallat 1989). Figure 4.22a shows the schemes of DWT and inverse DWT based on filter banks. The signal passes the filter banks and is decomposed into the approximation coefficients obtained from the low-pass filter and the detail (wavelet) coefficients obtained from the high-pass filters which are represented by $a_k^{(n)}$ and $d_k^{(n)}$, respectively. The decomposition levels and the filter coefficients are determined by the mother wavelet function. It is important to select an appropriate mother wavelet function according to the property of signals. Figure 4.22b shows that the low-frequency component in the signal is analyzed by a narrow band filter at high decomposition level, while the high-frequency component is analyzed by a broadband filter at the low decomposition level. And, wavelet shrinkage is a signal denoising method based on the idea of threshold wavelet coefficients of noisy signal.

4.2.3 Inertial Algorithm

The inertial algorithm is based on a simple law that states that the double integration of acceleration gives its displacement, and is one of the general methods used for rail corrugation measurement. The wavelet denoising method is proposed to eliminate trend term and interference signal. Figure 4.23 shows the scheme of the inertial algorithm. And, the wavelet denoising method for rail corrugation measurement system consists of three steps: ① apply the DWT on the raw noisy acceleration

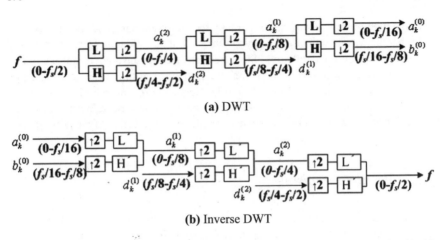

(a) DWT

(b) Inverse DWT

Fig. 4.22 Schemes of DWT and inverse DWT based on filter banks with decomposition level 3: L, L': low-pass filter; H, H': high-pass filter; f_s: sample frequency

signal due to the rail corrugation to obtain the wavelet coefficients; ② choose an appropriate threshold rule to the wavelet coefficients; ③ reconstruct the denoised signal with the inverse DWT. Through the threshold step, the wavelet coefficients of the noise, such as trend term and interference signal, can be eliminated since the noises embedded in the signal usually corresponds to the wavelet coefficients with small absolute value.

4.2.4 Test Scheme

We have developed a rail corrugation measurement system that uses high sensitivity fiber laser accelerometers as sensing element. The high-resolution demodulation system of FLA bases on interference demodulation via nonequilibrium interferometer to transform the variation of laser wavelength to phase variation, and applies PGC method for phase demodulation to demodulate the wavelength of fiber laser vibration sensor. Through inertial algorithm, the rail corrugation can be estimated

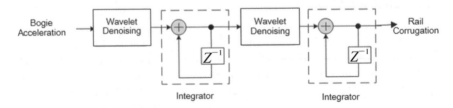

Fig. 4.23 Scheme of inertial algorithm

in LabVIEW. The system uses wavelength division multiplexing scheme, and the overall scheme is shown in Fig. 4.24.

4.2.5 Test Results

The field test was performed on Datong–Qinhuangdao Railway in North China. Figure 4.25a shows the installation of one FLA on the bogie. The four sensors are installed on the bogie using special fixed device, which is shown in Fig. 4.25b. The demodulation system is installed in the carriage and connected with the FLAs with the optical fiber cable. When the train passes the rail with corrugation, the corrugation induced acceleration of the bogie will be detected by the FLAs. The digital inertial algorithm is employed to calculate the rail corrugation from acceleration of FLAs.

Figure 4.26 shows acceleration signals recorded by rail inspection car and FLA. And Fig. 4.27 shows rail corrugations estimated by rail inspection car and inertial algorithm. We can see that the experimental results of acceleration recorded by one of FLA and rail corrugation estimated by inertial algorithm are in good agreement with those of rail inspection car. And the details of acceleration and rail corrugation are clearer due to high sensitivity FLA and wavelet denoising method.

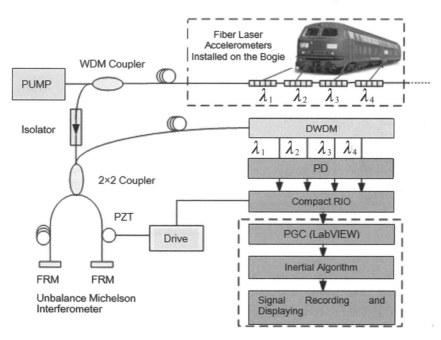

Fig. 4.24 Scheme of rail corrugation measurement

(a) Installation site

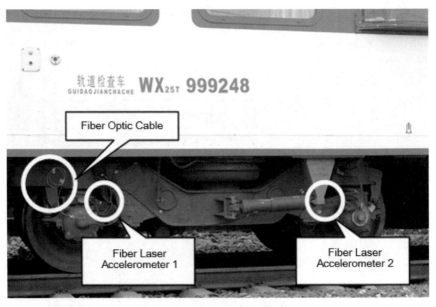

(b) General layout of installation

Fig. 4.25 Installation of FLAs

4.3 Summary

In this chapter, an application of DFB fiber lasers for AE source localization, based on a method of digital signal analysis, is presented and discussed. The proposed location algorithm has higher strain resolution for AE detection compared with the

4.3 Summary

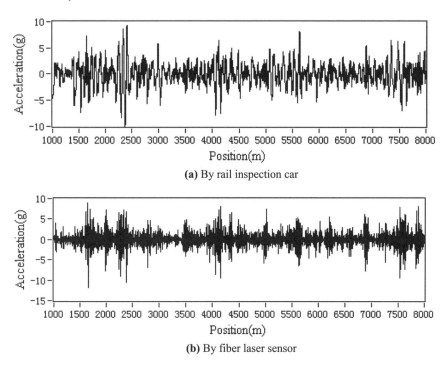

Fig. 4.26 Acceleration signals

traditional methods based on FBG and can eliminate the influence of distance and intensity of AE source. Especially, we take into account the two different types of AE source for location in this algorithm. This is the first demonstration, to the best of our knowledge that a periodic AE signal or a burst AE source can be located by one location algorithm at the same time, providing a powerful technique for AE source location in health monitoring of civil structures.

A novel all-fiber-optic rail corrugation detection technology based on fiber laser accelerometers (FLAs) is proposed. Due to the high sensitivity, larger dynamic range, and immunity to EMI, FLAs are suited to detect the vertical axle box acceleration of the train caused by rail corrugation. A flexible inertial algorithm based on double integration and wavelet denoising method is proposed to accurately estimate the rail

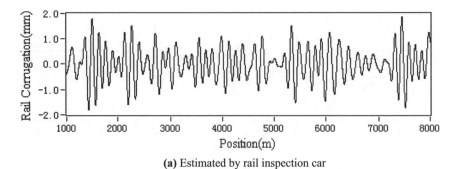

Fig. 4.27 Rail corrugations

corrugation. And, these FLAs based on rail corrugation detection system shows good performance in the field test on Datong–Qinhuangdao Railway, which shows that it's a practical method to use fiber laser accelerometer as a corrugation detector.

References

Baldwin CS, Vizzini AJ (2003) Acoustic emission crack detection with FBG. In: Smart structures and materials, 2003. SPIE, pp 133–143. https://doi.org/10.1117/12.484254

Berggren EG, Li MXD, Spännar J (2008) A new approach to the analysis and presentation of vertical track geometry quality and rail roughness. Wear 265(9):1488–1496. https://doi.org/10.1016/j.wear.2008.01.029

Carpinteri A, Lacidogna G, Pugno N (2007) Structural damage diagnosis and life-time assessment by acoustic emission monitoring. Eng Fract Mech 74(1):273–289. https://doi.org/10.1016/j.engfracmech.2006.01.036

Du Y, Zhang W (2009) A novel heavy haul train on road safety status monitoring technique based on fiber laser sensors. Eng Sci 32(6):61–66. https://doi.org/10.1016/j.engstruct.2010.02.020

Filograno ML, Rodríguez-Barrios A, González-Herraez M et al (2010) Real time monitoring of railway traffic using fiber Bragg Grating sensors. IEEE Sens J 2:493–500

Finkel P, Miller R, Finlayson RD et al (2001) Development of fiber optic acoustic emission sensors. In: AIP conference proceedings, 2001. p 1884. https://doi.org/10.1063/1.1373977

Hill DJ, Hodder B, Freitas JD et al (2005) DFB fibre-laser sensor developments. In: 17th international conference on optical fibre sensors, 2005. SPIE, p 4. https://doi.org/10.1117/12.624303

Huang W, Du Y, Zhang W et al (2012a) Rail corrugation measurement using fiber laser accelerometers. In: OFS2012 22nd international conference on optical fiber sensor. SPIE, p 4. https://doi.org/10.1117/12.968594

Huang W, Ma H, Zhang W et al (2012b) Rock mass acoustic emission detection using DFB fiber lasers. In: Photonics Asia. SPIE, p 8. https://doi.org/10.1117/12.999771

Huang W, Zhang W, Ma H et al (2013) Distributed feedback fiber laser rosette for acoustic emission detection. Appl Mech Mater 330:412–417. https://doi.org/10.4028/www.scientific.net/AMM.330.412

Ian R, Peter F, Stuart M (2002) Optical fibre acoustic emission sensor for damage detection in carbon fibre composite structures. Meas Sci Technol 13(1):N5–N9

Jiang M-S, Sui Q-M, Jia L et al (2012) FBG-based ultrasonic wave detection and acoustic emission linear location system. Optoelectron Lett 8(3):220–223. https://doi.org/10.1007/s11801-012-1190-4

Kazuro K, Hideaki M, Isamu O et al (2005) Acoustic emission monitoring of a reinforced concrete structure by applying new fiber-optic sensors. Smart Mater Struct 14(3):S52–S59

Kobayashi M, Naganuma Y, Nakagawa M et al (2008) Digital inertial algorithm for recording track geometry on commercial shinkansen trains. In: Eleventh international conference on computer system design and operation in the railway and other transit systems (COMPRAIL08), 2008. Computers in Railways XI, pp 683–692. https://doi.org/10.2495/cr080661

Kojima T, Tsunashima H, Matsumoto A (2006) Fault detection of railway track by multi-resolution analysis. Nihon Kikai Gakkai Nenji Taikai Koen Ronbunshu 88(7):321–322. https://doi.org/10.2495/CR060931

Lee Jun S, Choi S, Kim S-S et al (2012) A mixed filtering approach for track condition monitoring using accelerometers on the axle box and bogie. IEEE Trans Instrum Meas 61(3):749–758. https://doi.org/10.1109/TIM.2011.2170377

Lee J-R, Tsuda H (2005) A novel fiber Bragg grating acoustic emission sensor head for mechanical tests. Scripta Mater 53(10):1181–1186. https://doi.org/10.1016/j.scriptamat.2005.07.018

Liang Y-j Mu, L-l Liu J-f et al (2008) Combined optical fiber interferometric sensors for the detection of acoustic emission. Optoelectron Lett 4(3):184–187. https://doi.org/10.1007/s11801-008-7158-8

Liang S, Zhang C, Lin W et al (2009) Fiber-optic intrinsic distributed acoustic emission sensor for large structure health monitoring. Opt Lett 34(12):1858–1860. https://doi.org/10.1364/OL.34.001858

Mallat SG (1989) A theory for multiresolution signal decomposition: the wavelet representation. IEEE Trans Pattern Anal Mach Intell 11(7):674–693. https://doi.org/10.1109/34.192463

Mandriota C, Nitti M, Ancona N et al (2004) Filter-based feature selection for rail defect detection. Mach Vision Appl 15(4):179–185. https://doi.org/10.1017/s00138-004-0148-3

Matsuo T, Cho H, Takemoto M (2006) Optical fiber acoustic emission system for monitoring molten salt attack. Sci Technol Adv Mater 7(1):104–110. https://doi.org/10.1016/j.stam.2005.11.010

Minardo A, Cusano A, Bernini R et al (2005) Response of fiber Bragg gratings to longitudinal ultrasonic waves. Trans Ultrason Ferroelectr Freq Control 52(2):304–312. https://doi.org/10.1109/TUFFC.2005.1406556

Molodova M, Li Z, Dollevoet R (2011) Axle box acceleration: measurement and simulation for detection of short track defects. Wear 271(1):349–356. https://doi.org/10.1016/j.wear.2010.10.003

Nair A, Cai CS (2010) Acoustic emission monitoring of bridges: review and case studies. Eng Struct 32(6):1704–1714. https://doi.org/10.1016/j.engstruct.2010.02.020

Qin Y, Liang Y, Zhang Y et al (2010) Experimental study on an optical fiber acoustic emission sensor array. In: 2010 Academic symposium on optoelectronics and microelectronics technology and 10th Chinese-Russian symposium on laser physics and laser technology optoelectronics technology (ASOT), 2010. IEEE, pp 299–302. https://doi.org/10.1109/rcslplt.2010.5615293

Suda Y, Komine H, Iwasa T et al (2002) Experimental study on mechanism of rail corrugation using corrugation simulator. Wear 253(1):162–171. https://doi.org/10.1016/S0043-1648(02)00095-9

Takeda S, Okabe Y, Yamamoto T et al (2003) Detection of edge delamination in CFRP laminates under cyclic loading using small-diameter FBG sensors. Comp Sci Technol 63(13):1885–1894. https://doi.org/10.1016/S0266-3538(03)00159-3

Takeo S (2008) A development of accurate axle box accelerometer for railway vehicle. In: IEEE, SICE annual conference, pp 2642–2645. https://doi.org/10.1109/sice.2008.4655113

Tam HY, Liu SY, Guan BO et al (2005) Fiber Bragg grating sensors for structural and railway applications. In: Photonics Asia. SPIE, p 13. https://doi.org/10.1117/12.580803

Tan AC, Kaphle M, Thambiratnam D (2009) Design and performance of a high temperature/high pressure, hydrogen tolerant, bend insensitive single-mode fiber for downhole seismic systems and applications. In: ICRMS 8th international conference on reliability, maintainability and safety, p 7. https://doi.org/10.1109/icrms.2009.5269952

Thursby G, Culshaw B, Betz DC et al (2008) Multifunctional fibre optic sensors monitoring strain and ultrasound. Fatigue Fracture Eng Mater Struct 31(8):660–673. https://doi.org/10.1111/j.1460-2695.2008.01250.x

Tsuda H, Takahashi J, Urabe K et al (1999) Damage monitoring of carbon fiber-reinforced plastics with Michelson interferometric fiber-optic sensors. J Mater Sci 34(17):4163–4172. https://doi.org/10.1023/A:1004626129871

Tsuda H, Kumakura K, Ogihara S (2010) Ultrasonic sensitivity of strain-insensitive fiber Bragg grating sensors and evaluation of ultrasound-induced strain. Sens (Basel) 10(12):11248–11258

Wei C, Xin Q, Chung WH et al (2012) Real-time train wheel condition monitoring by fiber Bragg grating sensors. Int J Distrib Sens Netw 8(1):1–7. https://doi.org/10.1155/2012/409048

Wild G, Hinckley S (2007) Fiber Bragg grating sensors for acoustic emission and transmission detection applied to robotic NDE in structural health monitoring. In: Sensors applications symposium IEEE, pp 1–6. https://doi.org/10.1109/sas.2007.374388

Ye CC, Tatam RP (2005) Ultrasonic sensing using Yb 3+/Er 3+-codoped distributed feedback fibre grating lasers. Smart Mater Struct 14(1):170–176

Zumpano G, Meo M (2006) A new damage detection technique based on wave propagation for rails. Int J Solids Struct 43(5):1023–1046. https://doi.org/10.1016/j.ijsolstr.2005.05.006

Chapter 5
Fully Distributed Optical Fiber Sensor

The fully distributed optical fiber sensor takes the whole fiber as the sensing unit and the sensing point is continuously distributed. It can measure the information at any location along the fiber-optic fiber called as the mass sensing head. Therefore, the distributed sensor, with its unique advantages of a wide range of sensing and continuous nonpoint continuous distribution, can effectively extract the distribution of information in a wide range of fields, to solve many problems in current measurement. It has become the most promising research direction of fiber-optic sensing technology. This chapter will describe the basic principle of fully distributed optical fiber sensor. Based on Rayleigh scattering and Brillouin scattering principle, the helical fiber strain sensor and Brillouin fully distributed large displacement sensor based on fiber Bragg grating are designed. Its structure design, theoretical analysis, experimental verification, and other aspects are discussed in detail.

5.1 Spontaneous Scattering Spectrum in Optical Fiber

Light waves are electromagnetic (EM) waves. When an electromagnetic wave hits a medium such as optical fiber, the incident electromagnetic wave interacts with the molecules or atoms that make up the material, creating a scattering spectrum. Spontaneous scattering can be observed when the incident light intensity is relatively low. When the light with angular frequency ω_0 is incident into the fiber, its scattering spectrum diagram is shown in Fig. 5.1.

Among them, Rayleigh scattered light has the same angular frequency as the incident light, see ω_0. In other words, the photon energy is conserved before and after the whole scattering process, so Rayleigh scattering is also known as elastic scattering. The scattering of other frequencies different from incident photon frequencies is inelastic scattering. When the frequency of the scattered light is higher than the frequency of the incident light, it becomes the anti-stokes light; on the contrary, it is called stokes light. The inelastic scattering process can be further divided into

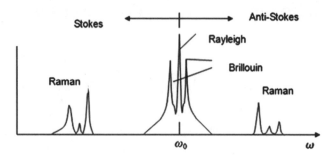

Fig. 5.1 Sketch diagram of typical spontaneous scattering on solid material

Brillouin scattering and Raman scattering. Brillouin scattering describes the energy conversion of photons and acoustic phonons. In form, the phonons are a kind of collective vibrations that include the corresponding nucleon motion in the scattering material. Raman scattering is caused by the energy conversion of the incident light to the electron structure of the independent molecule or atom. In condensed matter physics, Raman scattering is described as the light scattering of optical phonons. In particular, the molecular structure has two important characteristics: one is that the rotation of the molecule has several waves (cm^{-1}); the second is the vibration of molecules with greater energy. However, the rotational energy of the molecules is rarely observed in the optical fiber, which is caused by limited rotation freedom due to densely packed neighboring molecules. The molecules are excited in the process of reconstruction. However, because the energy range of the reconstructed molecule in the excitation state is smaller, the main vibration spectrum associated with it is not uniformly spread. Therefore, the Raman scattering spectrum contains many narrow bands. The interval of bands corresponds to electronic vibrations. Its bandwidth is derived from the excited state of molecular rotation or reconstruction. Sometimes, it is thought that Raman scattering is caused by optical phonons in solid matter. What needs to be raised is that the above spontaneous scattering refers to Raman scattering when the incident light intensity is weak. The scattering spectrum obtained by extremely high-intensity lasers is entirely different.

5.2 Application of Spontaneous Scattering in Fully Distributed Optical Fiber Sensing Technology

When the light wave travels forward in the optical fiber, the backscattered light will be continuously generated along the fiber, which mainly includes Rayleigh-scattered light, Brillouin-scattered light, and Raman-scattered light. Therefore, the scattered

along the optical fiber. In this way, the related parameters such as pressure, temperature, and bending are measured. Therefore, based on the different scattering signals, the fully distributed optical fiber sensing technology is divided into fully distributed optical fiber sensing technology based on Rayleigh scattering, fully distributed optical fiber sensing technology based on Raman scattering, and fully distributed optical fiber sensing technology based on Brillouin scattering. This chapter will mainly discuss the spiral fiber strain sensors based on Rayleigh scattering and Brillouin scattering principle as well as Brillouin fully distributed large displacement sensor based on optical fiber grating.

5.3 Winding Optical Fiber Strain Sensor

The prestressed tendon has the characteristics of small space and concealment (Jin et al. 2007). The traditional detection method is difficult to meet the needs of prestressed reinforcement stress distribution in prestressed concrete bridge monitoring requirements. Therefore, this chapter proposes a prestressed steel monitoring method with the fully distributed optical fiber sensor. Relative to Brillouin scattering and Raman scattering, Rayleigh scattering has the highest energy and can be more easily detected. Therefore, the distributed strain sensor based on Rayleigh scattering is proposed as a solution for distributed stress monitoring of prestressed reinforcement (Du et al. 2004).

5.3.1 Theoretical Basis and Analysis

The main principle of intensity-modulated fiber-optic sensor is fiber loss theory. Optical fiber loss includes absorption loss, scattering loss, and perturbation loss. The testing principle of backscattering method is the basis of realizing fiber loss distribution test and has a wide application prospect in the field of distributed sensing.

5.3.1.1 Principle of Backscattering Sensing

1. Rayleigh Scattering Loss Theory

Rayleigh scattering is the most important loss in scattering, which is a linear scattering loss with unchanged light frequency. Rayleigh scattering is caused by structural inhomogeneities such as tiny particles or pores of a core material that are much smaller in size than the wavelength of incident light (typically less than $\lambda/10$). Rayleigh scattering loss exists in all directions of fiber and its loss value is proportional to $1/\lambda^4$.

The Rayleigh scattering coefficient is usually calculated by the following equation (Zhang et al. 1988).

$$\alpha_r = \frac{8\pi^3}{3\lambda^4} n^8 p^2 k T \beta_{co} \qquad (5.1)$$

where λ is the wavelength of incident light; n is the refractive index of fiber core; p is the photoelastic coefficient; k is the Boltzmann constant; T is the curing temperature; and β_{co} is the isothermal compression rate.

Rayleigh scattering is very sensitive to material composition. For example, the incorporation of a small amount of P_2O_5 into SiO_2 can significantly reduce Rayleigh scattering. If P_2O_5 is incorporated while GeO_2 is reduced to maintain the original relative refractive index difference Δ, it will further reduce the Rayleigh scattering. In addition, Δ value of optical fiber is higher, the Rayleigh scattering loss is greater.

The refractive index of the scattering fiber core material is fluctuated, which is caused by the density and structure change of the fiber in the cooling process. Structural changes in the optical fiber can be eliminated by improving the manufacturing process, but the resulted uneven density during cooling is unavoidable. Therefore, the scattering in the optical fiber cannot be eliminated, especially Rayleigh scattering. Although Rayleigh scattering leads to the loss of transmission optical power, the Rayleigh scattering phenomenon in the optical fiber can be used to measure the loss characteristics of the whole fiber through the backscattering method, so that it can play an important role in the field of fiber distributed sensing.

2. Backscattering

Scattering loss causes the dispersion of light power, so that the energy is distributed in all directions and the backscattering is usually scattered along the axial direction of the fiber axis. Analyzing the loss characteristics of the whole fiber by measuring the backscattering intensity is called the backscattering method (Sun et al. 1999). The method is to inject a high-power and narrow-pulsed light into the fiber under test and then monitor the power of the scattered light back in the axial direction of the fiber at the same end. The loss of each point along the fiber should be characterized by the distance from the starting point of the optical fiber to scattering and the light power at the scattering point.

When the backscattering method is used to measure the fiber loss, the optical signal received by the photodetection system will have not only the Rayleigh backscattering light but also the front and back end faces of the fiber and the Fresnel-reflected light at the fiber-optic connector.

Fresnel reflection usually occurs in fiber splice and fiber breakage, where the interface between the fiber core and the air is existing. Fresnel reflection principle is shown in Fig. 5.2. Assuming that the end face of the optical fiber is perpendicular to the axis of the optical fiber and has a mirror-like shape, that is, the interface 1 is parallel to the interface 2, the light will be reflected and transmitted in the interface 1 and 2, respectively.

5.3 Winding Optical Fiber Strain Sensor

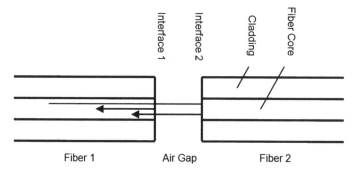

Fig. 5.2 Principle of Fresnel reflection

Fresnel reflection coefficient γ can be calculated by the following formula:

$$\gamma = P_{\text{ref}}/P_{\text{in}} = (n_1 - n_0)^2/(n_1 + n_0)^2 \tag{5.2}$$

where P_{ref} and P_{in} are the reflected light power and incident light power, respectively; n_1 is the refractive index of optical fiber core, 1.5; n_0 is the air refraction index, which is 1.0. If n_1 and n_0 are substituted into Eq. (5.2), $P_{\text{ref}}/P_{\text{in}} = 4\%$.

It can be seen that the Fresnel reflection signal is much more than the reverse signal of Rayleigh scattering. Because the Fresnel reflection signal is too strong, it will make the receiving signal circuit become saturated and it will take some time to recover to a normal state. During this time, a section of back Rayleigh scattering signals will be lost, which must be paid attention to in the practical application.

5.3.1.2 Theory of Fiber Bending Loss

Optical fiber loss does not occur in the ideal straightening condition. While the optical fiber is bent, the energy radiating along the radius of the fiber will be generated, and the conduction mode of the original optical waveguide will become the radiation mode causing the bending loss. The bending loss can be divided into two types: micro-bending loss and macro-bending loss. The bending loss is generated by three factors: spatial filtering, pattern leakage, and pattern coupling. Different bending types have different mechanisms: the main cause of micro-bend loss is pattern coupling; the main reason for macro-bend loss is spatial filtering.

The micro-bend loss of optical fiber (see Fig. 5.3a) mainly refers to the loss caused by light wave conduction relative to the random small amplitude deviation occurring in the flat state mainly caused by mechanical coupling between modes. Pattern coupling refers to the coupling of energy from one mode to another, which enables the distribution of light energy in various modes tending to be equal to a dynamic equilibrium. Due to the fiber bending, the fiber structure is changed, and the transmission mode in the fiber core is coupled with the radiation mode in the cladding

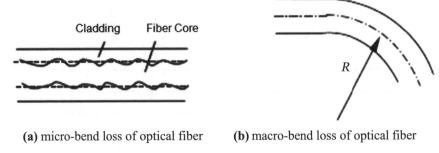

(a) micro-bend loss of optical fiber   (b) macro-bend loss of optical fiber

Fig. 5.3 Sketch diagram of optical fiber bending

layer resulting in energy loss of conduction mode. This loss increases dramatically as the spatial period of mechanical perturbations along the fiber coincides with the wave number difference between adjacent modes in the fiber.

The macro-bending loss of optical fiber (Fig. 5.3b) is the loss due to bending with a much larger radius of curvature than the diameter of the fiber, which is mainly caused by the spatial filtering effect. Spatial filtering is a physical effect during which the total reflection condition of light wave propagation is destroyed by the optical fiber bending, the high-order guided mode that enters the bending part of optical fiber is refracted to cladding, and the energy carried is radiated out of optical fiber. This is the ray theory of optical fiber bending loss. Due to the existence of spatial filtering effect, the mode number of the fiber bending part is less than that of the linear part. The greater the degree of fiber bending is, the more significant the spatial filtering effect is, the less the mode number of fiber transmission is, thus resulting in fiber transmission loss. This is the mode theory of fiber bending loss.

5.3.1.3 Optical Loss of Winding Fiber

1. Influence of Fiber Curvature Change on Optical Loss.

The optical fiber is spirally wound around the round cross-section rod. When the rod is pulled out, the fiber winding curvature will also change with it due to stress resulting in the change of bending loss. The curvature of the fiber is analyzed below.

When the rod is stretched by force, the changes of screw pitch h and radius r are shown in Formulas (5.3) and (5.4).

$$\Delta h = \varepsilon h, \tag{5.3}$$

$$\Delta r = -\mu \varepsilon r \tag{5.4}$$

The relationship between fiber curvature ρ and rod tensile strain ε is

5.3 Winding Optical Fiber Strain Sensor

$$\rho = \frac{(1-\mu\varepsilon)r}{(1-\mu\varepsilon)^2 a^2 + \frac{(1+\varepsilon)^2 h^2}{4\pi^2}} \tag{5.5}$$

From the view of horizontal and vertical strains of rod, the ρ-ε expression related to Poisson's ratio of rod is obtained. In addition, the ρ-ε expression unrelated to Poisson's ratio of rod can also be obtained based on the principle of constant volume before and after stretching.

Before rod stretching, the volume V_1 within a screw pitch length is

$$V_1 = \pi \times r^2 \times h \tag{5.6}$$

The stretched volume V_2 becomes

$$V_2 = \pi(r + \Delta r)^2 (h + \Delta h) \tag{5.7}$$

From $V_1 = V_2$

$$\pi r^2 h = \pi(r + \Delta r)^2 (h + \Delta h) \tag{5.8}$$

If the longitudinal strain of rod ε is

$$h + \Delta h = (1+\varepsilon)h \tag{5.9}$$

Formula (5.8) becomes

$$\pi r^2 h = \pi(r + \Delta r)^2 (1+\varepsilon)h \tag{5.10}$$

To arrive at

$$r^2 = (r + \Delta r)^2 (1+\varepsilon) \tag{5.11}$$

$$r + \Delta r = \frac{r}{\sqrt{1+\varepsilon}} \tag{5.12}$$

The optical fiber curvature ρ after stretching is

$$\rho = \frac{r + \Delta r}{(r + \Delta r)^2 + \frac{(h+\Delta h)^2}{4\pi^2}} = \frac{\frac{r}{\sqrt{1+\varepsilon}}}{\frac{r^2}{1+\varepsilon} + \frac{(1+\varepsilon)^2 h^2}{4\pi^2}} \tag{5.13}$$

If $r=5$ mm and $h=40$ mm, the curve shown in Eq. (5.13) is shown in Fig. 5.4. In the figure, the symbol "•" means the derivation from vertical and horizontal strains and "■" means the derivation from constant volume. Thus, the results obtained from the two derivation methods are much closer. The two derivation methods are different because of the influence of specific Poisson's ratio. When the rod is stretched and deformed, the curvature of the winding fiber will decrease with the increase of tensile

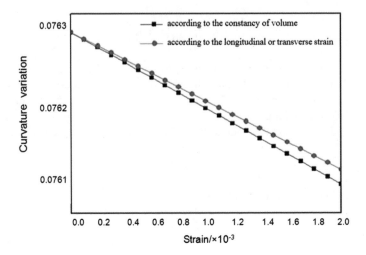

Fig. 5.4 Relation between curvature and stress

strain. The bending loss affected by the curvature is reduced accordingly. However, because the object of study is steel reinforcement and other materials, considering its Poisson's ratio and elasticity modulus, when the rod with $r=5$ mm and $h=40$ mm produces 2,000 $\mu\varepsilon$, the curvature ρ of the fiber changes only 0.0002. Therefore, the change of fiber curvature is not the main factor affecting the loss.

2. Influence of Stress on Fiber Loss.

The fiber is mainly affected by axial tension and normal lateral pressure. When the fiber is subjected to longitudinal (axial) mechanical stress, the three main physical effects that cause the change of the optical fiber parameters will be generated.

(1) Fiber Length Changes–Strain Effect

$$\varepsilon_{\text{axial}} = \frac{\Delta L}{L} \tag{5.14}$$

(2) Diameter Variation of Fiber Core–Poisson Effect

It is known that the Poisson's ratio of optical fiber μ is 0.17 so the transverse strain of the core $\varepsilon'_{\text{core}}$ is

$$\varepsilon'_{\text{core}} = \frac{\Delta r}{r} = -\mu\varepsilon_{\text{axial}} \tag{5.15}$$

That is, the change of fiber core diameter is

$$\Delta r = -\mu\varepsilon_{\text{axial}} r \tag{5.16}$$

5.3 Winding Optical Fiber Strain Sensor

(3) Refractive Index Variation of Optical Fiber Core–photoelastic Effect

When the fiber is subjected to the longitudinal mechanical stress, the optical fiber not only produces the strain, but also has the variation of the refractive index because of photoelastic effect. The relationship between photoelastic effect and strain is

$$\Delta\left(\frac{1}{n^2}\right)_t = \sum_{j=1}^{6} p_{ij} s_j \tag{5.17}$$

where, $i, j = 1, 2, \ldots, 6$; s_j is the fiber strain tensor; the p_{ij} is the component of the photoelastic tensor, or the photoelastic coefficient.

For homogeneous and isotropic optical fiber media, the photoelastic tensor p is

$$p = \begin{Bmatrix} p_{11} & p_{12} & p_{12} & 0 & 0 & 0 \\ p_{12} & p_{11} & p_{12} & 0 & 0 & 0 \\ p_{12} & p_{12} & p_{11} & 0 & 0 & 0 \\ 0 & 0 & 0 & p_{44} & 0 & 0 \\ 0 & 0 & 0 & 0 & p_{44} & 0 \\ 0 & 0 & 0 & 0 & 0 & p_{44} \end{Bmatrix} \tag{5.18}$$

where

$$p_{44} = (p_{11} - p_{12})/2 \tag{5.19}$$

So, this photoelastic tensor has only two independent components p_{11} and p_{12}. Also because

$$\Delta\left(\frac{1}{n^2}\right)_t = -2n^{-3}\Delta n_i \tag{5.20}$$

So, Formula (5.17) becomes

$$\Delta n_i = -\frac{n^3}{2} \sum_{j=1}^{6} p_{ij} s_j \tag{5.21}$$

where n is the initial refractive index of the fiber core. In the absence of shear strain, $s_4 = s_5 = s_6 = 0$, and $s_1 = s_2$ in each isotropic fiber medium. Now, there are

$$\Delta\left(\frac{1}{n^2}\right)_1 = (p_{11} + p_{12})s_1 + p_{12} s_3 \tag{5.22}$$

$$\Delta\left(\frac{1}{n^2}\right)_2 = (p_{12} + p_{11})s_1 + p_{12} s_3 \tag{5.23}$$

$$\Delta\left(\frac{1}{n^2}\right)_3 = 2p_{12}s_1 + p_{11}s_3 \tag{5.24}$$

Therefore, the refractive index change Δn of optical fiber core is

$$\Delta n = -\left(\frac{n^3}{2}\right)[(p_{11} + p_2)s_r + p_{12}s_1] \tag{5.25}$$

where $s_1 = s_3$, s_1 is the fiber axial strain, s_3 is the longitudinal strain.

When the rod is pulled and deformed, the axial strain of the optical fiber $\varepsilon_{\text{axial}} = \frac{s_2 - s_1}{s_1}$. The quartz fiber is known to have $n = 1.456$, $p_{11} = 0.121$, $p_{12} = 0.270$, and $\mu = 0.17$, which can be substituted in Formula (5.25) to obtain

$$\Delta n = -0.314 \varepsilon_{\text{axial}} \tag{5.26}$$

From Formulas (5.16) and (5.26), the axial strain of the optical fiber decreases the refractive index n of the core and the core diameter r, so that the guiding conditions in some high-order modes are destroyed. The total communication mode is reduced, and the optical power loss increases.

The tightening degree of the fiber in the production process has a great influence on the initial loss of the fiber. The larger the winding tension, the more closely the fiber is attached to the rod, and the greater the initial loss of the fiber. Similarly, the longitudinal strain of the rod will be increased, and the fiber will be attached to the rod. Therefore, the influence of normal side pressure of optical fiber in the stretching process is similar to tension in the winding. The normal lateral pressure increases with the increase of the axial strain of the optical fiber and the longitudinal strain of the rod, which leads to the increase of the optical power loss.

Therefore, when the rod is stretched and deformed, the fiber curvature in the sensitive segment decreases, which leads to the decrease of optical loss. However, the variable quantity is small when the rod deforms distinctly. The axial force and lateral pressure increase the optical loss. Therefore, due to the combined action of many factors, the optical power loss is increased, namely that it is increased when the wound rod is stretched.

5.3.2 Structure and Parameters of Winding Optical Fiber Strain Sensor

In the measurement of optical fiber loss, the light intensity received by the photoelectric detection system mainly depends on the power change of the optical fiber transmission. To achieve the sensing objective, it is necessary to make a significant change in strain.

Flat fiber is not sensitive to structural strain. When strain changes, the power of transmitting light in the fiber is small and it is difficult to be used as the sensor. To

5.3 Winding Optical Fiber Strain Sensor

Fig. 5.5 Optical sensing fiber winded on rods

aim at the sensing objective using a single optical fiber, the structure design should be reasonable. By changing the form of optical fiber to form one sensitive segment after another on the optical fiber, a fiber-optic sensor array is formed on a single fiber. Such design can greatly improve the response sensitivity of the fiber to structural strain and can realize the aim of quasi-distribution test of structural strain.

The structure design of the fiber-sensitive segment must be suitable for the application object. For example, if the prestressed tendons in large reinforced concrete structures are taken as the research object, the optical fiber spiral is wound on the circular section rod (see Fig. 5.5), named as winding fiber strain sensor. It has multiple winding segments, which constitutes a distributed sensor array as shown in Fig. 5.6.

In this form, the optical fiber is wound on the rod, which will first form an initial bending loss. When the rod deformation is stretched, not only the bending rate of the fiber is changed, but also the optical power transmission in the optical fiber can be changed by the combined action of axial tension and normal lateral pressure. Thus, the strain of the structure is obtained by the optical loss. In addition, it is easy to attach the fiber on the tested member, thus increasing the length of the fiber and improving the spatial resolution.

5.3.2.1 Influence of Sensing Fiber

When the optical fiber sensor is designed, the sensitivity of fiber sensor must be considered first. The sensitivity is affected by not only the fiber properties but also structural deformation. The influence of fiber-optic parameters on its sensing function is analyzed below.

1. Influence of Sensing Fiber Type

When the structure of fiber-optic sensor is designed, it is necessary to select the type of optical fiber according to the sensing principle. For example, the optical fiber sensor with phase modulation and interference theory is generally used, with high birefringence single-mode polarization-maintaining optical fiber. The optical fiber temperature sensor adopts the twinkling fiber with doped fluorescent material. Special grating fiber is used as the Bragg sensor.

Fig. 5.6 Schematic diagram of winding fiber sensing array

The backscattering method can be used to measure the backscattering of the normal single-mode fiber or multi-mode fiber. Thus, the loss information of the fiber is obtained. Compared with single-mode fiber, the multi-mode fiber core has large diameter, more propagation modes, and larger light transmission. Therefore, under the same external disturbance, the intensity of light intensity is much larger than that of single-mode fiber, and its backside Rayleigh scattering coefficient is much larger than that of single-mode fiber. In addition, the optical coupling efficiency of multi-mode fiber is in the range from 3 to 8 dB larger than single-mode fiber. Therefore, under the same light power, the backscattering power of multi-mode fiber is 10~15 dB larger than that of single-mode fiber.

To sum up, the optical intensity loss value of multi-mode fiber measured by the photoelectric detection system is much larger than that of single-mode fiber in the same external disturbance. Therefore, the use of ordinary multi-mode fiber as a sensing element can get higher sensitivity, accuracy, and signal-to-noise ratio than ordinary single-mode fiber.

2. Influence of Sensing Optical fiber Characteristic Parameters

Because the relative refractive index difference is small, the power coupling coefficient of the adjacent model group is proportional to the core diameter. That is to say, the greater the coupling degree of the conduction mode and the radiation mode, the greater the optical power loss in the fiber core. Therefore, the selection of certain relative refractive index difference and core diameter can increase or decrease the optical loss.

When the backscattering method is used to test the free flat optical fiber, the signal changes detected by the optical detection system are caused by the attenuation of the optical fiber or the variation of the scattering coefficient. In the demodulation of the sensing signal, the attenuation signal is a useful signal reflecting the measured physical information. Therefore, to ensure that the measured signals are truly indicators of the measured physical quantities, signals detected by the optical detection system should be only caused by structural strain changes. Therefore, optical fiber used must have the same scattering coefficient to remove the interference caused by the different initial scattering coefficients of the fiber.

5.3.2.2 Selection of Sensor Structure Parameters

When the fiber is wound around a round rod, an initial loss is generated on the fiber. Considering the principle of backscattering, the initial loss of sensor must be controlled in a certain range. The extremely small initial loss could be mistaken as a natural change along the optical fiber loss. The extremely large initial loss will reduce the signal-to-noise ratio at the measuring point due to too small light intensity of backscattering. Therefore, it must be selected properly.

The optical fiber spirally wound on the round rod can be taken as the cylindrical helix. Its parametric equation is

5.3 Winding Optical Fiber Strain Sensor

$$\begin{cases} x = r\cos\theta \\ y = r\sin\theta \\ z = \pm b\theta = \pm\frac{h}{2\pi}\theta = \pm r\theta\cot\beta \end{cases} \quad (5.27)$$

where $\theta = \omega t$ and ω is the angular velocity; h is pitch; β is the helix angle. The curvature of this cylindrical helix is

$$\rho = \frac{r}{r^2 + \frac{h^2}{4\pi^2}} \quad (5.28)$$

Fiber loss is mainly caused by bending when the optical fiber is spirally wound on the round rod and the rod is not stressed. The curvature is determined by the radius of the rod and the winding pitch. The number of winding turns is also a major factor in the determination of fiber loss. Therefore, the influence of structural parameters on the initial loss of fiber can be analyzed indirectly by analyzing the relationship between the structural parameters and the curvature of the fiber.

The test results show that the smaller the pitch, the faster the initial loss changes with the winding radius. The larger the pitch, the slower the initial loss changes with the winding radius. The initial loss and the pitch are approximately the power function under the winding radius.

The experimental results show the initial loss of optical fiber and winding number are simple proportional relation for several winding schemes with different winding radius, different screw pitch, and different winding radius and screw pitch. That is to say, the winding radius and screw pitch is dependent of the initial loss of the optical fiber in a single screw pitch, while the winding number plays the role of superposition.

In addition to the winding radius, screw pitch, and winding number, the looseness degree is also a key factor of the initial loss. The degree of tightness is the amount of tension applied to the optical fiber when the optical fiber is wound. The greater the tension when winding, the larger the initial loss; on the contrary, the smaller the tension, the smaller the initial loss. In the case of manual production, it cannot be quantitatively analyzed if the winding tension is not strictly controlled.

5.3.2.3 Determination of Sensor Distribution Parameters

1. Determination of Location of Measurement Points.

The winding fiber strain sensor is a quasi-distributed optical fiber sensor. Therefore, in addition to the design of single point (sensing unit), how to determine the location of test point is significant in the measurement of quasi-distribution. The determination of test point includes two parts: first, determining the structure signal position needs be used to analyze the structure correctly; second, determining the position of sensitive fiber along the fiber is also of significance depending on the test principle and various concrete influencing factors.

The determination of the locations of the first and final measurement points is mainly affected by the dead zone. To solve this problem, the following measures can be adopted: transition fiber method, masking method, and directional coupler method.

In the distribution test, the first measurement point on the optical fiber must be arranged outside the dead zone. Because of Fresnel reflection at the end of the whole optical fiber, it is necessary to make an optical fiber at the end to avoid the last sensitive segment being "flooded". The main limiting factor of other measurement points is the spatial resolution of the instrument. For every intermediate test points, it should also be noted that the distance between any two measuring points should not be less than the spatial resolution. To test the stress of the measuring distance between two points less than the spatial resolution, a time delay optical fiber between two fiber sensitive segments can be used to solve the above problem.

2. Number of Measurement Points

In practical application, the number of sensitive segments (i.e., measuring points) formed on single fiber is restricted by several factors. The number of measuring points in the distributed sensor system is restricted by three factors: the dynamic range of testing instruments, the initial loss of single measurement point, and the maximum variation amplitude of single point loss. Dynamic range is one of the indicators of test instruments. Therefore, the large dynamic range of instruments can be used to obtain more measurement points. The maximum amplitude of variation of single point loss is generally less than 0.1 dB, far less than the initial loss. Therefore, it can be considered as an auxiliary factor when the number of measurement points is estimated. After the dynamic range of the instrument and the initial loss and approximate range of each measuring point are known, a formula can be summarized to estimate the number of measurement points, that is

$$n = \frac{A - \alpha L - \beta}{k + \Delta k} \tag{5.29}$$

where A is the dynamic range of the instrument: α is the attenuation coefficient of the fiber; L is the length of fiber; β is the total value of other losses (such as connector loss, coupling loss, etc.); k is the single point initial loss; Δk is the variation amplitude of single point loss caused by the measured physical quantity.

5.3.3 Measurement System of Winding Optical Fiber Strain Sensor

For intensity modulated fiber optic sensors, the optical power meter is generally used for testing. The light source and test instrument must be placed at both ends of the sensor. Because of the large space length, the quasi-distributed optical fiber sensor of multi-winding section makes the test work difficult to be carried out. At the same

5.3 Winding Optical Fiber Strain Sensor

time, the optical power meter can only measure the total light intensity loss of the optical fiber from the light source to the optical power meter, which can mix the information of multiple sensing points and cannot attain accurate each point loss of multi-point quasi-distributed information.

For the purpose of quasi-distribution measurement, the optical time domain reflectometer (OTDR) is selected as the test instrument for the twined optical fiber strain sensor. The light source and photoelectric detection system are integrated in OTDR, so it is easy to conduct single end test for long-distance fiber. In addition, it can measure the light intensity loss at each sensitive segment (sensing point) along the fiber distribution. The principal block diagram of OTDR is shown in Fig. 5.7. The main clock control pulse generator modulates the laser light source with a narrow pulse light, which enters the optical fiber by directional coupler. Then, the information from optical fiber enters O/E unit through directional coupler, and then amplified, processed and reproduced on the display screen.

Light is attenuated exponentially in the transmission process. To show more intuitively, modern OTDR will take the signal through logarithmic processing. After logarithmic treatment, the backscattering signal of a uniform continuous fiber is shown as a strength indicator for linear distribution of time.

5.3.3.1 Measurement Principle of Five-Point Method

The light intensity received by the optical time domain reflectometer decreases exponentially with the length of the measured fiber. After the logarithmic transformation of the circuit, a straight-line waveform is obtained. However, and the signal-to-noise ratio S/N gets worse due to the large noise, the waveform curve often becomes the shape shown in Fig. 5.8 (nonlinear, and flapping up and down). If two points (M, R) are taken in the graph to measure the loss of the sensitive segment, there will

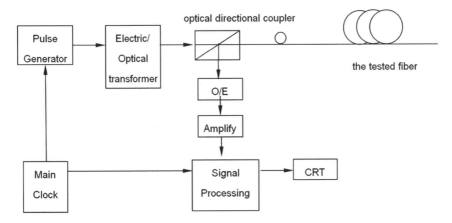

Fig. 5.7 Schematic diagram of OTDR measurement

Fig. 5.8 Linear approximation method using the least square method

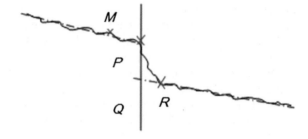

be a serious error in the measurement value due to the difference in the method of determining the location of P (mark point).

To solve this problem, statistical processing may be used to fit least squares of multiple data points within the desirable range. The calculation and processing of obtained fitting straight lines with this method can minimize the measurement error caused by the different positions of marker points.

The fiber loss of sensitive segment is shown as a "step" on the curve of backscattering light waveform. After the straight line is fit with the least square method, the optical loss value of sensitive segment fiber can be calculated with the five-point method. On the waveform curve at the sensitive section of the fiber, specify representative five points as shown in Fig. 5.9. On the two straight lines in front of and at the back of the "step", two points are specified, respectively (①, ② and ③, ④), and then the waveform linear equations in front of and at the back of the "step" are listed by two points, respectively.

The linear equation of point ② is

$$y = a_1 x + b_1 \tag{5.30}$$

The linear equation of point ④ is

Fig. 5.9 Schematic diagram of loss measuring based on five-spot method

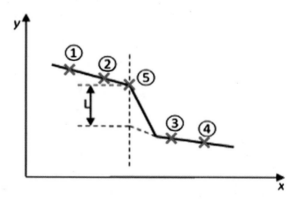

5.3 Winding Optical Fiber Strain Sensor

$$y = a_2 x + b_2 \tag{5.31}$$

If the position of the sensing point ⑤ is specified, the difference of optical power levels in front of and at the back of the "step" determined by the two equations can be used to calculate the optical power loss value of the sensitive section.

Suppose the coordinate of point ⑤ is $x=x_s$, the optical power loss of sensitive segment is

$$L = a_1 x_S + b_1 - (a_2 x_S + b_2) \tag{5.32}$$

In summary, the accurate optical power loss value can be measured. Above operation is automatically performed in the optical time domain reflectometer and the results are shown as insertion loss at point ⑤.

5.3.3.2 Noise Analysis and Information Processing

1. Noise Analysis

The backward Rayleigh scattering signal carrying the information along the optical fiber is very weak due to the existence of various noise sources. Therein, the main noise is optical detection system noise.

The optical detection system is actually a system for the conversion, transmission, and processing of optical signals. In addition to the photoelectric detector, it is also equipped with an electronic system. When the system works, it will always be interfered by useless signals. These non-signal components are collectively referred to as noise. The task of the detector and detection system is to extract the required optical signal and bring it to the required amplitude.

The noise of optical detection system is mainly thermal noise, shot noise, and $1/f$ noise (also called low-frequency noise). In most cases, the individual noise sources are essentially non-correlated noise independent of each other, so the total noise superposition becomes the sum of the individual noise powers.

All resistive elements with power consumption have thermal noise, which originates from the irregular thermal motion of free electrons or charge carriers inside the resistance. The thermal current noise power is

$$L_{nh}^2 = 4kT\Delta f / R \tag{5.33}$$

where k is a constant coefficient; R is the resistance value; T is the temperature of resistance; Δf is the electronic bandwidth of the measuring instrument. The spectral density of thermal noise power is of flatness, and the high-end limit frequency is 6×10^{12} Hz at room temperature. Since the operating frequency of general electronic system is much lower than this value, the white noise spectrum can be regarded as a typical form of thermal noise. From the relationship between the thermal noise power and the operating temperature of the optical detection system, the thermal noise of the optical detection system operating at low temperature will be reduced significantly.

Shot noise exists in all electron tubes and semiconductor devices, and light detectors are no exception. The shot noise of optical detector is a kind of noise caused by the particle property of electron or photo-induced carrier. If the operating bandwidth of the detector is Δf, the shot noise power is

$$I_n^2 = 2e\bar{I}\Delta f \tag{5.34}$$

where $\bar{I}$ is the average current of the detector; e is the electron charge. If the dark current of the optical detector I_d is included, the dark current noise power of the optical detector in the absence of light irradiation is

$$I_{nd}^2 = 2eI_d\Delta f \tag{5.35}$$

The shot noise of the optical radiation generated by the action of the light field can also be written directly

$$I_{np}^2 = 2eI_p\Delta f \tag{5.36}$$

where I_p is the average light current produced by the optical radiation field acting on the optical detector.

As can be seen from the above, the shot noise is determined by the current and the working bandwidth of the detector is independent of the operating frequency of the detector, and has the spectral characteristic of white noise.

As the noise of the detection system itself, there is also a $1/f$ noise related to the modulation frequency f, which is characterized in that the noise power spectral density is inversely proportional to the frequency. The mean square value of $1/f$ noise can be expressed as the following by empirical formula:

$$I_n = \left(\frac{AI^\alpha \Delta f}{f^\beta}\right) \tag{5.37}$$

where A is a proportional coefficient; f is the modulation frequency, $\alpha \approx 1$, $\beta \approx 2$. In general, such noise can be significantly reduced as long as the low-frequency modulation frequency is limited.

The three noise powers of detection system above can also be collectively expressed as

$$I_n^2 = S_n(f)\Delta f \tag{5.38}$$

To reduce the noise of detector or detection system, first, the noise power spectral density is the smaller, the better; second, the bandwidth of the system should be reduced as far as possible. In a word, the noise of detection system can be divided into low-frequency noise (i.e., $1/f$ noise) and white noise (thermal noise and shot

5.3 Winding Optical Fiber Strain Sensor

noise). $1/f$ noise can be reduced by limiting the modulation frequency, while the white noise can be limited by averaging method.

2. Information Processing

Because there are many kinds of noises in OTDR detection system, and the power of backward Rayleigh scattering signal is very weak, the signal-to-noise ratio of the signal received by OTDR is poor, which makes the useful signal often submerged by the noise. Therefore, it is necessary to extract the useful signal that is submerged by the noise through the weak signal processing technology.

The sampling integrator (BOXCAR) is a powerful tool for weak signal detection. The operating principle of sampling integrator is as shown in Fig. 5.10. The periodic signal submerged in the noise is discretized through sampling, and then sent to an integrator for accumulation, averaging and holding. Since there is a correlation between signal and sampling pulse, and the noise is random, the average value of noise will be smaller and smaller after a certain superposition average. Meanwhile, the periodic signal will be continuously increased by exponential law, the signal-to-noise ratio can be improved as a result, and the weak signal can be detected from the noise.

The sampling integrator usually has two operating modes, namely, fixed-point type and scanning type. A fixed-point sampling integrator is used to measure the average value of a certain transient of the periodic signal. The scanning sampling integrator can recover and record the whole waveform of the measured signal. This paper mainly analyzes the operating principle of scanning sampling integrator to help understand the working process of OTDR.

Since the sampling pulse of variable time delay changes gradually in the sampling process, the sampling process of the scanning sampling integrator is limited by the gate pulse width T_g. It can only be sampled in the width range of the gate pulse, and the gate pulse jumps at ΔT_g time interval. Since the slow scanning voltage varies very slowly with respect to the time base voltage, the movement of the sampling

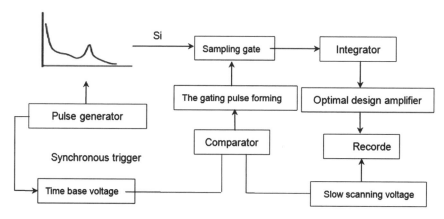

Fig. 5.10 Principle of sampling integrator

pulse with respect to the trigger pulse is also very slow. As a result, a plurality of sampling pulses with a gate width T_g can be successively swept by each point on the input signal waveform (see Fig. 5.11), so as to realize multiple accumulated averages for each sampling point p on the waveform.

If the gate width of the sampling pulse is T, and the pulse jumps to move (time delay) at interval of ΔT_g, the sampling times of any point on the measured signal waveform within gate width T_g is

$$n_s = \frac{T_g}{\Delta T}. \tag{5.39}$$

According to $\sqrt{N}$ law, the improvement in signal-to-noise ratio can be written as

$$\text{SNIR} = \sqrt{\frac{T_g}{\Delta T}}. \tag{5.40}$$

In slow scan time T, the amount of all sampling pulses sweeping one waveform of the sweep signal is

$$n_t = \frac{T_{SR}}{T_s}, \tag{5.41}$$

where T_s is the period of the measured signal. And the interval of sampling pulses is

$$\Delta T = \frac{T_b}{n_t} \tag{5.42}$$

where T_b is the time base voltage period. Substitute Formulas (5.41) and (5.42) into Formula (5.40), and obtain

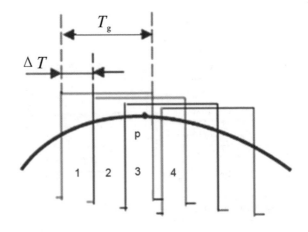

Fig. 5.11 Schematic diagram of pulse scanning

5.3 Winding Optical Fiber Strain Sensor

$$\text{SNIR} = \sqrt{\frac{T_g T_{SR}}{T_b T_{SR}}} \quad (5.43)$$

The above formula shows that the gate width T_g and slow scanning time T_{SR} are required to be larger to obtain larger SNIR. That is, more accumulated sampling times are required for the signal, and the measuring time is lengthened. Since the sampling interval of signal depends on the gate width T_g. The narrower the gate width, the smaller the sampling interval, and the higher the resolution of the signal after recovery. However, it can be seen from Formula (5.43) that when T_g is small, SNIR is small. The only way to maintain a certain SNIR is to reduce ΔT_g. Therefore, the resolution of waveform cannot be improved without limitation, and can only be considered according to the measurement requirements as appropriate.

3. Analysis on System Response Time

The response time of system, i.e., the acquisition time of measured value, is the whole time from pressing the test key, beginning to emit laser pulse to the acquisition of measured value. This time reflects the response speed of the system to the measured physical quantity, which adopts OTDR as a characteristic parameter of test instrument.

The repetition period T of the optical pulse emitted by OTDR varies. The larger the range, the longer the period. Since in the period that the backscattered light and Fresnel-reflected light of the probe optical pulse are transmitted to OTDR when the detected optical pulse arrives at the end of the measured optical fiber under test, only one optical pulse is allowed, the backscattered curve is the collection of continuous scattered or reflected signals of an optical pulse at the point along the line. If a second optical pulse is emitted during this period, the continuity of curve will surely be broken. To measure an optical fiber with a total length L, the optical pulse emitted by OTDR arrives at the optical fiber terminal, and then the time required for scattered or reflected light of the optical pulse at the terminal to be transmitted to the measurement output/input port of OTDR is

$$t = \frac{2Ln}{c} \quad (5.44)$$

Therefore, the repetition period of the optical pulse must be greater than t.

In addition, because the signal-to-noise ratio of the signal received by OTDR is poor, multiple averages are required to improve the signal-to-noise ratio. Therefore, the total response time of the system should consider not only once measurement time (i.e., a single test period T), but also the average number of times N, so that the total test time T_g of the system is always

$$T_a \geqslant \frac{2NLn}{c} \quad (5.45)$$

4. Selection for the Wavelength of the Sensing Test

The behavior of optical fiber sensing system based on scattering principle is directly related to the transmission wavelength, and different wavelengths have different optical fiber attenuation characteristics. Therefore, the application of different test wavelengths will have a significant impact on the test sensitivity. As for the same fiber category, the attenuation per unit length of 1,550 nm wavelength is smaller than that of 1,310 nm wavelength, but the sensitivity of 1,550 nm wavelength to bending loss is greater than that of 1,310 nm wavelength.

The following tests strongly verify the effect of test wavelength on bending loss sensitivity. It can be seen from Fig. 5.12 that when the optical wavelength is 1,550 and 1,310 nm, respectively, with the increase of bending degree, the measured loss value of 1,550 nm increases sharply, while the measured loss value of 1,330 nm changes slightly. Especially when the bending diameter is less than or equal to 20 mm, the bending sensitivity of 1,550 nm wavelength is more prominent. Therefore, 1,550 nm test wavelength can greatly improve the sensitivity of optical fiber sensing system.

5.3.4 Sensing Characteristic of Winding Optical Fiber Strain Sensor

5.3.4.1 Composition of the Measurement System

The performance testing system of the winding optical fiber strain sensor (see Fig. 5.13) consists of three parts: the winding optical fiber strain sensor to be tested, the test instrument, and the loading device. The test specimen was loaded with a certain step in the elastic range, the output of optical fiber sensor, and the strain of test specimen are recorded followed by analyzing and processing .

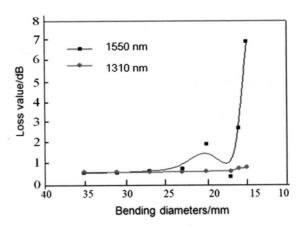

Fig. 5.12 Curve of bend sensitivity versus different wavelengths

5.3 Winding Optical Fiber Strain Sensor

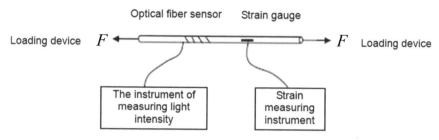

Fig. 5.13 Test system of twisted optical fiber sensor

1. Fabrication of Winding Optical Fiber Sensor

The used optical fiber is ordinary single-mode communication fiber (such as Lucent Technologies G.652 high-quality single-mode optical fiber). Its parameters are as follows: The core diameter is 8.3 μm, the cladding diameter is 125 μm, and the relative refractive index difference is 0.33%. Two types of test specimens are selected, namely, steel wire and steel bar. The ordinary single-mode optical fiber is spirally wound on the specimen with two diameters, and fixed with adhesive. The main parameters of test specimen are as shown in Table 5.1.

2. Selection of the Test Instrument

The test instrument includes an optical fiber sensor test instrument and strain test instrument. Model HP8147 OTDR is selected, and Model YE2536 resistance strain gauge is selected as the strain measuring instrument.

3. Selection of Test Loading Device

Because OTDR is selected as the test instrument of optical fiber sensor, and the principle of backscattering method determines that the obtained signal must be subjected to multiple averages, it takes a certain time to get the test result. While loading, to make the test result accurate and reliable, it is necessary to ensure that the load applied to the experimental machine is stable during OTDR test. Therefore, Model WJ-10A mechanical universal tester is selected as the test loading device.

Table 5.1 Specimen parameters of test

Diameter $2r$ (mm)	Pitch h (mm)	Winding number n
10	40	4
10	37	5
10	37	6
1.8	22	30

5.3.4.2 Design of Loading Device for Quasi-Distributed Test

After the optical fiber is spirally wound on the test specimen, a sensing zone is formed. Each continuous winding section constitutes a "point"-type sensor, and the test value of the "point"-type sensor is the average stress value of the continuous winding section; a plurality of winding section of delayed optical fibers at certain distance from a "point" sensor array, namely, a quasi-distributed optical fiber sensor as shown in Fig. 5.14.

It is known that the thickest specimen to be stretched is a 10 mm diameter circular section steel bar. Therefore, to ensure cyclic loading and unloading in the elastic range, the maximum tensile force shall be calculated as follows:

$$F = \sigma_p \frac{\pi d^2}{4} = 200 \times 10^9 \times \frac{\pi \times 0.01^2}{4} = 15,707.963 \text{ N} \qquad (5.46)$$

where the proportional limit $\sigma_p = 200$ MPa. Select No. 10 channel steel as raw material, weld 4 sections of channel steel into the form as shown in Fig. 5.15, and ensure a certain parallelism of two short sides. After strength analysis and stability check are carried out on the structure according to the maximum load to be applied to the test apparatus, it is known that the structure size design in the figure meets the test requirements.

In this device, an axial force can be applied to reinforcement by rotating a nut at one end of the frame. When the thread pitch is 1.5 mm, it can not only make the steel bar bear large force, but also control the load force accurately. However, if one end of that steel bar is fully fixed, torque will be generated on the steel bar when the nut is rotated to apply force. Therefore, it is necessary to install a movable jaw on one end (use in conjunction with retaining ring, aligning roller bearing and drill chuck). When the test specimen is subjected to torque, the bearing rotates to automatically eliminate the torque. The specific structure arrangement is as shown in Fig. 5.16.

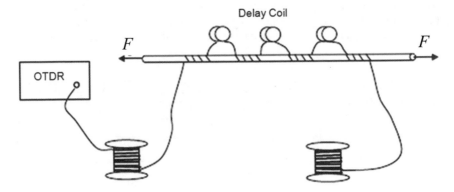

Fig. 5.14 Test system of quasi-distributing twisted optical fiber sensor

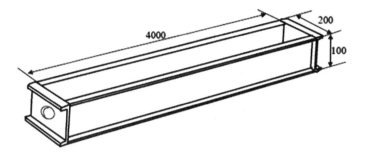

Fig. 5.15 Device used to apply on distributed sensor

5.3.4.3 Test and Analysis of Winding Optical Fiber Strain Sensor

In the test, the simplest axial force is applied to the specimen, and the loading and unloading tests are carried out step by step in the elastic range. By recording and summarizing the test data, the output–input relationship of the sensor and various static characteristics embodied by the static performance indexes are analyzed in detail. The static characteristic is an important basis to evaluate the static performance of sensors. The main indexes for the static characteristics of the sensor include linearity, sensitivity, hysteresis, and repeatability. The experimental study is conducted for ϕ 10 reinforced type specimen with 40 mm pitch and 4 turns.

Since the maximum tolerable tensile force of ϕ 10 specimen is 15 kN, the maximum load is taken as 8,000 N in consideration to material defects. If the loading step was determined as 500 N, the corresponding strain is as follows:

$$\varepsilon = \frac{\sigma}{E} = \frac{F}{EA} = \frac{F}{E\frac{\pi d^2}{4}} \approx 30 \times 10^{-4} \mu\varepsilon \tag{5.47}$$

Since there are 17 test points from 0 to 8,000 N, the step can also be modified to be 1,000 N, so that the entire test range can be divided into 8 levels of loading and unloading. By processing the test data, the fitting curve is as shown in Fig. 5.17.

Fig. 5.16 Local unit applied with loads

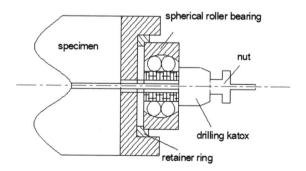

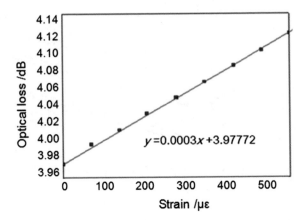

Fig. 5.17 Fitting curve of optical loss of steel bar specimen

1. Linearity

Linearity is usually expressed in terms of relative error, i.e.,

$$e_L = \frac{\Delta_{max}}{y_{FS}} \times 100\% \tag{5.48}$$

where Δ_{max} is the maximum deviation between the output average value and the fitted straight line, and y_{FS} is the full range output. The principle of least square method is used to obtain the fitting straight line, which can guarantee the minimum residual sum of squares for the calibration data of the sensor. As known from the calculation, the maximum deviation between the output average value and the fitted straight line $\Delta_{max} = 0.0043$ dB, and the full range output $y_{FS} = 0.1414$ dB after the constant component, namely the original loss value 3.9972 is eliminated. Substitute it into Formula (5.48), and obtain the linear error e_L 3.0%.

2. Sensitivity

Sensitivity refers to the ratio of the output increment of the sensor to the input increment, i.e., the slope of its fitting line expressed by $k = y/x$. Thus, the sensitivity of the sensor is 0.3×10^{-3} dB/με.

3. Hysteresis

Hysteresis is usually expressed by the ratio percentage of the maximum deviation of the output in the inverse and reverse strokes to the full range output, i.e.,

$$e_H = \frac{\Delta H_{max}}{y_{FS}} \times 100\% \tag{5.49}$$

Through the analysis on the test data, it is found that the full range output $y_{FS} = 0.1414$ dB, and maximum deviation of the output in the inverse and reverse strokes $\Delta H_{max} = 0.008$ dB. The hysteresis of 5.7% is obtained by substituting then into Formula (5.49).

5.3 Winding Optical Fiber Strain Sensor

4. Repeatability

The repeatability error reflects the dispersion degree of the calibration data and is a random error, so it can be calculated according to the standard deviation, i.e.,

$$e_R = \frac{a\Delta_{max}}{y_{FS}} \times 100\% \tag{5.50}$$

In that formula, Δ_{max} is the maximum value in the standard deviation of the output in the inverse and reverse strokes at each calibration point; a is the confidence factor usually taken as 2 or 3. When $a=2$, the confidence probability is 95.4%; When $a=3$, the confidence probability is 99.73%.

When the number of measurements in the test is less than 10, the range method can be used to calculate the deviation. The formula for calculating the deviation is:

$$\sigma = \frac{W_n}{d_n} \tag{5.51}$$

where W_n is the range; d_n is the range coefficient related to the number of measurements n, the value of which can be searched from Table 5.2.

By calculating the range $W_n = 0.007$, and searching $d_n = 1.91$ on the basis of $n=3$, take $a=2$, and substitute it into Formula (5.50), and obtain the repeatability error e_R 5.2%.

5. Resolution

There is a close relationship between strain resolution and sensitivity. By defining the sensitivity, it can be seen that the sensitivity k is the loss variation corresponding to the unit strain. The strain resolution is the minimum strain variation value that can be "felt" by the sensor. At present, if the minimum component of OTDR test loss value is known to be 0.001 dB, the strain resolution is the size of the strain variation value corresponding to 0.001 dB. Therefore, the strain resolution δ_ε can be calculated by the following formula:

$$\delta_\varepsilon = \frac{0.001}{k} \tag{5.52}$$

The sensitivity of the sensor k is known to be 0.3×10^{-3} dB/με. Substitute it into the formula above, and obtain the strain resolution of the sensor 3.3 με.

Table 5.2 Range coefficient

n	2	3	4	5	6
d_n	1.41	1.91	2.24	2.48	2.67
n	7	8	9	10	
d_n	2.88	2.96	3.08	3.18	

6. Measurement Range

The maximum measured value (i.e., input) that a sensor can measure is referred to as the upper measurement limit, the minimum measured value is referred to as the lower measurement limit, and the measurement interval from the lower measurement limit to the upper measurement limit is defined as the measurement range, the sensor was selected to be correctly used according to the measurement lower limit and the measurement upper limit of the sensor; the full input range of the sensor can be known by measuring range, and the corresponding full output value range is a significant parameter indicating the sensor performance.

In this test, the maximum input of the sensor is corresponding to the elastic limitation of the test specimen used. But according to the test data for the initial loss of the optical fiber sensor in Chap. 3, its loss is in the range of 1–7 dB, so the loss either caused by initial bending or caused by strain change can be measured by the test instrument. In addition, the amplitude of light intensity loss change is much less than its initial loss (generally less than 0.1 dB) due to the strain variation Therefore, according to the principle of the test instrument and the winding optical fiber strain sensor, the winding optical fiber strain sensor has a larger theoretical measuring range, which is enough for the strain range of the engineering structure.

5.3.5 Distributed Sensing Characteristics of Winding Optical Fiber Strain Sensor

The optical fiber is spirally wound on a circular bar to form a stress/strain sensitive segment, thus forming a single point winding optical fiber strain sensor. The multi-point quasi-distributed sensor array on a single optical fiber can be formed by spiral winding at different positions of the optical fiber, so as to form a quasi-distributed optical fiber sensor. When performing multi-point measurement, the typical test curve of OTDR is as shown in Fig. 5.18. In the figure, the reading of ordinate A is the level value, which is the attenuation value of optical power subjected to attenuation in inverse and reverse directions; abscissa L is the fiber length. Each "step" on the curve represents a single point winding optical fiber strain sensor. When different positions of the structure are in different stress states, each "step" will be reduced to different degrees on the original basis so as to reflect the change of the structure, and realize the test of the distributed stress.

In the distributed measurement, the measurement accuracy of each point is subjected to the combined influence of instrument pulse width, dynamic range, and distance to the test end and signal-to-noise ratio.

For measuring points at the same position, when measuring fiber loss, using narrow pulse can get higher spatial resolution than using wide pulse. There is a sharp falling edge at the "step" waveform displayed on OTDR, which is easy to be recognized in display and can read out more accurate measurement values. However, the transmission power of the narrow pulse is lower than the dynamic range of the

5.3 Winding Optical Fiber Strain Sensor

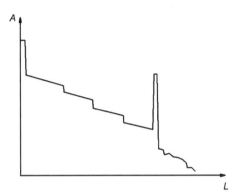

Fig. 5.18 Result of OTDR test

wide pulse, so the measurement accuracy decreases with the decrease of the signal-to-noise ratio at the measuring point at the same measuring distance.

Under the condition of the same pulse width, because of the optical power loss along the optical fiber, the transmission optical power of measuring point at far distance is decreased, the signal-to-noise ratio is decreased, and the test accuracy is lower than that of the near end. Therefore, when it is arranged at the measuring point, the stress state of the structure to be measured should be preliminarily estimated, and the most critical point should be set at the near test end of the optical fiber.

Although the abscissa of the OTDR measurement curve is shown as the distance, the true measurement is the time delay, i.e., the physical distance L displayed by OTDR is calculated by time delay t, the vacuum speed c, and the refractive index n of the optical fiber.

The distance accuracy depends on the refractive index error, the time base error (variation in the range of $10^{-5} \sim 10^{-4}$) and the sampling resolution S_r. When the refractive index error is not taken into consideration, the distance accuracy P_D can be expressed as below by the following empirical formula:

$$P_L = \pm 1 \text{ m} \pm 5 \times 10^{-5} \times L \pm S_r \tag{5.53}$$

When using OTDR to test, if there is an error between the refractive index setting and its true value, the positioning accuracy of the far-end sensing point on the optical fiber will be greatly affected. Therefore, only when the refractive index of the optical fiber is set correctly, can the obtained distance be guaranteed to be accurate and reliable. In a word, to improve the accuracy of distance measurement, an accurate method must be adapted to measure the refractive index of the optical fiber.

5.4 Large Displacement Sensor Based on Fully Distributed Optical Fiber Sensor

Displacement monitoring is an important method to evaluate the deep underground deformation of landslide, and has been widely used in the stability prediction of dam, rock mass, and slope. However, the monitoring technology with large measuring range, high precision, and good stability cannot keep up with the development of national economic construction. Especially in recent years, construction investment in railways, highways, and water conservancy projects has been increased, and it appears to be important to monitor the deformation and other stability problems of dams and slopes in real time and accurately. In order to monitor the slope and dam, much attention has been paid to FBG displacement sensors (Sun et al. 1999; Zhang et al. 2014; Zhao et al. 2015) and fully distributed fiber displacement sensors (Ding et al. 2015; Li et al. 2014; Nishio et al. 2010; Wylie et al. 2011; Tong et al. 2014) in recent years. The fiber grating displacement sensor monitors the displacement through the wavelength shift reflected by fiber grating—the internal sensitive element, but the structure design is very complicated (He et al. 2010; Hiroyuki et al. 2012; Zhang et al. 2014; Zhao et al. 2015), has small range, and cannot obtain the displacement directly. It is difficult to use in deep geotechnical engineering. Since the fiber grating quasi-distributed displacement sensor (Duan et al. 2013; Li et al. 2014) and fully distributed fiber displacement sensor (Behrad et al. 2016; Ding et al. 2015; Li et al. 2014; Nishio et al. 2010; Tong et al. 2014; Wylie et al. 2011; Ricardo et al. 2015) obtain structural displacement (Li et al. 2014; Nishio et al. 2010) through structural distributed strain integral, the displacement measurement error is extremely great resulting in the measurement of distortion. Therefore, this book presents a principle of optical time domain displacement measurement based on fiber grating positioning, and designs two kinds of displacement sensors characterized by direct displacement measurement without integral, large measuring range, simple structure, easy to be used in deep geotechnical engineering structure and so on.

5.4.1 Principle of Fully Distributed Displacement Sensing Based on Fiber Bragg Grating

The optical time domain reflection technique is an effective method to evaluate the loss and location of long-distance optical fiber, and it is also an important method to detect the intensity modulated optical fiber sensor. In this chapter, the principle of optical time domain reflection displacement measurement based on fiber grating positioning is proposed. In accordance with the characteristics of the same direction and greater optical power difference for fiber grating backward reflected light and the backward Brillouin-scattered light (see Figs. 5.19, 5.22 and 5.25, respectively), the optical grating positioning can be realized by optical time domain in such principle. Therefore, there will be a corresponding reflection event at the optical grating. When

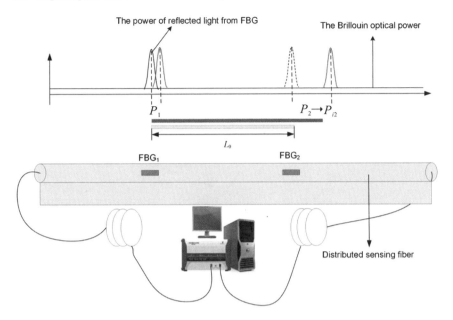

Fig. 5.19 Mechanism of displacement measuring based on FBG

the external displacement changes, the spatial distance of two reflection events of two optical gratings cascaded on the distributed sensing fiber, called as the position indicator, will change relatively via optical time domain spectrum. When the distance between two gratings changes, the spatial position of FBG$_1$ changes from P_1 to P_{i1}, and the spatial position of FBG$_2$ changes from P_2 to P_{i2}. Therefore, the structural displacement can be reflected by the relative position change $(P_{i2} - P_{i1}) - (P_2 - P_1)$ between gratings as shown in Fig. 5.19 (Li et al. 2015).

The strain limitation of the optical fiber is about 10,000 με, and the optical fiber will be broken and damaged if it is beyond the limited strain of the optical fiber. Therefore, according to the strain-displacement theory, the measuring range of the displacement sensor is related to the loaded length of the structure, namely

$$D = (P_{i2} - P_{i1}) - (P_2 - P_1)/n = \varepsilon_{max} L_0 \tag{5.54}$$

where D is the measurable displacement range of the sensor; L_0 is the length of the tested structure; n is the number of loaded sections of optical fiber laid around the structure.

As can be seen from Formula (5.54), the output of the displacement sensor is directly the length or displacement and there is no need to be computed by conversion or integration. The measurable displacement range of the sensor can also be designed according to the actual engineering needs. If a large displacement is desired, the loaded length of the displacement sensor will correspondingly be increased. Thus, the displacement sensor can measure a wide range of displacement variations through

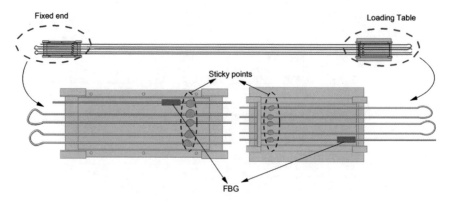

Fig. 5.20 Experimental setup used to apply strain on the distributed sensor

a flexible structural design. For example, the sensor may be applied to the large length loaded structures such as prestressed tendons.

The displacement sensitivity coefficient of the displacement sensor is

$$K_\mathrm{D} = \frac{(P_{i2} - P_{i1}) - (P_2 - P_1)}{\Delta L_0} \tag{5.55}$$

where K_D is the strain sensitivity coefficient of the displacement sensor, ΔL_0 is the displacement of the loaded structure.

The range and strain sensitivity of the displacement sensor can be freely designed according to the actual engineering needs. From Formula (5.54), it is known that increasing the measuring range of the displacement sensor requires increasing the length of the loaded structure; From Formula (5.55), it is noted that increasing the displacement sensitivity coefficient requires increasing the distance between two fiber gratings on the premise that the displacement of the loaded structure remains unchanged. Therefore, in practical engineering, when the structure displacement remains unchanged, the distance of between gratings can be elongated by increasing the number of loaded optical fiber segments followed by improving the sensitivity coefficient of the sensor.

5.4.2 Displacement Loading Test

To verify the principle and characteristics of fully distributed displacement sensing based on fiber grating positioning, displacement loading experiments was conducted (Figs. 5.20 and 5.21). The performance parameters of the fiber grating used in the test are shown in Table 5.3. Displacement sensor 1# and displacement sensor 2# fabricated with different Bragg gratings. Therein, displacement sensor 1# has 3 fiber gratings, and displacement sensor 2# has 2 fiber gratings. The arrangement of the fiber gratings

5.4 Large Displacement Sensor Based on Fully Distributed ...

Fig. 5.21 Photo of displacement measuring

in the displacement sensor is as shown in Fig. 5.20. In the experiment of displacement loading, the length of loaded optical fiber is limited by the experimental space, so it loaded optical fiber cannot satisfy the desired length. Therefore, to improve the range and the sensitivity coefficient of the displacement sensor, multiple loaded optical fiber sections are designed in the experiment. Among them, displacement sensor 1# is arranged with 4 sections tensile optical fiber of 3.96 m for each, and loaded with 1, 2, and 3 cm displacements (corresponding strains are 2,525.3, 5,050.5, 7,575.8 $\mu\varepsilon$), to ensure that the loading strain is less than the limit strain; 5 sections optical fiber was elongated to 5.0 m in displacement sensor 2, loaded with 0.5, 1, 1.5, 2, and 2.5 cm displacements, and followed by reducing the applied strains (1,000, 2,000, 3,000, 4,000, 5,000 $\mu\varepsilon$), the debonding of optical fiber from the holer does not occur. During the experiment, the fixed position at one end of the fiber is not changed, and displacement is applied at the other end by the displacement sliding table, so as to change the distance between the fiber gratings. In the test, NBX-7020

Table 5.3 Parameters of FBG

Sensor	Wavelength λ (nm)	Reflectivity R (%)	Bandwidth W (nm)	Distance L (m)	Fiber length L (m)
1#	1,509.930	90.98	0.23	12	8
	1,514.92	90.68	0.20	12	
	1,519.90	89.50	0.20		
2#	1,525.60	88.63	0.22	35	5
	1,535.80	90.27	0.19		

optical fiber stress analyzer of Neubrex Company is used to measure the position of fiber grating. The spatial resolution, the sampling interval and the room temperature is 2 cm (pulse width 0.2 ns), 1 cm, and 23.4 °C in the test procedure, respectively.

5.4.3 Analysis of the Displacement Sensing Characteristics

1. Displacement Sensing Characteristics of 1# Displacement Sensor

Figure 5.22 is the optical time domain spectrogram of displacement sensor 1# (Li et al. 2016). This figure shows that 1# displacement sensor can indicate the spatial positions of three fiber gratings by the optical time domain method. The reflection peaks of three fiber gratings can be clearly seen from Fig. 5.22a, which coincides with the location of preset three fiber gratings in 1# displacement sensor. From Fig. 5.22b, it can be seen that the position of FBG_1 remains basically unchanged as the displacement of the structure is 1, 2, and 3 cm. Meanwhile, the repeatability of the positioning data is good, and there is about an error of 1 cm (Li et al. 2014). It can be seen from Fig. 5.22c that when the loading displacement is 1 cm, the position of FBG_2 is 20.236 m, which varies by about 2 cm compared with the reference spatial position (20.215 m) indicating that the spatial position of FBG_2 is shifted due to the displacement, and that the measured displacement of 2 cm is less than the theoretical value 3.2 cm. Since the minimum sampling interval is 1 cm, there is 1 cm error in the system. When the loading displacement is 2 and 3 cm, the spatial position of FBG_2 basically remain unchanged, and the spatial position of FBG_2 does not vary by displacement. The reason is that ① the first section optical fiber has relaxed, and so the strain is small, although the strain of the second and the fourth optical fibers increases (Fig. 5.23), the displacements produced by both offset each other, and the total displacement basically remains unchanged; ② due to the minimum sampling interval of 1 cm. As can be seen in Fig. 5.22, when the loading displacement is 1 cm, the position of FBG_3 is 32.32 m, which varies by 3.1 cm compared with that of the reference spatial position (32.289 m). Meanwhile, it indicates that the spatial position of FBG_2 occurs the shift to longer distance due to the displacement. However, the position variation of 3.1 cm is less than the theoretical displacement, it is inferred that the cause of the measurement error is that the degumming of the

5.4 Large Displacement Sensor Based on Fully Distributed …

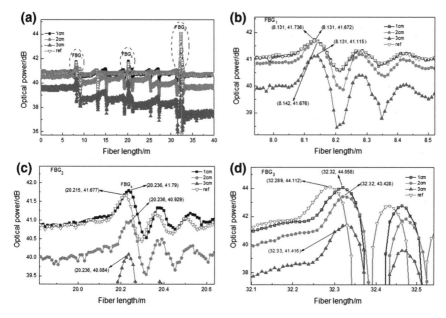

Fig. 5.22 Time domain spectrum of 1# DOS. **a** Overall time domain spectrum of 2#DOS; **b** time domain spectrum of FBG_1; **c** time domain spectrum of FBG_2; and **d** time domain spectrum of FBG_3

fiber in the bonding area at both ends results in fiber relaxation due to the great applied strain. For example, in Fig. 5.23 and Table 5.4, it can be seen that the loading displacement is 1, 2, and 3 cm, the strain of the first section fiber decreases from 9,015.478 to 6,358.872 $\mu\varepsilon$, and even 5,105.708 $\mu\varepsilon$. It indicates that the first section fiber has been relaxed, and the second section fiber and the fourth section fiber are well bonded under the applied displacement of 2 cm. However, the optical fiber has been degummed under the applied displacement of 3 cm. The third section is well bonded on both ends under the applied displacement of 1 cm, but the bonding points on both ends are degummed under the applied displacement of 2 and 3 cm.

As can be seen from the above, due to the double reasons of fiber relaxation and the limitation of sampling point spacing, the measured displacement error compared with the theoretical calculated displacement of the fully distributed fiber displacement sensor results in the lower linear correlation of the fitting curve between the theoretical displacement and the measured displacement, which is only 0.67 (see Fig. 5.24). At the same time, the measured strain sensitivity of the displacement sensor cannot be obtained due to the debonding at both ends of the loaded optical fiber.

2. Displacement Sensing Characteristics of 2# Displacement Sensor

To enlarge the measured displacement range, improve the displacement sensitivity and prevent the degumming phenomenon of the optical fiber in the bonding zone, the number of tensile sections and the load length (5 m) of the optical fiber are increased. Figure 5.25 is the positioning spectrum of 1# displacement sensor. From

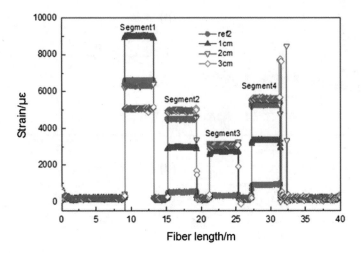

Fig. 5.23 Strain spectrum of 2# DOS

Fig. 5.25c, it can be clearly seen that there are 2 reflection peaks of fiber gratings, which are consistent with the actual location of 2 preset fiber gratings in 2# displacement sensor. From Fig. 5.25a, it can be seen that when the loading displacements are 0.5, 1, 1.5, 2, and 2.5 cm, the position of FBG_1 remains basically unchanged and is basically consistent with the reference spatial position. It can be seen from Fig. 5.25b that when the loading displacement increases, the position of FBG_2 (40.041 m) also shifts to 40.174 m, which is increased by 13.3 cm, and is basically consistent with the theoretically calculated displacement (12.7 cm). From Fig. 5.26 and Table 5.5, it can be seen that when the loading displacements are 0.5, 1, 1.5, 2, and 2.5 cm, the strain of the 5 loaded optical fiber sections is relatively uniform, and the loading strain is consistent with the actually calculated stain. Meanwhile, it also indicates that optical fiber degumming has not occurred in the bonding area on both ends of

Table 5.4 Strain data of sample 1#

Fiber section	Displacement (cm)			
	0	1	2	3
The first section fiber	6,613.312	9,015.478	6,358.872	5,105.708
The second section fiber	520.996	2,934.686	4,510.664	5,032.992
The third section fiber	314.669	2,714.147	3,083.99	3,083.99
The fourth section fiber	937.439	3,348.755	5,295.45	5,639.475
Interval between FBGs (m)	0	0.06431	0.044023	0.047454

5.4 Large Displacement Sensor Based on Fully Distributed …

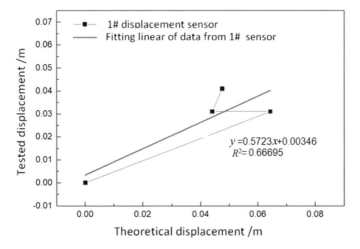

Fig. 5.24 Comparison between theoretical displacement and tested displacement from 1# DOS

2# displacement sensor. The measured displacement from the displacement sensor is consistent with the calculated displacement. For example, when the actual displacement of the structure is 2.5 cm, the theoretically calculated value of grating spacing is 12.5 cm (2.5 cm × 5 loaded optical fiber sections), while the measured displacement from the 2# displacement sensor based on grating intervals is 12.7 cm. So the above two displacements (12.5 and 12.7 cm) are highly consistent with each other. It can be seen that the slope of the curve is 1.078 (see Fig. 5.27). The displacement sensitivity coefficient of the fully distributed displacement sensor is 5.21 (see Fig. 5.28), and the linearity is better. According to the formula calculation, it is found that the theoretical sensitivity coefficient is 5.32, that is, it is feasible to improve the displacement sensitivity by installing five loaded optical fiber sections. Due to the maximum error of the sensor is 2 cm, if the testing accuracy of the displacement sensor was improved, it is necessary to increase the number of loaded optical fiber segments.

5.5 Summary

According to the principle of Rayleigh scattering and Brillouin scattering, two kinds of fully distributed optical fiber sensors are designed and verified by experiments. The main contents are as follows:

(1) Based on the theory of optical fiber bending loss, the winding optical fiber strain sensor is fabricated with the technology of backscattering OTDR, which is characterized by simple structure, low cost, good compatibility with the structure to be measured and flexible layout, and can realize quasi-distributed measurement.

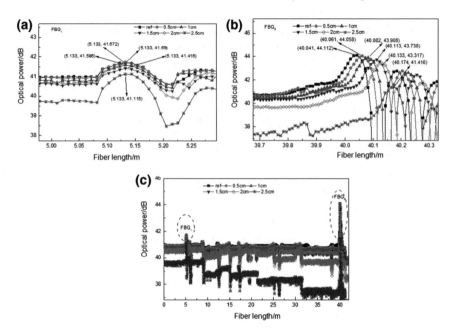

Fig. 5.25 Time domain spectrum of 2# DOS. **a** Time domain spectrum of FBG$_1$; **b** time domain spectrum of FBG$_2$; and **c** overall time domain spectrum of 2# DOS

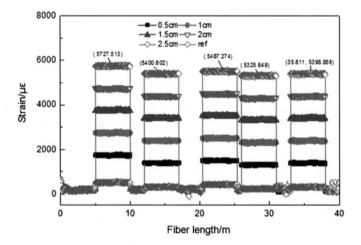

Fig. 5.26 Strain spectrum of 2# DOS

(2) Due to the measuring difficulty in large-range displacements such as deep rock mass and slope slip, a large-range fully distributed optical fiber displacement sensor based on fiber grating is designed and developed. The sensor has the

5.5 Summary

Table 5.5 Strain data of sample 2#

Fiber section	Displacement (cm)					
	0	0.5	1	1.5	2	2.5
The first section fiber	531.273	1,729.36	2,777.916	3,761.938	4,762.592	5,727.513
The second section fiber	318.578	1,452.04	2,430.681	3,426.12	4,412.82	5,400.602
The third section fiber	470.482	1,527.351	2,544.246	3,558.745	4,489.249	5,487.274
The fourth section fiber	239.801	1,285.168	2,325.649	3,314.122	4,320.92	5,325.649
The fifth section fiber	302.92	1,445.745	2,403.624	3,417.728	4,403.624	5,398.649
Interval between FBGs (m)	0	0.022169	0.042592	0.06516	0.082127	0.127383

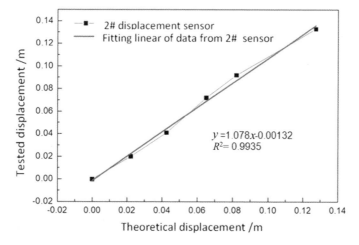

Fig. 5.27 Comparison between theoretical displacement and tested displacement from 2# DOS

remarkable advantage that the output data can be obtained without parameter conversion or integration, and the test accuracy is high. The experimental results show that the principle of fully distributed displacement sensing based on fiber

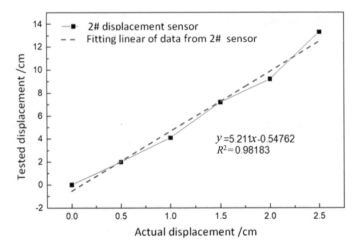

Fig. 5.28 Displacement sensitivity of 2# DOS

grating is feasible. All the developed fully distributed displacement sensors can realize displacement measurement, and the displacement sensitivity coefficient is high up to 5.21.

References

Behrad M, Valley B, Dusseault MB (2016) Experimental evaluation of a distributed Brillouin sensing system for measuring extensional and shear deformation in rock. Measurement 77. https://doi.org/10.1016/j.measurement.2015.08.040

Ding Y, Wang P, Yu S (2015) A new method for deformation monitoring on H-pile in SMW based on BOTDA. Measurement 70:156–168. https://doi.org/10.1016/j.measurement.2015.02.027

Du Y, Jin X, Sun B et al (2004) The spiral winding quasi-distributed optic fiber sensor based on general optic fibers. Eng Mech 21(1):48–51. https://doi.org/10.3969/j.issn.1000-4750.2004.01.009

Duan K, Zhang Q, Zhu H (2013) Application of fiber Bragg grating displacement sensors to geotechnical model test of underground salt rock gas storages. Rock Soil Mech 34(S2):471–477. https://doi.org/10.16285/j.rsm.2013.s2.034

He J, Dong H, Zhou Z (2010) A kind of new displacement sensor based on FBG for engineering. J Harerbin Univ Sci Techno 15(5):61–65. https://doi.org/10.15938/j.jhust.2010.05.026

Hiroyuki S, Yutaka S, Yoshio K (2012) Development of a multi-interval displacement sensor using Fiber Bragg Grating technology. Int J Rock Mech Min 54:27–36. https://doi.org/10.1016/j.ijrmms.2012.05.020

Jin XM, Du YL, Sun BC (2007) Research on the detection of fiber Bragg grating array using optical time domain reflection technology. Chin Sci Instrum 238(1):62–66. https://doi.org/10.19650/j.cnki.cjsi.2007.01.013

Li S, Wang K, Li LP (2014) Development and application of an extendable model test system for water inrush simulation in Subsea tunnel. Chin J Rock Mech Eng 33(12):2409–2418. https://doi.org/10.13722/j.cnki.jrme.2014.12.006

References

Li JZ, Sun BC, Kinzo K (2015) The influence of FBG on Brillouin distributed sensor. J Civil Struct Health Monit 5(5):629–643. https://doi.org/10.1007/s13349-015-0104-0

Li JZ, Xu LX, Sun BC (2016) Distributed displacement sensor based on fiber Bragg grating. J Tianjin Univ Sci Technol 49(6):653–658. https://doi.org/10.11784/tdxbz201601093

Nishio M, Mizutani T, Takeda N (2010) Structural shape reconstruction with consideration of the reliability of distributed strain data from a Brillouin-scattering-based optical fiber sensor. Smart Mater Struct 19(3):0350111–0350112. https://doi.org/10.1088/0964-1726/19/3/035011

Ricardo M, Javier S, Felipe B (2015) Estimating tunnel wall displacements using a simple sensor based on a Brillouin optical time domain reflectometer apparatus. Int J Rock Mech Min 75:233–242. https://doi.org/10.1016/j.ijrmms.2014.10.013

Sun S, Wang T, Xu Y (1999) Optical fiber measurement and sensing technology. Harbin Institute of Technology Press, Harbin

Tong H, Shi B, Wei G (2014) Study on distributed measurement of PHC pile deflection based on BOTDA. J Distrib Prev Mitig Eng 36(6):693–699. https://doi.org/10.13409/j.cnki.jdpme.2014.06.005

Wylie MT, Colpitts BG, Brown AW (2011) Fiber optic distributed differential displacement sensor. J Lightwave Technol 29(18):2847–2852. https://doi.org/10.1117/12.882743

Zhang G, He J, Xiao G (1988) Optical fiber sensing technology. Water Resources and Electric Power Press, Beijing

Zhang Y, Zhang Y, Wang Q (2014) Improved design of slow light interferometer and its application in FBG displacement sensor. Sensor Actuat A-Phys 214(4):168–174. https://doi.org/10.1016/j.sna.2014.04.034

Zhao Z, Zhang Y, Li C (2015) Monitoring of coal mine roadway roof separation based on fiber Bragg grating displacement sensors. Int J Rock Mech Min 74:128–131. https://doi.org/10.1016/j.ijrmms.2015.01.002

Chapter 6
Monitoring Technology for Prestressing Tendons Using Fiber Bragg Grating

Along with the rapid development of infrastructure construction such as expressway and high-speed railway in China, the construction of large-scale concrete bridge springs up exuberantly. Generally speaking, prestressed concrete structure is the main body of large-scale building structure and it is also a structural form with the largest material consumption. Therein, the prestressing tendon is the critical bearing area. Through long-term, real-time, and online monitoring for the stress of prestressing tendons, the safety and reliability of large-scale concrete structures can be scientifically evaluated so as to find out the problems in time and take corresponding technical measures to strengthen and repair the structure or discard the structure and prevent crises before they emerge. It is of very important practical significance and uses value to ensure the safety, reliability, and durability of large-scale building. Through the prestress loss theory analysis, this chapter mainly designs the FBG prestress sensor at the anchor head, proposes stress monitoring technology for prestressing tendon based on the optical fiber grating sensor array, and provides the theoretical foundation for the stress monitoring of prestressing tendon, which is the key component of the prestressed structure.

6.1 Theoretical Analysis on Prestress Loss of Concrete Structure

6.1.1 Calculation of Prestress Loss

In the process of tensioning and force transfer anchorage (prestressing stage) for prestressing tendon as well as in the long-term operation, the prestress may be gradually reduced due to material properties, tensioning process, anchoring, and other reasons, that is, the prestressing loss. There are many reasons for the loss of prestress and the forming time varies differently. Taking the tensile process as an example, the prestress losses caused by the two different processes of the pre-tensioning method and the post-tensioning method (Xue 2003) are not exactly the same.

Specifically, the prestress loss mainly includes friction loss, anchorage loss, elastic compression loss of concrete, temperature stress loss, shrinkage, creep loss and stress relaxation loss of concrete, etc. Among them, friction loss, anchorage loss, concrete elastic compression loss, and temperature stress loss are called instantaneous loss, and shrinkage, creep loss, and stress relaxation loss of concrete are called long-term loss because they take a long time to complete. The following sub-item discusses the reasons for the loss of prestress and the calculation method.

6.1.1.1 Friction Loss

Friction losses between prestressing tendons and pipe walls occur in post-tensioned members, while tensioning the prestressing tendon, due to the position deviation of the reserved pipeline or the rough pipeline wall, friction is caused by the contact between the prestressing tendon and the reserved pipeline wall, and the pre-tensile stress of the prestressing tendon is gradually reduced after leaving the jacking end. The value of the friction loss from the tensioned end to the calculated section is expressed by σ_{l1}. The friction loss consists of two parts: one is the loss of pipeline deviation, which is mainly the frictional resistance loss caused by the length of prestressing tendon, the friction coefficient between contacting materials and the quality of pipeline construction; the other part is the pipe bending loss, which is mainly the frictional resistance loss caused by the radial compressive stress of the prestressing tendon on the inner wall of the tunnel as a result of pipe bending. Deriving from the classical Coulomb friction theory, the formula for calculating the prestress loss is (Chen and Chen 2006)

$$\sigma_{l1} = \sigma_{con}\left[1 - e^{-(\kappa x + \mu\theta)}\right] \tag{6.1}$$

when $\kappa x + \mu\theta \leq 0.2$, σ_{l1} can be calculated according to the following approximate formula, that is:

$$\sigma_{l1} = \sigma_k(\kappa x + \mu\theta) \tag{6.2}$$

where σ_k is the control stress under the anchor when the reinforcing bar is tensioned; κ is the friction coefficient for local deviation of pore length per meter (see Table 6.1); x is the channel length (m) from the tensioned end to the calculated cross section, and the projection length of the channel on the longitudinal axis can be approximated; μ is the friction coefficient between the prestressing tendon and the tunnel wall (see Table 6.1); and θ is the angle from the tensioned end to the tangent of the channel section of the calculated section curve.

Formula (6.1) also applies to spatial prestressing tendons (Zhao 2002). Assuming that the curve equation of a prestressing tendon with a space curve is as follows:

6.1 Theoretical Analysis on Prestress Loss of Concrete Structure

Table 6.1 Value of friction coefficient κ and μ

Molding mode of channel		κ	μ	
			Strand wire, steel strand, plain face reinforcement	Deformed steel bar
Iron tube embedded in the hole		0.0030	0.35	0.4
Pre-laid bellow		0.0015	0.25	–
Extrusion forming		0.0015	0.35	0.60
Unbonded tendon	7ϕ5 steel wire	0.0035	0.10	–
	ϕ15 steel	0.0040	0.12	–

$$\begin{cases} x = x(t) \\ y = y(t) \\ z = z(t) \end{cases} \tag{6.3}$$

Assuming that t is a point on the prestressing tendon, the calculation formula of the prestressing friction loss is as follows:

$$\sigma_{1(t)} = \sigma_{\text{con}}\left[1 - e^{-(\mu \theta_t + \kappa X_t)}\right] \tag{6.4}$$

where X_t is the space curve length from the calculated section to the tension end, which is calculated according to Formula (6.5) and for the plane curve, calculated according to Formula (6.6); θ_t is the space angle of the curve, which is calculated according to Formula (6.7) and for the plane curve, calculated according to Formula (6.8).

$$X_t = \int_0^t \sqrt{x'^2 + y'^2 + z'^2}\, dt \tag{6.5}$$

$$X_t = \int_0^t \sqrt{1 + y'^2}\, dt \tag{6.6}$$

$$\theta_t = \int_0^t \frac{\sqrt{\left|\begin{matrix} x' & y' \\ x'' & y'' \end{matrix}\right|^2 + \left|\begin{matrix} y' & z' \\ y'' & z'' \end{matrix}\right|^2 + \left|\begin{matrix} z' & x' \\ z'' & x'' \end{matrix}\right|^2}}{(x'^2 + y'^2 + z'^2)^{3/2}}\, dt \tag{6.7}$$

$$\theta_t = \int_0^t \frac{y''}{1 + y'^2}\, dt \tag{6.8}$$

Prestressing tendon tensioning includes two modes, namely, single-end tensioning and tensioning at both ends. Therein, tensioning at both ends is composed of two modes: simultaneous tensioning at both ends, tensioning at one end and supplemental tensioning at the other end. In the following, different tension modes are analyzed and the corresponding frictional resistance loss calculation formula is established.

1. Calculation of frictional resistance loss in tensioning at one end

Before calculating the friction losses, the prestressing tendons are first segmented (Zhao et al. 2003). The number of segment points n is related to the linear category, and the segment points are located at the transition points. When the linear function is continuous and the first order is derivable, the distribution of friction losses is continuous. However, when a folding point is encountered, the friction losses at this point will be discontinuous since the transition of actual engineering is always made with a certain length of arc (the length of arc section is generally short). Therefore, a small arc section can be added to the bending point during calculation, and it is approximately considered that the prestress frictional resistance loss of the arc section appears mainly at the turning point (Yang 2003) (see $\overline{2, 3}$ and $\overline{2k-1, 2k}$ in Fig. 6.1) so that the friction loss distribution remains continuous.

Assuming that prestressing tendon is tensioned at the left end, the frictional resistance loss of each segment point is calculated. Namely, the effective prestress value after deducting the frictional resistance loss of the latter segment point in the previous segment is taken as the "tension control force" of the next segment so as to calculate the friction loss increment of the two segment points and determine the frictional resistance loss of the next segment point. The frictional resistance loss of each segment point can be gradually calculated by the following formula:

$$\begin{cases} \sigma_{11}(x_{2k}) = \sigma_{11}(x_{2k-1})\left[1 - e^{-(\mu\theta_{2k,2k-1} + \kappa x_{2k,2k-1})}\right] (k = 1, 2, \ldots, N) \\ \sigma_{11}(x_{2k+1}) = \sigma_{11}(x_{2k})\left(1 - e^{-\mu\theta_{2k+1,2k}}\right) (k = 1, 2, \ldots, N-1) \end{cases} \quad (6.9)$$

where x_{2k}, x_{2k-1} is the channel length from the segment point $2k$ to the segment point $2k - 1$, that is, $x_{2k}, x_{2k-1} = x_{2k} - x_{2k-1}$; $\theta_{2k}, \theta_{2k-1}$ is the space angle from segment point $2k$ to segment point $2k - 1$, i.e., $\theta_{2k}, \theta_{2k-1} = \theta_{2k} - \theta_{2k-1}$; $\theta_{2k+1}, \theta_{2k}$ is the space angle of the approximate circular arc added at the k-th turning point, that is, the complementary angle to the angle between $\overline{2k-1, 2k}$ and $\overline{2k+1, 2(k+1)}$.

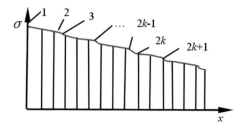

Fig. 6.1 Calculation segments of prestress friction loss

6.1 Theoretical Analysis on Prestress Loss of Concrete Structure

2. Calculation of frictional resistance loss in tensioning at both ends

The calculation method of frictional resistance loss for tensioning at both ends is as follows:

(1) According to the above method, the distribution of frictional resistance loss in tensioning at one end of left and right ends is calculated, respectively. As shown in Fig. 6.2, the curves SS' and EE' are, respectively, the effective prestress curves obtained under the assumption of tension at both ends with frictional resistance deducted. Obviously, $\sigma_{S'} = \sigma_{E'}$.

(2) According to the frictional resistance loss value of each segment point, the same point of tension friction loss is determined on both ends, which is called "fixed point", that is, the intersection point C shown in Fig. 6.2. Finally, the effective prestress curve after deducting friction loss takes SCE.

(3) Position and frictional resistance loss of fixed point C.

Step 1 Compare curves SS' and EE' to obtain that the fixed point C is between the point $2i - 1$ and $2i$.

Step 2 Calculate the difference between the two stress curves at points $2i - 1$ and $2i$, which should be as follows:

$$\begin{cases} h_1 = \sigma^Z_{2i-1} - \sigma^Y_{2i-1} \\ h_2 = \sigma^Y_{2i} - \sigma^Z_{2i} \end{cases} \quad (6.10)$$

Step 3 Approximate the distance from the fixed point C to the points $2i - 1$ and $2i$, which should be as follows:

$$l_1 = \frac{h_1}{h_1 + h_2} L \quad (6.11)$$

$$l_2 = \frac{h_2}{h_1 + h_2} L \quad (6.12)$$

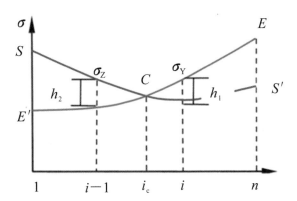

Fig. 6.2 Curve of effective prestress subtracted to σ_{11}

where L is the distance from point $2i-1$ to point $2i$.

Step 4 Calculate the position of point C by interpolation from Formulas (6.4)–(6.8) as given below:

$$x_C = \frac{x_{2i-1}\left(\sigma_{2i}^Y - \sigma_{2i}^Z\right) - x_{2i}\left(\sigma_{2i-2}^Z - \sigma_{2i-2}^Y\right)}{\sigma_{2i-1}^Z - \sigma_{2i-1}^Y + \sigma_{2i}^Y - \sigma_{2i}^Z} \tag{6.13}$$

$$\sigma_C = \frac{\sigma_{2i-1}^Z \sigma_{2i}^Y - \sigma_{2i}^Z \sigma_{2i-1}^Y}{\sigma_{2i-1}^Z - \sigma_{2i-1}^Y + \sigma_{2i}^Y - \sigma_{2i}^Z} \tag{6.14}$$

Step 5 Take the fixed point C as a new segment point i_C.

3. Frictional resistance loss at arbitrary point

According to the previous calculation formula and method, the frictional resistance loss at each segment point can be obtained. The frictional resistance loss at arbitrary point is

$$\begin{aligned}
\sigma_{11}(x_{2k+1}) &= \sigma_{11}(x_{2k})\left(1 - e^{-\mu\theta_{2k+1,2k}}\right) \quad (k = 1, 2, \ldots, N-1) \\
\sigma_{11}(x) &= \sigma_{11}(x_{2k-1})\left\{1 - e^{-[\mu\theta_{x,2k-1} + \kappa(x - x_{2k-1})]}\right\} \quad (k = 1, 2, \ldots, i-1, i+1, \ldots, N) \\
\sigma_{11}(x) &= \sigma_{11}(x_{2i-1})\left\{1 - e^{-[\mu\theta_{x,2i-1} + \kappa(x - x_{i-1})]}\right\} \quad (x_{2i-1} < x < x_C) \\
\sigma_{11}(x) &= \sigma_C\left\{1 - e^{-[\mu\theta_{x,C} + \kappa(x - x_C)]}\right\} \quad (x_C < x < x_{2i})
\end{aligned} \tag{6.15}$$

The linear approximation of the frictional resistance loss at any point can be obtained by the interpolation function as

$$\sigma_n(x) = \begin{cases} \frac{x - x_{2k-1}}{x_{2k} - x_{2k-1}}\sigma_n(x_{2k-1}) + \frac{x_{2k} - x}{x_{2k} - x_{2k-1}}\sigma_n(x_{2k}) & \begin{pmatrix} x_{2k-1} < x < x_{2k}, \\ k = 1, 2, \ldots, i-1, i+1, \ldots, N \end{pmatrix} \\ \frac{x - x_{2i-1}}{x_C - x_{2i-1}}\sigma_n(x_{2i-1}) + \frac{x_C - x}{x_C - x_{2i-1}}\sigma_C & (x_{2i-1} < x < x_C) \\ \frac{x - x_C}{x_{2i} - x_{C1}}\sigma_C + \frac{x_{2i} - x}{x_{2i} - x_C}\sigma_n(x_{2i}) & (x_C < x < x_{2i}) \end{cases} \tag{6.16}$$

Formulas (6.15) and (6.16) are also applicable to spatial prestressing tendons. Except that x is the position vector of spatial prestressing tendons and the above calculation process can be calculated according to the vector.

6.1.1.2 Anchorage Loss

Stress loss caused by anchor deformation, rebar retraction, and joint compression can be calculated according to the following formula:

6.1 Theoretical Analysis on Prestress Loss of Concrete Structure

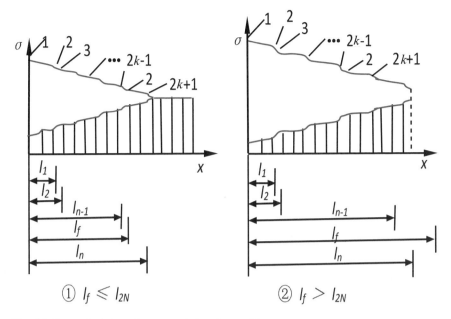

Fig. 6.3 Loss calculation of prestressed at anchor position

$$\sigma_{l2} = \frac{a}{l} E_s \qquad (6.17)$$

where a is the anchor deformation value and the prestressing retraction value, which can be determined according to the measured data. When there is no reliable data, it can be determined by the specification value; l is the length from tensioning end to anchor end; E_s is the elasticity modulus of the prestressing tendon.

Formula (6.17) assumes that this loss is equal over the full length of the prestressing tendon, so it is only suitable for the pre-tensioning method. For the post-tensioning method, the reverse frictional resistance action between the prestressing tendon and the pipe wall should be considered in order to better reflect the actual situation of the change of anchorage loss. Assuming that the reverse frictional resistance action is the same as the frictional resistance action when the prestressing tendon is tensioned, the influence length of the retraction of the prestressing tendon can be obtained and the value of the anchorage loss can be obtained.

1. Calculating the length affected by inverse frictional resistance l_f

As shown in Fig. 6.3, the total linear length of prestressing tendons is l_{2n}, and there are n segments of prestressing tendons in total. Assume that the anchorage loss disappears in paragraph n and the following formula is used for trial calculation paragraph by paragraph, and the first one satisfying $l_{n-1} < l_f \leq l_n$ is the length affected by reverse friction.

$$l_f = \frac{\sqrt{\frac{aE_P(x_{n+1}-X_n)}{1000[\sigma_n(x_{n+1})-\sigma_n(x_n)]}+(l_{n+1})^2}}{\frac{l_1^2\frac{\sigma_n x_2-\sigma_n x_1}{x_2-x_1}+\sum_{i-1}^{m-1}(l_i^2-l_{i-1}^2)\frac{\sigma_n x_{i+1}+\sigma_n x_i}{x_{i+1}-X_i}}{\frac{\sigma_n x_{n+1}+\sigma_n x_n}{x_{n+1}-X_n}}} \quad (6.18)$$

2. Calculation of anchorage loss in two cases

In the first case, when $l_f \leq l_{2n}$, $l_{n-1} < l_f \leq l_{2n}$, only add a segment point between $n-1$ and n, calculate the frictional resistance loss of the segment point according to Formulas (6.15) and (6.16), and then calculate the anchorage loss of each point by inverse calculation.

$$\sigma_{12}(x) = 0 \quad (x_{lf} < x < x_{2n})$$
$$\sigma_{12}(x_i) = 2\sigma_{12}(x_{lf}) - \sigma_{11}(x_i) \quad (0 < x < x_{lf}) \quad (6.19)$$

In the second case, when $l_f > l_{2n}$, and all l_f calculated by Formula (6.19) do not satisfy $l_{n-1} < l_f \leq l_{2n}$, then $l_f = l_{2n}$, and the anchorage loss of each point is calculated by means of inverse calculation.

$$\sigma_{i2}(x_{2n}) = \frac{1}{l_{2n}}\left[\frac{aE_P}{1000}+l_1^2\frac{\sigma_n(x_2)-\sigma_n(x_1)}{x_2-x_1}\right]+\sum_{i-1}^{2N}(l_i^2-l_{i-1}^2)\frac{\sigma_n(x_{i+1})-\sigma_n(x_i)}{x_{i+1}-x_i}$$
$$\sigma_{i2}(x_i) = \sigma_n(x_{2n}) + \sigma_{i2}(x_{2n}) - \sigma_n(x_i) \quad (0 \leq x < x_{2n}) \quad (6.20)$$

6.1.1.3 Stress Relaxation Loss

Under the action of constantly unchanged stress, the creep will increase with the extension of continuous loading time. If the prestressing tendon is tensioned to a certain stress value and its length is kept constant, the stress in the prestressing tendon will gradually decrease with the extension of time, which is called stress relaxation. Its characteristics are as follows: the higher the initial tensile stress, the greater the stress relaxation. When the initial tensile stress is less than 50% of the ultimate strength of the steel bundle, the relaxation is small and can be ignored. The amount of relaxation of prestressing tendon is closely related to its material quality. Generally, the relaxation of hot rolled steel bar is smaller than that of carbon steel wire, and the relaxation of steel strand is larger than that of original single steel wire. The relaxation of prestressing tendon develops the fastest at the initial stage of bearing initial tensile stress. The relaxation amount is the largest in the first hour, more than 50% amount is completed in 24 h, and then it will tend to stabilize gradually. Stress relaxation is also related to temperature. The amount of relaxation will increase with the increase of temperature. In addition, the relaxation stress loss is also related to the load-holding time, so different relaxation values should be adopted according to the load-holding time at different loading stages of the component. In the post-tensioning method, it can be considered that the relaxation stress losses are all completed in the use stage. The concrete calculation can be started since the prestress is established,

6.1 Theoretical Analysis on Prestress Loss of Concrete Structure

which should be determined by the fact that 50% of the final value of the relaxation loss can be completed within 2 d and 100% is completed within 40 d.

Code for Design of Highway Reinforced Concrete and Prestressed Concrete Bridge and Culverts (JTG D62-2004) stipulates that the final value of prestress loss caused by reinforcement stress relaxation of prestressing tendons can be calculated according to the following provisions:

1. Prestressing steel wire and steel strand

$$\sigma_{l3} = \psi \xi \left(0.52 \frac{\sigma_{pe}}{f_{pk}} - 0.26 \right) \sigma_{pe} \quad (6.21)$$

where ψ is the overtension coefficient and $\psi = 1.0$ in case of a single tension, $\psi = 0.9$ in case of over tensioning; ξ is the relaxation coefficient of steel bar, $\xi = 1$ in case of Grade I relaxation (ordinary relaxation), $\xi = 0.3$ in case of Grade relaxation (low relaxation); and σ_{pe} is the reinforcement stresses during force transfer anchorage. For post-tensioned components, $\sigma_{pe} = \sigma_{con} - \sigma_{l1} - \sigma_{l2} - \sigma_{l4}$ and for pre-tensioned components, $\sigma_{pe} = \sigma_{con} - \sigma_{l2}$; f_{pk} is the tensile standard value of prestressing tendon.

2. Finished deformed bar

$$\sigma_{l3} = A\sigma_{con} \quad (6.22)$$

where $A = 0.05$ in case of a single tension and $A = 0.035$ in case of over tensioning. When the relaxation loss of reinforcement should be calculated in segments for prestressing steel wire and steel strand, the ratio of its intermediate value to ultimate value shall be determined according to the time of establishing prestress as Table 6.2.

6.1.1.4 Elastic Compression Loss of Concrete

When prestressed concrete members are subjected to pre-pressure, elastic compressive strain ε_c will be produced immediately. At this time, the prestressing tendons bonded with concrete or tensioned and anchored will produce the compressive strain same as that of the concrete at the corresponding position $\varepsilon_p = \varepsilon_c$, thus causing prestress loss. Such loss is called elastic compression loss of concrete. The elastic compression amount of concrete causing stress loss is related to the way of prestressing.

The pre-tensioned concrete structural tendon tension and the compressive prestress applied to the concrete are two completely separated procedures. When the

Table 6.2 Ratio of intermediate stress loss and ultimate stress loss of tendon

Time (d)	2	10	20	30	40
Ratio	0.5	0.61	0.74	0.87	1.00

tendon relaxes to apply pressure to the concrete, the total elastic compressive strain produced by the concrete will cause the stress loss of the tendon. The value is as follows:

$$\sigma_{l4} = \varepsilon_c E_p = \frac{\sigma_{pc}}{E_c} e_P = N_P \sigma_C \tag{6.23}$$

where N_p is the ratio of the elasticity modulus of the prestressing tendon to the elasticity modulus of the concrete; E_p is the elasticity modulus of concrete; and σ_c is the normal force of concrete generated by all the pre-applied force of reinforcement at the center of gravity of the tendon for the section calculation of the pre-tensioned concrete structure, which can be calculated by $\sigma_c = \frac{N_{p0}}{A_0} + \frac{N_{p0}^2}{J_0}$; N_{p0} is the prestressing force of the prestressing tendon when the concrete stress is zero (minus the prestress loss at the corresponding stage); A_0 and J_{0f} are, respectively, the converted section area and converted section moment of inertia of the prestressed concrete structure; and e_{p0} is the distance from the gravity center of prestressing tendon section to the gravity center of converted section.

In post-tensioned prestressing concrete members, the elastic compression of concrete takes place in the tension process and the elastic compression of concrete is completed immediately after the tension is completed. Therefore, there is no need to consider stress loss for the post-tensioned members those have completed a single tension. However, in general, due to the large number of prestressing tendons in post-tensioned members, the prestressing tendons are usually tensioned and anchored in batches. In this case, the prestressing tendons that have been tensioned and anchored will be elastically compressed and deformed when the prestressed tendons are subsequently tensioned in batches, resulting in stress losses. Therefore, this stress loss in post-tensioned members, commonly referred to as batch tension stress loss, can be calculated according to the following formula:

$$\sigma_{l4} = \frac{m-1}{2m} n_p \sigma_c \tag{6.24}$$

where m is the number of tensioning batches of prestressing tendons, and the number and the prestressing force of reinforcing bars per batch are the same; σ_c is the positive stress of the concrete produced by the resultant force N_p of the m batches of tendons at its point of action (the center of gravity of all tendons), which is equal to the sum of the $\Delta\sigma_{pc}$ positive stresses of the concrete produced by tensioning each batch of tendons, i.e., $\sigma_{pc} = \sum \Delta\sigma_{pc} = m\sigma_{pc}$, where σ_{pc} is the positive stress of concrete generated by subsequent tensioning of each batch of tendons at the center of gravity of the tensioned tendon in the previous batch of the calculated section, which can be calculated by $\sigma_{pc} = \frac{N_p}{m}\left(\frac{1}{A_n} + \frac{e_{pn} y_n}{J_n}\right)$. Therein, N_p is the resultant force of prestress of all tendons (minus stress loss at the corresponding stage); e_{pn} is the distance from the center of gravity of the previous batch of tensioned tendons to the gravity axis of the net section of concrete, so $y_{ne} \approx e_{pn}$; and A_n and J_n are net section area and net section moment of inertia of prestressed concrete structure, respectively.

6.1.1.5 Shrinkage and Creep Loss

Shrinkage and creep are inherent characteristics of concrete. Due to shrinkage and creep of concrete, the prestressing concrete members are shortened and the prestressed tendons are also retracted, resulting in prestress loss. The corresponding values are as follows (Yang 2005):

$$\sigma_{l5} = \frac{0.9 n_p \sigma_h \varphi(t_m, \tau) + E_p \varepsilon(t_m, \tau)}{1 + 15\rho} \tag{6.25}$$

where σ_h is the normal stress of concrete generated by prestress (minus the loss in corresponding stage) at the center of gravity of all the stressed steel bars on the calculated cross section when the steel bars of the pre-tensioned concrete structure are relaxed or the steel bars of the post-tensioned concrete structure are anchored; ρ is the reinforcement ratio, which is the ratio of the cross-sectional area of prestressing and non-prestressing tendons in tensile area and compressed area to the cross-sectional area of the concrete structure; and $\varphi(t_m, \tau)$ is the final value of concrete creep coefficient when the loading age is τ. If there is no reliable test data, the specification can be checked; $\varepsilon(t_m, \tau)$ is the final value of shrinkage strain calculated from the concrete age τ. If there is no reliable test data, the specification can be checked.

6.1.1.6 Temperature Stress Loss

Temperature stress loss refers to the stress loss generated when the pre-tensioned component is heated and cured on the fixed pedestal by steam or other methods. If the natural temperature of the manufacturing site during tensioning of the prestressing tendon is t_1, the maximum temperature during curing is t_2, and the temperature difference is $\Delta t = t_2 - t_1$, then the calculation formula of stress loss σ_{l6} is as follows:

$$\sigma_{l6} = \alpha \Delta t E_s \tag{6.26}$$

where α is the linear expansion coefficient of the reinforcing steel bar, which is generally $\alpha = 1 \times 10^{-5}/°C$.

6.1.2 Calculation of Effective Prestress

The effective prestress of the prestressing tendon is the pre-tensile stress that existed in the prestressing tendon after the deduction of corresponding prestressing loss from tension control stress under the anchor. The combination of prestress losses shall generally be divided into prestressing and post-tensioning by the prestressing stage and the reaction stage of applied load according to the occurrence sequence of stress loss and the time required for full completion. The combination of prestress loss values at each stage can be calculated as per Table 6.3.

In the prestress stage, the effective prestress in the prestressing tendon is

$$\sigma_y = \sigma_k - \sigma_l^I \qquad (6.27)$$

where σ_l^I is the sum of the stress losses that occurred after the completion of prestress tension and the force transfer anchorage; σ_k is the control stress under the prestressing tendon anchor.

In the reaction stage of applied load, the effective prestress in the prestressing tendon, i.e., the permanent prestress is

$$\sigma_y = \sigma_k - \sigma_l^I - \sigma_l^{II} \qquad (6.28)$$

where σ_l^I is the sum of the stress losses that occurred after the completion of the prestressing tendon force transfer anchorage.

6.2 Design of FBG Prestress Sensor at Anchor Head

The prestress decrease of prestressing tendon will lead to the decrease of structural bearing capacity, and even lead to the failure of the whole engineering structure. Therefore, the long-term and real-time safety monitoring of prestressing tendons must be carried out. However, the traditional monitoring method based on electric measurement (Chen and Huang 2005) has the following disadvantages: the signal cannot be transmitted at a long distance, the distributed measurement cannot be realized, and it is especially difficult to survive in the prestressed structure and meet the practical needs of engineering application. Therefore, the monitoring scheme at anchor head of prestressed structure is put forward based on optical fiber grating, and the research on fiber grating prestress sensor is carried out. The results show that the sensor is characterized by high measurement accuracy, non-electromagnetic interference, prefabrication, strong practicality, and can meet the practical needs of long-term monitoring at the anchor of prestressed structure (Jin et al. 2008a, b).

The FBG prestress sensor at the anchor end has excellent sensing performance, overcoming the disadvantages of poor long-term stability of resistance-type and

Table 6.3 Prestress loss at various stages

Stage	Prestress method	
	Pre-tensioned	Post-tensioned
Stage I (prestress stage)	$\sigma_l^I = 0.5\sigma_{13} + \sigma_{14} + \sigma_{16}$	$\sigma_l^I = \sigma_{11} + \sigma_{12} + \sigma_{13}$
Stage II (load stage)	$\sigma_l^{II} = 0.5\sigma_{13} + \sigma_{15}$	$\sigma_l^{II} = \sigma_{13} + \sigma_{15}$

steel-string-type sensors, and inability of distributed measurement and signal transmission at long distance. It is especially suitable for long-term, long-distance, health and safety monitoring of prestressed structure and has great application value and development prospect.

6.2.1 Prestress Monitoring Principle at Anchor Head of Prestressed Concrete Structure

For the post-tensioned prestressing concrete structure, during force transfer anchorage, the stress σ_y of the prestressed tendon is

$$\sigma_y = \sigma_k - (\sigma_{S3} + \sigma_{S4} + \sigma_{S6}) \tag{6.29}$$

where σ_k is the minimum control stress of prestressing under the prestressed anchor; σ_{S3}, σ_{S4}, and σ_{S6} are three parts of prestress loss caused by anchorage loss, frictional resistance loss, and concrete elastic compression loss, respectively. In the operation stage, the prestressing stress of the prestressed tendon σ_y is

$$\sigma_y = \sigma_k - (\sigma_{S1} + \sigma_{S2} + \sigma_{S3} + \sigma_{S4} + \sigma_{S6}) + N_P \frac{M_g}{I_j} y_{jy} + N_P \frac{M_d + M_h}{I_0} y_{0y} \tag{6.30}$$

where σ_{S1} and σ_{S2} are, respectively, the creep loss of prestressed concrete and the stress relaxation loss of reinforcing bar; N_p is the ratio for the elasticity modulus of the prestressing tendon and the concrete; M_g, M_d, and M_h are the bending moments produced by the beam dead weight, and other constant loads and live loads in the calculated section, respectively; y_{jy} and y_{0y} refer to the distance from the prestressing tendon to the center of gravity axes of net section and the converted section, respectively; and I_j and I_0 are the net section moment of inertia and the converted section moment of inertia of the beam, respectively.

The external diameter of the sensor elastomer is D, the internal diameter is d, the prestress is F, and the strain is ε_T. The following is obtained through force analysis for the elastomer:

$$F = \frac{\pi}{4}(D^2 - d^2)\varepsilon_T E \tag{6.31}$$

Because the pressure on the anchor-end sensor F is equal to the tension force at the anchor end of the prestressing tendon in opposite direction (Du et al. 2007). If A_p is the cross-sectional area of the prestressing tendon, E_p is the effective elasticity modulus of the prestressing tendon, ε_p is the strain of the prestressing tendon, the relationship between the minimum control stress of the prestressing anchor σ_k and the stress of the prestressing tendon σ_y and the pressure on the anchor sensor F can

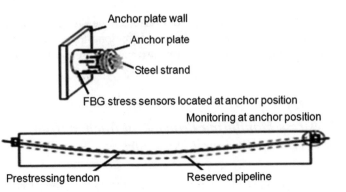

Fig. 6.4 Anchor stress test of prestressed steel strand

be obtained. The relational expression for the strain of the prestressing tendon ε_p and the sensor strain ε_T is, respectively, as follows:

$$\sigma_k = F/A_p = \frac{\pi(D^2 - d^2)\varepsilon_T E}{4A_p} \tag{6.32}$$

$$\sigma_y = \begin{cases} \frac{\pi(D^2-d^2)\varepsilon_T E}{4A_p} - (\sigma_{S3} + \sigma_{S4} + \sigma_{S6}) \\ \frac{\pi(D^2-d^2)\varepsilon_T E}{4A_p} - (\sigma_{S1} + \sigma_{S2} + \sigma_{S3} + \sigma_{S4} + \sigma_{S6}) + N_p \frac{M_g}{I_j} y_{ij} + N_P \frac{M_d+M_h}{I_0} y_{0j} \end{cases} \tag{6.33}$$

$$\varepsilon_p = \frac{\sigma_y}{E_p} = \begin{cases} \frac{\pi(D^2-d^2)\varepsilon_i E - (\sigma_{S3}+\sigma_{S4}+\sigma_{S6})}{4A_p E_p} \\ \frac{\pi(D^2-d^2)\varepsilon_i E - (\sigma_{S1}+\sigma_{S2}+\sigma_{S3}+\sigma_{S4}+\sigma_{S6})}{4A_p E_p} + N_P \left(\frac{M_g}{I_j E_p} y_{ij} + \frac{M_d+M_h}{I_0 E_p} y_{0j} \right) \end{cases} \tag{6.34}$$

For post-tensioned prestressed structures, the installation of FBG prestress sensor at the anchor head is feasible due to the normative algorithm of the above prestress loss under known constant and live loads (see Fig. 6.4). It will not only realize the whole stress monitoring of anchorage during tension and operation but also can estimate the stress of prestressing tendon and thereby complete the indirect measurement of stress for the whole prestressing tendon. On this basis, structure safety can be assessed objectively during tension and operation. At the same time, the prestressing tendon always keeps in large strain state during tension and operation, and it is difficult to monitor large strain of prestressing tendon directly. According to Formulas (6.33) and (6.34), the dimension of sensor elastomer can be designed, that is, adjusting the cross-sectional area of the elastomer, to measure an equivalent smaller strain at the anchor head, solving the difficulty of large strain measurement for prestressing tendon.

6.2.2 Structure Design and Principle of FBG Prestress Sensor at Anchor Head

6.2.2.1 Structure of FBG Prestress Sensor at Anchor Head

The structure of FBG prestress sensor at the anchor head is that a plurality of fiber gratings are uniformly arranged on the annular elastomer wall and the laying direction is consistent with the axial direction of the elastomer. A plurality of FBG is connected in series with the same conducting fiber through which the wavelength signal is transmitted. The wavelength variation is detected by the fiber grating demodulator and converted into an electric signal. The total load acting on the sensor and the load distribution are calculated by the secondary instrument to obtain the stress of the prestressing tendon.

6.2.2.2 Measuring Principle of FBG Prestress Sensor at Anchor Head

When the grating is in a uniform strain field, the relationship between strain and wavelength variation is

$$\varepsilon = \frac{1}{1 - P_e}\left[\frac{\Delta\lambda}{\lambda} - (\alpha + \zeta)\Delta T\right] \tag{6.35}$$

The elastomer is divided into the corresponding number of equal parts according to the number of pasting gratings and the average strain of the ith part of the elastomer can be approximately equal to Formula (6.31). As a result, the average pressure of this part can be expressed as

$$F_i = -A_i \varepsilon_i E = -\frac{A_i E}{1 - P_e}\left[\frac{\Delta\lambda}{\lambda} - (\alpha + \zeta)\Delta T\right] \tag{6.36}$$

where E is the elasticity modulus of the elastomer and A_i is the cross-sectional area of the corresponding portion of the elastomer. In this case, the total load to which the elastomer is subjected is the sum of the pressures to which the parts are subjected, i.e.,

$$F = \sum F_i = -\sum \frac{A_i E}{1 - P_e}\left[\frac{\Delta\lambda}{\lambda} - (\alpha + \zeta)\Delta T\right] \tag{6.37}$$

If that elastomer is only subjected to uniform loading, the above formula can be simplified to

$$F_i = -A\varepsilon_k E = -\frac{A_i E}{1 - P_e}\left[\frac{\Delta\lambda_k}{\lambda_k} - (\alpha + \zeta)\Delta T\right] \tag{6.38}$$

where A is the total cross-sectional area of the elastomer; ε_k is the strain measured from any part of the elastomer wall; and λ_k is the Bragg center wavelength of any part of pasting fiber grating on the wall of the elastomer.

It can be seen from Formulas (6.36)~(6.38) that if the drift amount of the center wavelength of the fiber grating and the variation of its working temperature are known, the total load acting on the elastomer and the load distribution can be calculated and the stress condition of the prestressing tendon can be obtained. This is the operating principle of FBG prestress sensor at the anchor end.

6.2.3 Design of FBG Prestress Sensor at Anchor Head

The design principle of FBG prestress sensor at the anchor end is as follows: ① the linearity is good and the sensitivity is high; ② the signal can be transmitted at long distance to realize distributed measurement; ③ anti-lateral pressure and simple structure; and ④ good stability and durability are available for long-term monitoring.

6.2.3.1 Selection and Heat Treatment for Elastomer Materials of Sensors

The elastomer of FBG prestress sensor at the anchor head must have high mechanical strength and elastic limit. Therefore, 40Cr steel material featured by wide source and low price is selected (refer to Table 6.4 for performance parameters). In order to make it have stable elasticity and low creep, it is required that its hardness can reach HRC40–HRC45 after heat treatment, and some aging treatment should be conducted to improve the repeatability and accuracy of the sensor.

6.2.3.2 Structure Design of the Elastomer

In structure design, firstly, the uniformity of strain working area of elastomer should be ensured. Therefore, the height should be large, the influence of Poisson effect should be reduced, and the structure should have good rigidity so as to reduce the disturbance of external vibration. Second, the space size of the elastomer and even the whole sensor should not only meet the mechanical requirements but also should not adversely affect the tensile process and normal operation of the prestressed structure.

During the prestress tension, according to Clause 3.4.2 of the current *Code for Design of Concrete Structures* (GB/T 13096—2008), the prestressing tendon shall be made of steel strands with the tension control stress value $\sigma_{con} = 0.70 f_{ptk}$. During structure operation, the unfavorable action on prestressing tendon from beam dead weight and various loads should be fully considered, and the stress of the prestressing tendon at the anchor end $\sigma_{max} = f_{ptk}$. Therefore, during elastomer design, if A_p is the

6.2 Design of FBG Prestress Sensor at Anchor Head

Table 6.4 Performance parameter of different steels

Material trademark	Elastic modulus (GPa)	Linear expansion coefficient (10^{-6} K^{-1})	Strength limit (GPa)	Yield limit (GPa)	Temperature coefficient of elastic modulus (10^{-4} K^{-1})	Remarks
40Cr	218	11.0	1.00	0.8	−3	General sensor
35CrMnSi2A	200	11.0	1.65	1.3	−3	High-precision sensor
60Si2MnA	200	11.5	1.60	1.4		Small thickness planar elastic element, high fatigue limit
50CrVA	212	11.3	1.30	1.1		Elastic element under alternating load
30CrMnSiA	210		1.08	0.834	–	–
40CrNiMoA	210		1.08	0.932	–	–

cross-sectional area of the prestressing tendon, the designed pressure on the sensor at the anchor end F_{design} is as follows:

$$F_{\text{design}} = \frac{\pi}{4}(D^2 - d^2)\varepsilon_t E_t = A_P \sigma_{\max} = A_P f_{\text{ptk}} \qquad (6.39)$$

Based on the mechanical analysis of the sensor and the optimum design for the structure size, it is possible to measure the small strain of the sensor at the anchor end in order to solve the problem of measurement difficulty for the strain–stress of the prestressing tendon under the condition of large strain during tension and operation.

On the basis of satisfying the mechanical requirements, the size of the elastomer is mainly determined by the size of the anchor structure and the size of the anchor-end zone of the prestressed structure. According to the dimension parameters of the anchor, the internal diameter d of the elastomer is selected, initially, which is generally not smaller than the central diameter of the anchor pad. Because the strain value of elastomer should be $(8\sim12) \times 10^{-4}$ under the maximum load, the sensor allows 20% overload. The strain is 9×10^{-4} under the rated load and the elasticity modulus of 40Cr steel is 218 GPa. The outer diameter D of elastomer can be calculated by substituting into Formula (6.39), while determining the elastomer height H, and in order to ensure sufficient uniformity of the elastomer strain working area, it is required that H should be at least greater than D. At the same time, when selecting the value of sensor height H, the influence of the elastomer size on prestressing tension and structure operation should be fully considered. See Table 6.5 for structural parameters of elastomer designed according to different anchor sizes.

Table 6.5 Elastomer parameter of FBG prestress sensor at anchor head

Sensor number	Inner diameter (10^{-3} m)	Outer diameters (10^{-3} m)	Height (10^{-3} m)	Suitable anchorage type
GC1	65	95	150	ϕ15-2, ϕ15-3
GC2	85	117	170	ϕ15-4, ϕ15-5
GC3	96	135	170	ϕ15-6, ϕ15-7
GC4	115	150	180	ϕ15-8, ϕ15-9
GC5	132	174	220	ϕ15-10, ϕ15-11, ϕ15-12

6.2 Design of FBG Prestress Sensor at Anchor Head

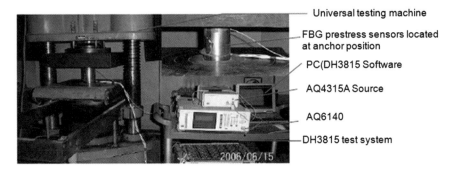

Fig. 6.5 Experiment diagram of FBG prestress sensor at anchor head

6.2.4 Calibration Experiment for FBG Prestress Sensor at the Anchor Head

6.2.4.1 Experimental Devices and Instruments

The FBG prestress sensor at the anchor end is calibrated by universal material tester (see Fig. 6.5). An optical fiber grating is stuck on the elastic wall of the FBG prestress sensor at the anchor end and a high-precision strain gauge is arranged near the sensor (Jin et al. 2005).

The light source is ASEAQ4315A. The central wavelength of FBG is identified by AQ6140 multi-wavelength meter. Its main technical indexes include: 900–1,680 nm wavelength range; 0.1 pm wavelength resolution; 5 GHz spacing resolution; up to 512 channels, etc.

6.2.4.2 Experimental Scheme

In experiment, the FBG prestress sensor at the anchor head is subjected to step-by-step loading and unloading tests on the universal material tester, which is conducted in its elastic range step by step. The output–input relation and various characteristics of the sensor are analyzed through the test data. The specific experimental scheme is as follows:

(1) Divide the full range (measuring range) of the sensor into a plurality of equal interval points, load from small to large, and record the corresponding output values.
(2) Gradually lower the load from large to small while recording the corresponding output value.
(3) Carry out a plurality of tests of a positive stroke cycle and a reverse stroke cycle for the sensor according to that steps described in Steps (1) and (2), and list a table or draw a curve based on the obtained output–input test data.

(4) The test data are processed to determine the static characteristic indexes such as linearity, sensitivity, hysteresis, and repeatability of the sensor.

6.2.4.3 Sensor Performance Analysis

The FBG prestress sensors at the anchor head are loaded and unloaded step by step on the universal material tester, and the readings of the universal material tester, resistance strain gauge, and the multi-wavelength meter are recorded, respectively, during the experiment. As a result, a curve for the relationship between the wavelength shift of the FBG prestress sensors at the anchor head during loading and unloading, and the load applied on it as well as the strain of reference strain gauge can be obtained (refer to Fig. 6.6).

As shown in Table 6.6, the performance parameters of FBG prestress sensor are better in linearity and repeatability, and the correlation coefficient is more than 0.99. Among them, the load sensitivity coefficient and strain sensitivity coefficient are in good consistency with the theoretical calculation. This shows that FBG prestress sensor has better repeatability, linearity, and sensitivity.

6.3 Prestress Monitoring Technology Using Fiber Bragg Grating Sensor Arrays

6.3.1 Structure and Performance Parameters of Steel Strands

Taking seven steel strands as an example, seven steel strands are twisted from seven high-strength steel wires (3–5). The six surrounding spiral steel wires are called outer

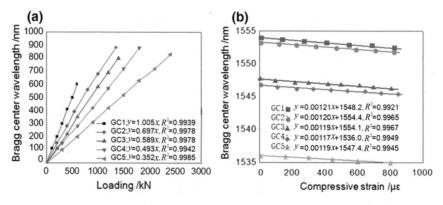

Fig. 6.6 Wavelength shift of FBG prestress sensor at anchor head versus load and compression strain of gauge

6.3 Prestress Monitoring Technology Using Fiber Bragg …

Table 6.6 Performance parameters of FBG prestress sensors at anchor head

Sensor number	GC1	GC2	GC3	GC4	GC5
Linearity (relative error) e_L (%)	2.6	0.4	0.6	0.6	0.6
Load sensitivity (pm/10^6 kN)	1.005	0.697	0.589	0.493	0.352
Strain sensitivity (pm/με)	1.21	1.20	1.19	1.17	1.19
Hysteresis e_H (%)	2.5	0.8	1.1	1.0	0.8
Repeatability error e_R (%)	2.6	0.6	1.0	0.7	0.7
Strain resolution (δ_ε/με)	0.83	0.83	0.84	0.85	0.84
Range (MPa)	700	1,200	1,400	1,800	2,400

wire, the straight wire in the center is called inner wire, and the overall dimension of circumcircle is called nominal diameter. The lay length is the length of the steel strand corresponding to 360° rotation of the spiral steel wire. Table 6.7 is the main performance parameters of common steel strands and carbon steel wires.

Table 6.7 Main performance parameters of steel strands and carbon steel wires

Performance index	Carbon steel wire	Steel strand
Nominal diameter (10^{-3} m)	$\phi 5$	15.0 (7 $\phi 5$)
The standard value of tensile strength (MPa)	1,570	1,860
Elongation (%)	4.0	3.5
Section area 10^{-6} m^2	19.63	139.98
Nominal weight (kg/m)	0.154	1.09
Elastic modulus (GPa)	200	190

6.3.2 Combination of Fiber Bragg Grating and Steel Strand and Stress Measurement Principle for Steel Strand

6.3.2.1 Composite Technology of Fiber Bragg Grating and Steel Strand

Prestressing steel strands are formed by spiral twisting of a plurality of steel wires. In the prestressed structure, the surplus space of prestressing tendon in the working environment is limited. The following geometric and mechanical factors should be mainly considered in the composite technology of fiber grating and steel strands: ① the installation space of the sensor carrier is limited by the size limitation of the reserved hole, and the optical fiber may be damaged by the mechanical shear and extrusion, so proper protection measures should be taken; ② the deformation of sensor and steel strand shall be synchronized; and ③ simple, practical, and easy operation should be realized by the bonding of sensor and steel strand as far as possible. Therefore, the composite technology of fiber grating and steel strand can adopt the following procedure to manufacture intelligent prestressing tendon with FBG (Luo 2004; Sun 2002):

(1) The fiber grating should be made into FBG sensor array. During manufacturing, FBG with small central wavelength should be placed at the position with maximum tensile stress according to the theory analysis of prestress.
(2) The semi-finished high-strength steel wire shall be drawn on the surface to form a shallow groove parallel to its axis. To prevent stress concentration caused by sharp corners, the shallow groove shall be a circular arc boundary.
(3) It is required that the fiber grating stuck in the shallow groove on the surface should be parallel to the axis of the high-strength steel wire, and the optical grating can be slightly pulled by hand, so as to make the steel wire have better sensing characteristics and higher sensitivity, then the shallow groove should be filled with 703 waterproof glue (the hardened 703 glue will become flexible material) in order to protect the fiber grating sensor system and realize water and moisture resistance.
(4) Place $\phi 5$ high-strength straight steel wires with FBG sensors in the center of the steel strand and twist the six surrounding outer steel wires to form a bundle of steel strands (see Fig. 6.7).

6.3.2.2 Stress Monitoring Principle of Steel Strand

In the local position of steel strand, it is considered that there is no significant difference in stress and strain of each steel wire. Therefore, the strain ε measured by FBG sensor in intelligent prestressing tendon can be approximated as the strain of steel strand at this position, so as to obtain the stress of steel strand at this position.

$$\sigma = E_p \varepsilon \tag{6.40}$$

Fig. 6.7 Photo of steel strands with FBG sensors

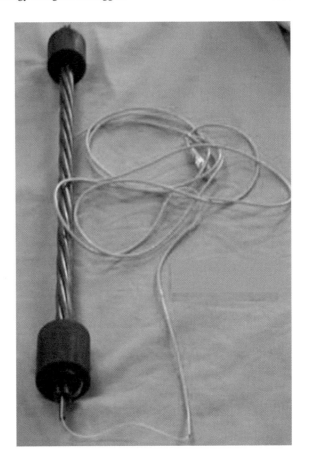

where E_p is the elasticity modulus of the steel strand.

When the fiber grating is in a uniform strain field, the stress of the steel strand at a certain position is obtained as follows:

$$\sigma = \frac{E_P}{1-p_e}\left[\frac{\Delta\lambda}{\lambda} - (\alpha+\xi)\Delta T\right] \quad (6.41)$$

6.3.3 Quasi-Distributed Stress Monitoring of Prestressing Steel Strand Based on Fiber Bragg Grating

In the monitoring system, the combination of optical frequency domain multiplexing and wavelength division multiplexing is adopted to realize quasi-distributed measurement of prestressing steel strand based on FBG sensors (Jin et al. 2008b, 2010; Jin and Du 2010).

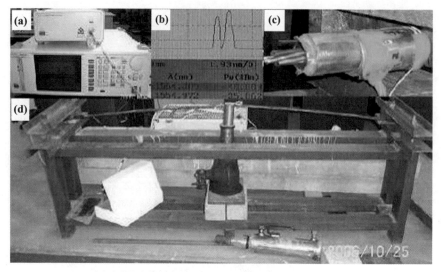

(a) ASE AQ4315A Source, (b) FBG spectrum,
(c) Detection on prestress sensors measurement,
(d) Distributed stress testing of prestressing tendons

Fig. 6.8 Diagram of measurement system for quasi-distributed stress based on intelligent prestressing tendon

In order to verify the feasibility of the quasi-distributed stress–strain monitoring method of prestressing tendon based on fiber grating, a traditional resistance strain gauge is stuck at the corresponding position of fiber grating for strain measurement. The quasi-distribution stress monitoring system of prestressing tendon based on fiber grating is shown in Fig. 6.8.

6.3.3.1 Experimental Devices and Instruments

Figure 6.8d shows the experimental device designed for transverse prestress loading of prestressing tendon. In the loading test, a jack is used for transverse loading and tensioning of the steel strand. The fiber grating is manufactured by Beijing Pi-optics Co., Ltd. The adopted connector is a universal optical fiber FC/APC jumper head. The ASE light source (AQ4315) is used. The FBG wavelength is measured by AQ6140 multi-wavelength meter.

6.3 Prestress Monitoring Technology Using Fiber Bragg ...

Table 6.8 Performance parameters of intelligent steel strand with FBG sensor

Linearity (relative error) e_L (%)	Load sensitivity (nm/10^9 kN)	Strain sensitivity (pm/$\mu\varepsilon$)	Hysteresis e_H (%)	Repeatability error e_R (%)	Strain resolution $\delta_\varepsilon/\mu\varepsilon$	Range (MPa)
1.1	41.6	1.21	2.5	1.20	0.83	252

6.3.3.2 Arrangement of Sensors and Attachment of Strain Gauges

As shown in Fig. 6.8c, an FBG prestress sensor is arranged at the anchoring end of the transverse loading test device of the prestressing tendons and a traditional resistance strain gauge is pasted near the FBG; the steel strand with FBG is prepared by adopting the composite technology of the FBG and the steel strand, and a resistance strain gauge is pasted on the surface of the corresponding position of the FBG.

6.3.3.3 Experimental Scheme

In the test, the steel strand is firstly tensioned in a state of slight stress through the transverse loading of jack after the testing device is assembled, then a step-by-step loading and unloading test is carried out on the prestressing reinforcement. The readings of strain gauge and the multi-wavelength meter are recorded during the loading and unloading process, respectively.

(1) Divide the measuring range into several equal interval points, load from small to large, and record the corresponding output values.
(2) Gradually lower the load from large to small while recording the corresponding output value.
(3) Carry out a plurality of tests of the forward stroke cycle and the reverse stroke cycle according to steps described in the (1) and (2) and list a table or draw a curve based on the obtained output–input test data.

6.3.3.4 Experimental Results and Analysis

The relationship between the center wavelength of FBG and the load is shown in Fig. 6.9, and the relationship between the wavelength of FBG and the strain of the referenced strain gauge is as shown in Fig. 6.10. The performance parameters of the intelligent steel strand with FBG sensor are shown in Table 6.8.

Test result shows that intelligent steel strand with FBG has good sensing performance, high linearity, good repeatability, and high sensitivity. During the test, the load is increased by the step of 10 kN from 0 to 252 kN and the FBG is broken when the strain of steel strand reaches 8,715 $\mu\varepsilon$. According to GB/T5224-2003, the yield load of Grade 1,860 steel strand is 234 kN, and the minimum breaking load

Fig. 6.9 Relation between FBG wavelength and load

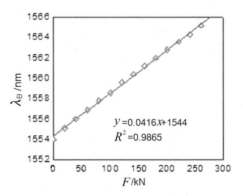

is 260 kN, so the measuring range of FBG basically meets the requirements of steel strand strain measurement. It is shown that stress measurement of steel strand at local position where FBG locating in the same cross section can be realized by the combination of FBG sensor and steel strand.

The strain curve between the FBG wavelength of the anchor-end prestress sensor and referenced strain gauge is shown in Fig. 6.11. The strain curve between the center wavelength of FBG attached on the steel strand inner wire after repeated cyclic loading and unloading, and referenced strain gauge is shown in Fig. 6.12. The strain curve between the anchor-end prestress sensor and FBG sensor attached on the steel strand inner wire is shown in Fig. 6.13. The relationship between the wavelength increment of the anchor-end prestress sensor and the wavelength increment of intelligent prestressing steel strand is shown in Fig. 6.14.

The following conclusions can be obtained from linear fitting of the test data:

(1) The relationship between FBG wavelength and strain gauge strain near the corresponding position is $y = 0.0012x + 1{,}554.3$ and $y = 0.0012x + 1{,}554.5$, respectively. It is shown that the linearity and repeatability are both good with a corre-

Fig. 6.10 Shift of FBG wavelength versus strain from gauge

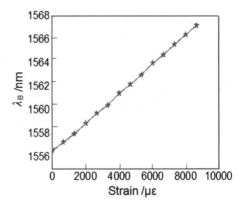

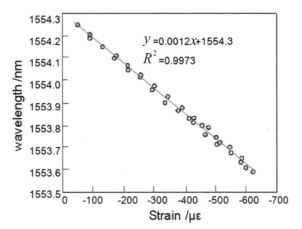

Fig. 6.11 Relation between wavelength shift from anchor sensor and strain from gauges

lation coefficient of 0.997 and a strain sensitivity coefficient of 1.2 pm/με. Since the wavelength resolution of spectrometer can reach 1 pm, the strain resolution of 0.83 με can be obtained. The central wavelengths of the fiber grating used in the test are 1,554.230 and 1,554.460 nm. According to the parameters of the typical quartz fiber $n = 1.46$, $\mu = 0.17$, $P_{11} = 0.12$, and $P_{12} = 0.27$, the strain sensitivity coefficient of 1.212 pm/με can be calculated which is in good agreement with the test result. Moreover, the FBG prestress sensor at anchor head and intelligent steel strand with FBG sensor both provide strain measurement with high linearity (0.99) and good repeatability, indicating that FBG sensor can basically meet the technical requirements for prestress monitoring.

(2) The strain relationship measure by the FBG prestress sensor at anchor head and the prestressing tendon strain gauge is as follows:

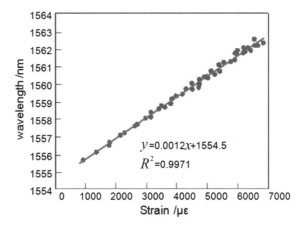

Fig. 6.12 Relation between wavelength shift of distributed FBGs and reference strain

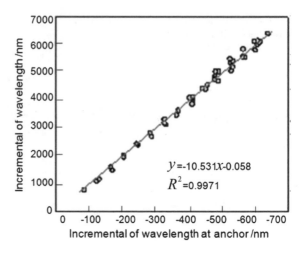

Fig. 6.13 Relation between wavelength shift from anchor sensor and distributed FBGs tendons

$$y = -10.435x - 5.562, R^2 = 0.9943.$$

The relationship between the wavelength increment of FBG prestress sensor at anchor head and that of intelligent prestressing tendon with FBG sensor is as follows:

$$y = -10.531x - 0.058, R^2 = 0.9971.$$

According to Formulas (6.39)~(6.41), it can be obtained that

$$\frac{\pi}{4}(D^2 - d^2)\varepsilon_t E_t = A_p \sigma_p = A_p E_p \varepsilon_p \tag{6.42}$$

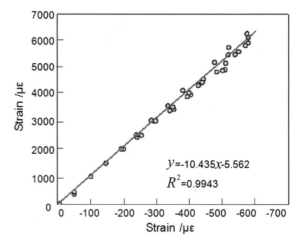

Fig. 6.14 Relation between strain shift from anchor sensors and distributed FBGs

$$\varepsilon_p = \frac{\pi(D^2 - d^2)E_t}{4A_p E_p}\varepsilon_t \tag{6.43}$$

$$\Delta\lambda_p = \frac{\pi(D^2 - d^2)\lambda_p E_t}{4A_p E_p \lambda_t}\Delta\lambda_t \tag{6.44}$$

where λ_p refers to the wavelength of intelligent prestressing tendon with FBG sensor; λ_t refers to the wavelength of FBG prestress sensor at anchor head; $\Delta\lambda_p$ refers to the wavelength increment of intelligent prestressing tendon with FBG sensor; and $\Delta\lambda_t$ refers to the wavelength increment of FBG prestress sensor at anchor head.

Substituting the parameters like $D = 47 \times 10^{-3}$ m, $d = 23 \times 10^{-3}$ m, $A_p = 1.4 \times 10^{-4}$ m^2, $E_t = 218$ GPa, $E_p = 195$ GPa, $\lambda_p = 1{,}554.460$ nm, and $\lambda_t = 1{,}554.230$ nm into Formulas (6.43) and (6.44), the results show $\varepsilon_p = -10.536\varepsilon_t$, $\lambda_p = -10.538\lambda_t$ indicating that the measured data of FBG is closer to the theoretical value and the linear correlation is high. It shows that the quasi-distributed measurement of prestress based on FBG sensor array by wavelength division multiplexing not only can monitor the prestress at anchor end but also can measure the prestress in a quasi-distributed and continuous way.

The test results that stress measurement of steel strand at local position where FBG locating in the same cross section can be realized by the combination of FBG sensor and steel strand.

6.4 Summary

According to the research on prestress loss theory and experiment, FBG prestress sensor at anchor head is designed in this chapter. Besides, a quasi-distributed monitoring technology of prestressing steel strand is proposed by combining FBG sensor and steel strand. The test results show that FBG prestress sensor at anchor head has good performance and can basically meet the technical requirements for prestress monitoring at anchor head. Moreover, stress measurement of steel strand at local position where FBG locating in the same cross section can be realized by the combination of FBG sensor and steel strand. Finally, the stress monitoring technology of prestressing steel strand based on FBG is established.

References

Chen Y, Chen Y (2006) Study on friction loss model of prestressed reinforcement in prestressed concrete structure. Ind Const 36(s1):250–252. https://doi.org/10.3321/j.issn:1000-8993.2006.z1.074

Chen H, Huang F (2005) The test-force ring of fiber optical gratings and its application. J Huazhong Univ Sci Technol Bat Sci Ed 33(5):58–60. https://doi.org/10.13245/j.hust.2005.05.019

Du YL, Li XY, Gui Y et al (2007) The fiber bragg grating sensors for monitoring prestress of the concrete located at anchor position. J Railw Sci Eng 4(1):82–86

Jin XM, Du YL (2010) Research on the FBG sensing technology for long-term measuring of the large strain of prestressed reinforcement. Piezoelectrics Acoustooptics 32(4):547–550. https://doi.org/10.3969/j.issn.1004-2474.2010.04.008

Jin XM, Du YL, Sun BC et al (2005) Study on optical fiber sensing technology for long-term real-time monitoring of prestressed reinforcement. Chin Safety Sci J 15(11):108–112. https://doi.org/10.16265/j.cnki.issn.1003-3033.2005.11.025

Jin XM, Du JH, Peng YB (2008a) Research on the strain transferring characteristics of coated optical fiber sensor embedded in measured structure. Chin J Sens Technol 21(9):1650–1653. https://doi.org/10.3969/j.issn.1004-1699.2008.09.036

Jin XM, Peng YB, Du YL (2008b) Research on stress measurement technology of prestressed tendons in large-scale prestressed concrete structures. Railw Eng (9):15–17. https://doi.org/10.3969/j.issn.1003-1995.2008.09.006

Jin XM, Lei Y, Du YL (2010) Influence of fiber structure on the error analysis in reference FBG temperature compensaiton. J Func Mater Dev 16(03):249–254. https://doi.org/10.3969/j.issn.1007-4252.2010.03.010

Luo S (2004) Research on key technology of strand production. Dissertation. Xi'an University of Architecture and Technology

Sun Y (2002) Research on optical fiber intelligent materials, devices and intelligent anchor cable structure system. Dissertation, Wuhan University of Technology

Xue W (2003) Design of modern prestressed structure. China Architecture & Building Press, Beijing

Yang T (2003) Study on prestress loss of long-span prestressed concrete continuous beam bridges. Dissertation, Wuhan University of Technology

Yang Z (2005) Three dimensional finite element analysis and research of prestressed concrete structure. Dissertation, Zhejiang University

Zhao Y (2002) Research on the friction loss and parameter of space prestress. Dissertation, Chang'an University

Zhao Y, Huang D, Li Y (2003) Analysis of transient prestress loss of prestressed tendons in slabs. Ind Const 33(3):40–44. https://doi.org/10.3321/j.issn:1000-8993.2003.03.013

Chapter 7
Cable Stress Monitoring Technology Based on Fiber Bragg Grating

Cable structure, as a common structural component in engineering, is widely used both in civil engineering structures (such as cable-stayed bridge, suspension bridge, cable dome, tower mast structure, etc.) and in aerospace structures (mesh antenna, etc.). The health status of cable structure directly determines the health status of cable-supported structure, so it is necessary to monitor the health status of cable structure.

This chapter mainly introduces the design idea of cable stress monitoring technology, pressure sensor structure, and cable state monitoring technology based on intelligent cable structure.

7.1 Current Status

The corrosion damage, vibration damage, tension relaxation, and quality problems of construction are common causes of cable damage. In particular, the cable is spanned across rivers and bays, and the steel wire rope and anchorage structure is vulnerable to corrosion damage due to long-term exposure to harsh environment of wind and rain, humidity, and air pollution. There have been some cases in China in which the stay cables have suffered serious corrosion damage and have to be replaced. Cable is a component with large length–diameter ratio and relatively small damping, which is easy to vibrate under the stimulation of wind and other loads. The vibration of cable produces fatigue damage caused by the repeated bending stress at the cable root, and also undermines the rigid casing and its anchorage. Therefore, the monitoring and control of cable vibration have always been a hot issue at home and abroad. In addition, the cable can be damaged by the incisions, scratches, misalignment of the sleeve, and wire connection failure during construction and installation. The loss of prestress caused by various reasons has always been the focus of attention in the field of engineering at home and abroad. Because the redistribution of the cable will leave a significant impact on the cable-supported structure. The stress relaxation caused by various reasons can be found in time by monitoring the cable tension; other damages, such as cable corrosion, should be monitored by means that can reflect its local health state.

For a long time, the manual monitoring of cable mainly includes checking whether the cable system is corroded and tilted, whether the fastener is loose. Through regular check on the number of corroded steel wires to determine the degree of corrosion, when the number of broken wires (equivalent to the superposition of corrosion number and corrosion level) is greater than 5% of the total number of wires, this cable should be replaced in a timely manner. Manual monitoring takes a lot of manpower and material resources, and the monitoring results cannot meet all the security needs, not to mention the real-time monitoring of emergencies.

To realize the real-time monitoring of the long-term working state of cable and ensure the safe operation during its service life, Swedish scholars have proposed the idea of intelligent anchor cable system based on the interferometric optical fiber sensing principle. However, due to the complexity of this measurement principle and the technical difficulty of system integration, no breakthrough results have been obtained. Sun Dongya et al. from Wuhan University of Technology have proposed an intelligent anchor cable structure (see Fig. 7.1) based on micro-bending strength modulation optical fiber sensing system. It can realize distributed measurement of anchor cable strain by using OTDR technology, which has the disadvantage of low precision. As shown in Fig. 7.2, Sun Dongya et al. have also proposed the design scheme of intelligent anchor cable structure based on FBG sensor, but it is difficult to realize in practice (Hu et al. 2001; Jiang 2000; Sun 2002).

The traditional method is restricted by factors like number of sensors, measurement conditions, and installation process, leading to insufficient data and difficulty

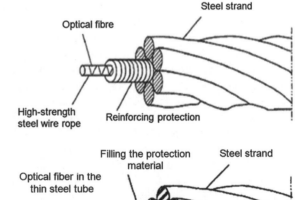

Fig. 7.1 Microbending intensity-modulated optical fiber sensors installed on steel strand

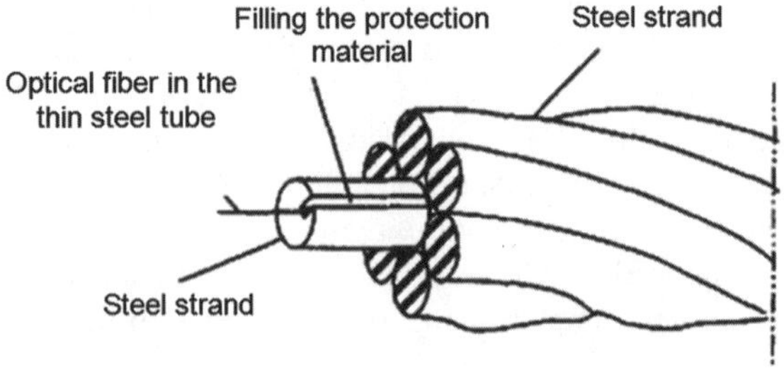

Fig. 7.2 Structural schematic of steel strand in intelligent anchor cable system

in giving a scientific and accurate conclusion of structural health assessment. And it is necessary to study new monitoring means and methods.

7.2 Cable Tension Monitoring System Based on FBG

The cable is the main force-bearing component of cable-supported structure. Through the monitoring of cable tension, cable damages such as prestress reduction caused by various reasons can be found in time. It not only can timely understand the health status of the cable itself, but also may find the safety status of the cable support structure system through the real-time monitoring of cable tension, so it is of great significance. However, it is difficult to measure the actual cable force quickly and accurately by the conventional cable force testing methods in engineering (Fang and Zhang 1997). Furthermore, for large structures, there may be multiple cables in the structure at the same time, and the distribution range is large. To realize the long-term and real-time monitoring of all cables, the traditional electrical measurement method is not competent.

To realize the long-term and real-time health monitoring of cable and its engineering structure, it is necessary to select reasonable state parameters which can truly and comprehensively reflect the cable damage. According to the structural characteristics of cable, this section proposes and designs two kinds of cable intelligent monitoring system based on FBG (Zhang et al. 2002, 2003a, b).

7.2.1 Composition and Working Principle of Cable Tension Monitoring System

The cable tension monitoring system is mainly used to monitor the change of cable tension caused by various reasons during service. It mainly consists of a sensing part, a data acquisition and processing part, and a health diagnosis and safety evaluation part, as shown in Fig. 7.3.

The sensing part is a pressure sensor composed of FBG and elastic element, the data acquisition is completed by wavelength monitor, and the data processing and health diagnosis and safety evaluation part is completed by corresponding software. Due to the incomplete function of the existing wavelength monitor, the virtual instrument and wavelength monitor are combined to complete the data aggregation and processing.

The pressure sensor based on fiber grating is composed of FBG and circular elastic body, and the FBG is rigidly adhered to the outer surface of elastic body circular ring to form a through-center pressure sensor. In operation, the pressure sensor is arranged between the anchorage device at the anchor end of cable and the cable hole backing plate, when the cable tension changes, the changed tension is transmitted to

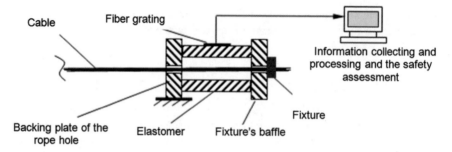

Fig. 7.3 Monitoring principle for cable tension

the elastic body through the clamp baffle plate, the elastic body is subjected to the same force as the cable tension, and the FBG adhered on the elastic body surface is correspondingly stretched and compressed, so that the central reflection wavelength of the FBG is changed, the change in central reflection wavelength of FBG is measured by a wavelength monitor, and the tension magnitude and the working state of cable can be judged through data processing and analysis. Therefore, as long as the relationship between the cable force and the FBG wavelength is established, the cable force sensing of FBG can be realized, thereby realizing the long-term monitoring of cable tension (Zhang et al. 2002).

7.2.2 Characteristics of Cable Tension Monitoring System

The pressure sensor installed at the anchor end converts cable tension into pressure to measure so that the monitoring structure is simpler, more reliable, and more practical. Meanwhile, by utilizing the characteristics that the FBG is easy to reuse and has a low transmission loss, all or groups of sensors of each cable can be multiplexed together to form a sensor array, namely, the cable force measurement of a plurality of cables can be simultaneously realized by a set of systems, which can greatly reduce the monitoring cost, and the FBG has the advantages of simple transmission lines and long-distance transmission. The outstanding advantage of this method is that the cable force changes caused by various loads can be accurately determined, and the linear shape of the whole engineering structure is ensured to be in the design state. The method is not only suitable for real-time monitoring of cable tension during construction, but also for long-term monitoring of cable tension after completion. Meanwhile, the vibration frequency of cable can be monitored in real time, and the health condition of cable as well to a certain extent. However, the defects of cable tension monitoring system are also obvious, that is, the stress distribution state of cable steel strand cannot be fully understood, and the frequency change is not sensitive to damage, so the damage degree of cable is often very serious when cable damage is found to exceed the natural frequency.

7.2.3 FBG Pressure Sensor

7.2.3.1 Design of Pressure Sensor

In a pressure sensor, an FBG is used as a sensitive element to constitute a sensor through an elastic element. The measured physical quantity is converted into a change in wavelength by the deformation of elastic element to lengthen (or shorten) the fiber grating adhered thereto. Therefore, the sensitivity and accuracy of the sensor depends to a large extent on the elastic element, so the elastic element is one of the key components of the sensor. In the design process of elastic element, it is usually necessary to select the materials of elastic element according to the measurement range, measurement accuracy, and sensitivity of sensor to determine the allowable stress, and then design its structural parameters (Shan 1999).

Taking the relationship among the stiffness, sensitivity, and natural frequency of the sensor as an example, if the sensor is expected to have higher sensitivity, the stiffness of elastic element cannot be too high; however, when measuring the dynamic signal, if the sensor is expected to have good dynamic response characteristics and a wide frequency response range, the elastic element is required to have a higher stiffness. Because of the contradiction between them, so it is required to repeat operation and comparison for comprehensive consideration. The following text describes the design process of the elastic element in the cable tension monitoring system.

1. Selection of Pressure Sensor Structure

The cable tension monitoring structure is installed at the anchor end of cable and requires a hollow center. Three pressure sensors can be chosen and their advantages and disadvantages are shown in Table 7.1.

Table 7.1 Comparison of three pressure sensors

Type	Advantages	Disadvantages
Cylinder	Compact and simple structure, easy to process, it can bear greater stress in equal volumes, it has been extensively used in large and medium sensors	Larger height, higher stress, larger nonlinear, lower output sensitivity, poor resistance to lateral load
Spoke-structure	Excellent linearity, high precision, low shape, strong resistance to lateral load, high ability in eccentric load	Complex structure, difficult to process
Plate-ring	Stability stress state, good temperature homogeneity, high output sensibility, low height, strong resistance to lateral load, high ability in eccentric load	Complex structure, difficult to process

Because of the complicated structure and processing difficulty of spoke-structure and plate-ring-structure, the cylindrical structure is selected according to the operating requirements.

2. Selection of Pressure Sensor Materials

The material of elastic element is the key to the high technical performance and stability performance of the sensor. The ideal materials for the elastic element should meet the following requirements (Liu 1999, 2001; Shan 1999; Yang 2001): ① it must have good linear elasticity and small elastic hysteresis and elastic aftereffect; ② the elastic modulus has good time stability and low-temperature coefficient; ③ the output sensitivity should be as high as possible; ④ it must have good impact toughness, so that the sensor has better performance to against impact, vibration, and fatigue; ⑤ simple structure, easy processing, and installation; ⑥ it has good stability and corrosion resistance, and can adapt to the harsh environment in long-term field monitoring, etc.

It is not possible for a material to have all of the above properties at the same time, but the metal material can be technically treated to have or approach to the above properties as much as possible.

The commonly used materials of elastic body are CrNiMo, CrMnSi alloy steel, aluminum alloy, beryllium bronze, etc. Under laboratory conditions, an important criterion for designing cable tension monitoring system is that both sensitivity and stiffness must be taken into account, that is to say, the sensor should have higher stiffness and natural frequency under the specified sensitivity. For this reason, it is necessary to select a material whose product of the elastic modulus E and strength σ is the smallest one. The $E\sigma$ product of aluminum alloy is only 1/9 of that of alloy steel, which is an ideal elastic element material for manufacturing cable tension monitoring system. As calibrated by tests, the elastic modulus of the aluminum alloy is $E = 72 \times 10^9$ Pa and ultimate strength is $[\sigma] = 203$ MPa.

3. Parameter Selection of Pressure Sensor

(1) Parameter Selection

According to the structural dimensions of the cable anchorage in the test model, the inner diameter $D = 15$ mm of the elastic body of the sensor is first selected.

The choice of outer diameter should be calculated according to the maximum pressure borne by the elastomer and the allowable stress $[\sigma]$ of the selected materials, and the calculation formula is (Shan 1999; Yang 2001)

$$D \geq \sqrt{\frac{4}{\pi} \frac{F}{[\sigma]} + d^2} \tag{7.1}$$

Assuming that the elastomer is subjected to a maximum pressure of 20 kN, $D = 20$ mm is obtained by the calculation according to the above formula.

The height of elastic body has influence on the precision and dynamic characteristics of the sensor. From the view of mechanics of materials, it is known that the height

7.2 Cable Tension Monitoring System Based on FBG

has influence on the deformation along its cross section. When $H > D$, the load properties acting along its end face are independent of the contact conditions. Therefore, the determination of H here, should mainly consider the elastic body under the action of cable force to make the strain field of FBG working area has enough uniformity, for the sensor structure selected herein, the following requirements should be met:

$$H \geq D - d + l \tag{7.2}$$

Considering the convenience of operation during processing and testing, $H = 45$ mm herein.

The dynamic characteristics and the hysteresis in change of the elastic sensitive element are related to its natural vibration frequency. Generally, it is always desirable to have a higher natural frequency. The value f_0 of the natural frequency of the cylindrical elastic element can be estimated by the following formula (Shan 1999; Yang 2001).

$$f_0 = 0.159 \frac{\pi}{2H} \sqrt{\frac{EA}{m_1}} \tag{7.3}$$

where m_1 refers to the mass of elastic element in unit length, calculated by the following formula:

$$m_1 = A\rho \tag{7.4}$$

where ρ refers to the density of elastic element.

For elastic elements designed in this section, it can be calculated that $f_0 = 28{,}574$ Hz.

The results show that the cable tension changes slowly and the vibration frequency are low. Therefore, the elastic element fully meets the requirements, and can realize real-time monitoring of cable tension and natural frequency.

(2) Influence of Temperature on Pressure Sensor Accuracy

When the FBG is adhered to the surface of tested structure or buried in the test material, the thermal expansion of the material structure will also change the grating cycle.

In general, the thermal expansion coefficient of measured structure is much larger than that of the optical fiber as the supporting material of the FBG. At this time, the Bragg wavelength shift caused by the thermal expansion effect of the optical fiber can be neglected, but the temperature sensitivity of the grating is greatly influenced by the supporting material or the packaging material. The greater the coefficient of thermal expansion of supporting material is, the higher the temperature sensitivity of the grating is.

In the cable tension monitoring system, the temperature sensitivity of pressure sensor is about 3.23 times as high as that of the original because the fiber grating is finally adhered to the aluminum alloy.

7.2.3.2 Performance Test of FBG Pressure Sensor

1. Static Characteristic Assessment of Pressure Sensor

Broadband light source BBS (1,525~1,575 nm) is selected for the test. The wavelength resolution of wavelength monitor (model: F200) is 0.001 nm. After pasting, the measured Bragg wavelength of fiber grating in free state is 1,547.8 nm, and the length is 10 mm. The test is carried out on a hydraulic universal testing machine. In the range of 0~25 kN, there are 25 test points in a circulation in total.

The experimental results show that the linear fitting is greater than 0.998, the sensitivity is 100.09 $\mu\varepsilon$/kN, the lag error is $\pm 0.249\%$, the repeatability error is $\pm 1.049\%$, and the tension resolution is 8.26 N. The lower limit of pressure measurement for the designed sensor is 8.26 N.

It should be noted that the sensor sensing characteristic index analyzed in the test is for the particular grating, the elastomer material, and the grating wavelength shift monitoring system selected in the book, and if any part of the sensor changes, the sensing characteristic index of the sensor will change accordingly. In addition, the maximum strain of fiber grating manufactured by different processes can be greatly different, which should be considered in the design.

2. Dynamic Characteristic Assessment of Pressure Sensor

The dynamic strain measurement test of fiber grating is carried out on a flexible beam (length × width × thickness = 600 × 30 × 2 mm, elasticity modulus = 196 GPa), and the test device is shown in Fig. 7.4. Various excitation forces are applied to the beam by the vibration exciter, and the optical signal induced by fiber grating is processed through a wavelength monitor and a computer.

The central wavelength of FBG used in the test is 1,550.45 nm, and the signal acquisition and processing are completed by F200 and virtual instrument.

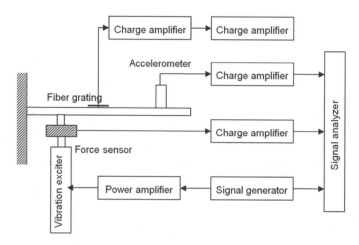

Fig. 7.4 Dynamic performance test device for FBG

7.2 Cable Tension Monitoring System Based on FBG

For ease of comparison, the acceleration response of the beam is obtained by adopting a piezoelectric acceleration sensor (4,374) and a charge amplifier (2,635) of Danish B&K Company. Its output end will pass through anti-aliasing filter. INV-306 software and hardware system signal acquisition and analytic processing (i.e., DASP system) of China Orient Institute of Noise & Vibration is adopted to realize the data acquisition, and data processing is conducted by a computer.

According to multiple measurements, the data measured by FBG sensor and the traditional accelerometer are highly identical with each other, and the dynamic response characteristics of both under the same condition have good consistency, which means FBG sensor can be used for dynamic real-time monitoring of the structure under low-frequency vibration condition.

7.2.4 Hardware Design

The hardware composition of the cable tension monitoring system based on fiber grating is shown in Fig. 7.5.

7.2.5 Software Design

Cable tensile analysis and processing software are based on the software platform LabVIEW for G language programming (graphic programming) of American NI Company, and the overall design of the cable tension analysis processing software developed by LabVIEW is shown in Fig. 7.6. It can complete self-monitoring, self-

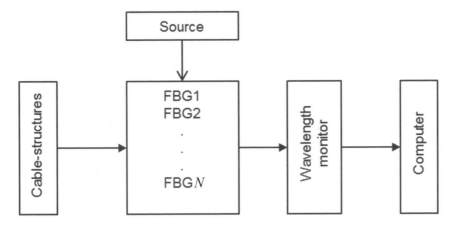

Fig. 7.5 Hardware for stress testing system

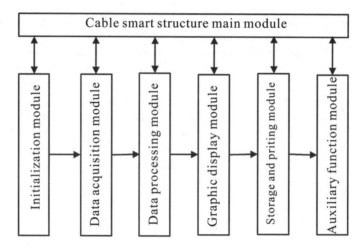

Fig. 7.6 Overall design of testing system

diagnosis, and other functions for cable structure damage. As can be seen from the figure, the monitoring system software mainly consists of six modules.

1. Initialization Module

The initialization module is mainly used to calibrate the sensor, specify the measurement object from the database, set various parameters of the hardware, read various initialization parameters, complete the basic configuration of the system, and provide the basis for subsequent data processing of the computer.

Since there are many sensors in the cable intelligent structure, and each sensor is not exactly the same, the sensors used should be calibrated one by one before use. The calibrated sensor can be used to measure the various parameters.

2. Data Acquisition Module

The module includes sampling parameters (such as channel number, sampling frequency, sampling points, sampling control mode, etc.) setting and sampling start control.

The data acquisition module of the wavelength monitor is mainly invoked among the data acquisition modules. On this basis, other functions of the software are completed.

In the system, two ways to read data are adopted. One is to read text files, and the other is to directly invoke serial data. In the real-time monitoring process, the method of invoking serial data is adopted, and the real-time information is obtained through the data processing function of the software. During data playback, the mode of text file reading is adopted.

3. Data Processing Module

The main function of the module is to analyze and process the wavelength signals acquired by the data acquisition module. The module comprises a plurality of analysis functions of amplitude domain, time domain, and frequency domain, of which, the amplitude domain includes amplitude domain statistical analysis (e.g., inclusive value, mean square values, effective values, variance, standard deviation, etc.) and probability distribution function analysis; the time domain includes auto-correlation analysis and cross-correlation analysis; the frequency domain includes self-power spectrum analysis, cross power spectrum analysis, amplitude spectrum analysis, phase spectrum analysis, frequency response function (e.g., amplitude frequency, phase frequency, etc.) analysis, etc. These data processing functions fully meet the requirements of test analysis.

In the design of the cable tension monitoring system, due to the influence of various noises, the signal-to-noise ratio of the optical fiber grating reflection signal itself is very low when broadband light source is used in the system, and the filtering processing must be performed before the signal analysis in order to improve the wavelength monitoring precision. This requires that the ripple amplitude of the passband should be narrow as far as possible, but the requirement is not very high for the width of the transition zone. Through comprehensive comparison for various low-pass filters, the Butterworth filter is selected (Wang 2003).

4. Graphic Display Module

In the process of signal acquisition, the signal can be displayed in real time and multichannel waveform. In the process of signal processing, the spectrum analysis and modal analysis of the signal can be displayed.

5. Storage and Printing Module

The displayed image, waveform, and statistical table are stored and replayed in the form of files, and can be printed out through the printer.

In order to facilitate the observation of the measurement result, the module has the functions of local stepless amplification, cursor tracking reading, curve splitview and curve overview and the like.

6. Auxiliary Function Module

Its main function is the real-time help and use instruction of "Cable Intelligent Structure Monitoring and Analysis System". Real-time help is a detailed explanation of all controls, settings, operational steps, etc. on the interface, which can be used in all interfaces, effectively improve the operability of the system, and reduce operating errors. The instructions for use can only be invoked in the main interface.

In summary, the intelligent structure monitoring and analysis system of cable can realize real-time acquisition and processing of test data in the test process, and has a plurality of analysis functions for amplitude domain, time domain, and frequency domain of the test result, and can meet the requirements of the intelligent structure of cable.

7.3 Distributed Stress Monitoring System for Cable Based on FBG

7.3.1 Composition and Working Principle of Distributed Stress Monitoring System for Cable

In order to realize the real-time monitoring of cable prestress distribution and change and its long-term working state and ensure safe operation of cables and their engineering structures in usage period, it is necessary to monitor their stress distribution. Through the monitoring the cable stress distribution, not only can the strain distribution information of the cable be known, but also the strain modal information of the cable vibrating under the excitation of the external environment can be obtained. Based on the character that the local damage on structure is relatively sensitive to the strain, the strain modal information can be used to judge whether there are any broken wire, crack and other specific damages on the cable, as well as damage location and damage degree, which is favorable for safety monitoring and integrity evaluation of the cable.

The cable distributed stress monitoring system is mainly used to monitor the strain distribution along the entire length of the cable and the modal parameters of the cable vibration. It is also composed of three parts, namely, the sensing part, the data acquisition and processing part, and the health diagnosis and safety evaluation part as shown in Fig. 7.7.

The sensing part is composed of optical fiber grating which is directly used as the sensing element to measure the stress and strain of the cable, and the data acquisition and processing is completed by the wavelength monitor and the virtual instrument together (Zhang et al. 2004, 2005).

In the monitoring system, optical fiber grating is adopted as a sensing element, and during the manufacturing process of the cable, the optical fiber grating is implanted into the cable, the dynamic strain inside the cable is monitored in real time through an FBG sensor. The signals such as structural damage and acting load are sensed, which will indicate the stress course and damage position of key parts after treatment, and be saved in the structural health status database of central processing unit. The database includes the use and maintenance history of the cable throughout the lifespan, as

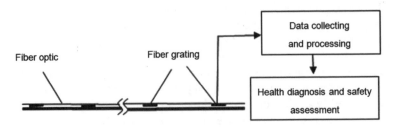

Fig. 7.7 Schematic diagram of monitoring principle for cable distributed stress

well as the stress and damage values required by the system to analyze the residual strength and remaining life of the structure on a regular basis. The analysis program can predict the durability and damage tolerance of the structure, including the specific conditions of fatigue, fracture and the like, and the analysis results are used as the basis for the expert to conduct system evaluation for structural health.

7.3.2 Characteristics of Cable Tension Monitoring System

The cable distribution stress monitoring system can comprehensively and timely understand the stress distribution state of the cable steel strand, so as to find out the strain distribution and vibration modal information of each part of the cable, and judge whether the cable is damaged or not on such basis. The design of the monitoring system not only provides a new method for the health monitoring of the cable and its structural system, but also greatly improves the reliability of their health monitoring, thereby providing the possibility for realizing the whole service life and integrated monitoring of the engineering structure from manufacturing to service.

7.3.3 System Design of Signal Acquisition Processing and Analysis

1. Hardware Design

The hardware composition of cable distribution stress monitoring system is the same as before. According to the results of finite element analysis, the first-order frequency of the test cable is less than 10 Hz. Therefore, the sampling frequency of the optical fiber grating signal demodulator adopted is 100 Hz. According to the sampling theorem, the requirement is fully met.

2. Software Design

The program flow of software in cable distributed stress monitoring system based on fiber grating is shown in Fig. 7.8. The basic function of the software is similar to that of the tension monitoring system. But its signal processing includes statistical analysis, power spectrum estimation, digital filtering, modal analysis, and other functions.

7.3.4 Realization of Remote Monitoring for Smart Structure in Cable

The traditional electric measuring method is easy to be influenced by the ambient temperature, which is characterized by poor stability and low reliability. It is difficult for

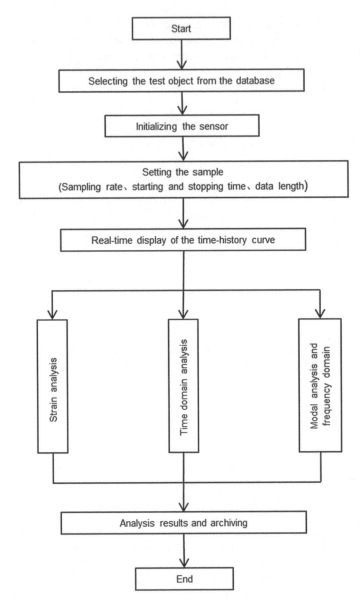

Fig. 7.8 Flowchart of distributed stress monitoring system of cable

the signal to realize long-distance transmission. The abovementioned disadvantages can be well overcome by using optical fiber sensor. The signal from the single-chip microcomputer (or optical transceiver) is transmitted to the main control computer through Ethernet, and the data receiving, processing, and analyzing are carried out with LabVIEW. In order to realize the remote fault diagnosis of cable structure, this design uses TCP/IP protocol programming to realize network communication.

The data is transmitted from the monitoring site to the remote monitoring center for real-time monitoring and analysis, which requires powerful network function of LabVIEW. The program reads the machine address and port of the data to be transmitted, and transmits the sampled data and length to the designated machine through TCP/IP protocol.

7.4 Test for Condition Monitoring of Cable Structure

In order to verify the correctness of the cable theoretical calculation model and the reliability of the designed cable intelligent structure, it is necessary to establish a complete cable test measurement model in the laboratory. The cable is subjected to static and dynamic characteristic tests by using the designed cable intelligent structure, which will be compared with the traditional test method and the theoretical calculation value.

7.4.1 Cable Model and Test System

The main purpose of the test is to verify the effectiveness and reliability of cable intelligent structure based on fiber grating for cable tension and modal parameter monitoring. It should meet the following requirements: ① The characteristics of cable structure can be simulated; ② The cable tension can be adjusted; ③ The structure is simple, economical, and practical.

Based on the design of cable intelligent structure, cable model and test system for cable tension and modal parameter measurement are established. The cable tension is the main parameter to simulate the dynamic characteristics of the cable, and the response characteristics of the cable with different tension are obviously different. In order to simulate the different response characteristics of the cable under different forces, a cable tension adjustment mechanism is designed (see Fig. 7.9).

The lower end of the cable is fixed with the slide block, the slide block is fixed on the guide rail of the base, and the position of the sliding block on the guide rail can be adjusted. Therein, the pressure sensor is the pressure sensor based on the fiber grating in the cable tension monitoring system. The size of cable tension can be adjusted by changing the counterweight, and after adjusted to certain cable force, the upper end of the cable can be fixed with the cable bracket through a clamp.

The cable tension measurement principle of cable tension monitoring system is shown in Fig. 7.10. The light emitted by the light source is incident on the optical fiber grating through a coupler. Under the action of the external environment, the cable tension changes and causes the drift of Bragg central wavelength, and the light

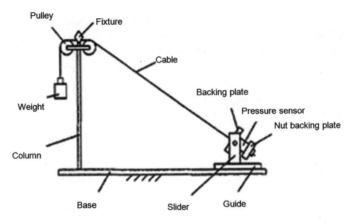

Fig. 7.9 Test device of cable tension

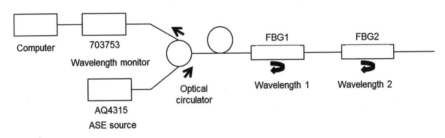

Fig. 7.10 Measuring principle of cable stress

reflected by FBG is detected by the wavelength monitor so as to obtain the size of the cable tension.

The cable force changes slowly during the stretching process, which can be regarded as a static loading mode, and after the stretching is completed, the cable in operation is always in the vibration state under the action of external loads such as rain-wind induced vibration, vehicle-mounted, and other external load. The two load states are simulated in the test, respectively, and the cable tension is measured by the cable tension monitoring system.

Considering convenience and cost, the cable used in the test consists of a single strand of fine copper wire, and its characteristic parameters are shown in Table 7.2.

7.4.2 Test of Cable Tension

1. Cable Tension Monitoring under Static Loading Conditions

First, the static loading mode is simulated, the size of the cable tension is changed by using different counterweights, and the cable tension is tested by the cable tension

7.4 Test for Condition Monitoring of Cable Structure

monitoring system. In the design of the cable tension monitoring system, the sensing characteristics of the cable tension monitoring system have been calibrated, so the value of the cable tension can be directly read out. A large number of tests show that the value of the cable tension and the counterweight are in good compliance with each other.

2. Cable Tension Monitoring under Dynamic State

The cable tension is adjusted to a certain state, and both ends of the cable are fixed, to make the cable deviating from the balance state, then suddenly release it, and measure the response during the free vibration of the cable. Simultaneously, a strain gage is glued on the elastic element of the pressure sensor. According to the comparison between the frequency values from spectrogram obtained from signal analysis for strain gage measured value and FFT analysis for value of cable tension monitoring system measured by visual instrument, the frequency value measured by both methods are consistent, which means the real-time monitoring for cable tension can be realized by the cable tension monitoring system of this design.

The cable tension monitoring system based on the fiber grating not only can measure the cable tension during the stretching process, but also can carry out real-time monitoring on the cable tension under the working state, even monitoring the vibration frequency in the cable vibration process, which is of great importance to the safety evaluation of the cable and its supporting structure.

7.4.3 Test of Cable Stress Distribution

This section studies the results of cable stress distribution measured by cable distribution stress measurement system under different excitation. In order to verify the validity and reliability of the test results, the strain gage and the fiber grating are pasted on the corresponding model cable adjacent to each other, and the two measurement methods are subject to test comparison and analysis. In the test, nine measuring points are uniformly distributed on the cable, and the selection of the strain gages and their test instruments is identical to that in Sect. 7.3.

Adjust cable tension, and fix it when $F = 50$ N. An initial displacement of the cable is made to measure the distribution stress of the cable during free vibration.

Table 7.2 Characteristic parameters of cable used in test

Parameter name	Value
Length (m)	3.24
Inclination angle (°)	20
Diameter (mm)	1.585
Elastic modulus (MPa)	1.96×10^5
Density (kg/m³)	7,900

The results show that the results measured by strain gage are highly consistent with those measured by cable distribution stress monitoring system.

Similarly, when $F = 50$ N, the swept sine is carried out on the cable, and the sweep frequency range is 8–9 Hz. The results measured by the strain gage and the results measured by the cable distribution stress monitoring system can well reflect the resonance phenomenon while the sweep frequency passes the first-order inherent frequency (8.74 Hz) of the cable.

Through contrastive analysis on the measurement results from both methods, it can be seen that the cable distribution stress monitoring system can realize the measurement of cable distribution stress well.

7.4.4 Test of Cable Modal Parameter

1. Cable Calculation Model

Whether the cable characteristic parameters will be affected by the cable tension monitoring system installed on the cable shall be illustrated by calculation. This section checks whether the installation of cable tension monitoring system will affect cable properties by simulating the change of natural vibration frequency before and after installing the cable tension monitoring system.

First, the natural vibration frequency of the cable under three different tensions is calculated by using the established finite element model. Then, a unit (elastic element) is added on the basis of the original model to form a new model, and calculate the natural vibration frequency of the cable in this model under different conditions. From the result of the finite element analysis, there is almost no change in the natural vibration frequency calculation value of the two models. Namely, there is no influence on the cable vibration characteristics after installing the cable tension monitoring system and its influence on the cable characteristic can be neglected.

When $F = 70$ N, the finite element calculation is conducted for the strain modal vibration mode of cable shown in Table 7.2 by using finite element model established in the previous text, and the calculation results in the first two-order vibration modes are shown in Fig. 7.11.

2. Composition of Test System

The method of single-point excitation and multi-point measurement can be selected from the following three aspects:

(1) Selection of Excitation Mode

The excitation mode used in the modal test mainly includes sinusoidal excitation, random excitation, transient excitation, etc. The sinusoidal excitation is adopted in the test.

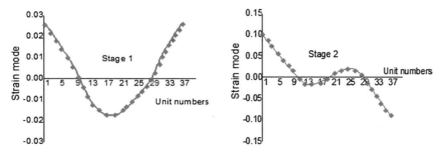

Fig. 7.11 Diagram of strain mode shape

(2) Selection of Excitation Position and Response Position

For a single-input multi-output method, the position of the beat point should first be determined, and then the number of response points and the position of each measuring point are determined. The position selection for the excitation point and the response point is important, and the vibration mode obtained can be analyzed with reference to the finite element theory. The selection at node in the previous orders of modal vibration mode should be avoided as far as possible, and the quality of the test should be ensured so as to avoid losing the modal information. While measuring, the sensor shall be pasted to the measuring point, and the response signal shall be monitored. According to the vibration mode calculation results, in order to obtain the previous two orders of modal vibration mode comprehensively and accurately, the nine uniformly distributed points in the distribution stress measurement can meet the requirements.

(3) Selection of Acquisition System and Modal Analysis Software

The acquisition system and modal analysis software for dynamic characteristic monitoring of fiber grating cable are completed by the self-designed cable intelligent structure monitoring and analysis. A block diagram of cable dynamic characteristics is shown in Fig. 7.12. When the optical fiber grating is used for measurement, various selected elements are formed according to the connection mode shown in the figure, and then the test can be carried out.

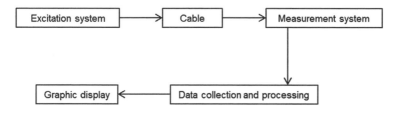

Fig. 7.12 Flowchart of monitoring for dynamic characteristics of stay cable

Table 7.3 Measured natural frequency

Tension (F/N)	The first-order frequency (Hz)	The second-order frequency (Hz)
$F=50$	8.742	17.501
$F=70$	10.341	20.703
$F=90$	11.730	23.479

(4) Test and Result Analysis

Adjust the size of the cable tension, measure the response characteristic of the cable in different situations, and process the response signal to obtain the inherent frequency and the vibration mode of the cable in different orders. In the case of three different tensions, the measurement of the inherent frequency of the cable is shown in Table 7.3.

The measured results are highly identical with the calculated frequency value and vibration mode, which means relatively clearer cable modal can be obtained by cable test in the cable distribution stress monitoring system based on optical fiber optic grating. It is proved that the method can be used for the dynamic test of cable structure. The experimental results also show that Link10 model established for the cable can better reflect the performance of cable structure, and further verify the accuracy of the model.

7.5 Summary

A cable tension monitoring system, including FBG pressure sensor, data acquisition and processing, health diagnosis and safety assessment, etc. is designed and developed based on fiber grating sensing. Through real-time data acquisition and processing analysis, the long-term monitoring of cable tension can be realized. By implanting FBG into the cable, a cable distributed stress monitoring system is established. Indoor test is carried out to obtain cable tension, stress distribution, and modal parameters. The results show that the cable tension monitoring system can monitor the cable tension and the vibration frequency in real time, and the cable distributed stress monitoring system can well realize the measurement of cable distributed stress. The modal of cable vibration can be clearly expressed.

References

Fang Z, Zhang Z (1997) Test of cable tension in cable-stayed bridges. China J Highw Transp 10(1):51–58

Hu S, Wang Q, Jiang Z (2001) Study and design of the digital intelligent cable pre-stress sensor. Explor Eng (Rock Soil Drilling Tunneling) (2):30–33. https://doi.org/10.3969/j.issn.1672-7428.2001.02.016

Jiang D (2000) Smart materials devices structures and their applications. Wuhan University of Technology Press, Wuhan

Liu G (1999) Modern detection technology and test system design. Xi'an Jiaotong University Press, Xi'an

Liu J (2001) Analysis and selection of load cell spring component metal material. Weighing Apparatus 30(5):10–16. https://doi.org/10.3969/j.issn.1003-5729.2001.05.003

Shan C (1999) Sensor theory and design basis and its application. National Defense Industry Press, Beijing

Sun Y (2002) Research on optical fiber intelligent materials, devices and intelligent anchor cable structure system. Dissertation, Wuhan University of Technology

Wang Y (2003) Digital filters and signal processing. Science Press, Beijing

Yang B (2001) Modern sensor technology foundation. China Railway Publishing House, Beijing

Zhang X, Du Y, Sun B (2002) Application of optical fiber grating pressure sensor on monitoring of cables' tension. J China Railway Soc 24(6):47–49. https://doi.org/10.3321/j.issn.1001-8360.2002.06.011

Zhang XS, Du YL, Jin XM et al (2003a) In suit monitoring for cables' tension with optical fiber grating sensors. J Traffic Trans Eng 3(4):22–24

Zhang XS, Du YL, Sun BC (2003b) Study in monitoring of cables' tension with fiber Bragg sensor. China Saf Sci J 13(7):80–82. https://doi.org/10.16265/j.cnki.issn.1003-3033.2003.07.021

Zhang XS, Du YL, Sun BC (2004) On-line monitoring method for stayed-cables dynamic response based on fiber Bragg grating. China Saf Sci J 14(7):101–103. https://doi.org/10.16265/j.cnki.issn.1003-3033.2004.07.025

Zhang XS, Ning CX, Zhou RF (2005) Fiber Bragg grating vibrating sensors and its signal processing system. J Hebei Univ Sci Technol 26(2):142–145

Chapter 8
Intelligent Monitoring Technology for Fiber Reinforced Polymer Composites Based on Fiber Bragg Grating

The fiber reinforced polymer/plastic (FRP) is a high-performance new material which is mixed in a certain proportion by the fiber and the base material and is integrated by a certain process. Such material has the advantages of lightweight, high-strength, corrosion resistance, good fatigue resistance, large damping, and low coefficient of thermal expansion, good molding processability and convenient construction (Wang 2007). Due to small size, routability and other significant advantages of optical fiber, it is very suitable to be implanted into fiber composite, forming fiber composite intelligent structure based on fiber grating. This intelligent structure shows very broad application prospect in structural health monitoring (Chen et al. 2016), especially for stay cable of long-span bridge and prestressing tendon in engineering structure (Wang and Zhou 2017). This chapter mainly analyzes and studies the preparation and properties of the fiber composite material, the interface bonding between FBG sensor and the composite material, as well as the sensing characteristics of the intelligent rod.

8.1 Preparation and Properties of Fiber Reinforced Polymer Composites

8.1.1 Selection and Proportioning of Component Materials

The composite material consists of three parts, namely, fiber, interface, and resin. Therein, the fiber is the reinforcement with reinforcing function, and the mechanical properties of the composite material are mainly determined by the fiber. Therefore, the selection of suitable fiber plays an important role in improving the strength of the composite material.

The tensile strength of ordinary carbon fiber composite material is 2,000 MPa, and the elasticity modulus is about 160 GPa. The tensile strength of the steel cable is generally 1,860 MPa, but the elasticity modulus is about 200 GPa. M40J carbon fiber produced by Toray Industries Inc. (Wang 2014) is a kind of high-strength and high-modulus fiber. Its monofilament tensile strength is 4,400 MPa, and elasticity modulus is 377 GPa. Such fiber is selected as the main reinforcement material, which provides a strong guarantee for the production of high-modulus, high-strength fiber composite material. Producing fiber composite material with large damping and good vibration reduction effect is also an important study content. The research shows that through mixing several kinds of fibers together, the damping of the material can be improved to a certain extent, the complementary advantages can be realized among them, and the problem that the performance of a single fiber composite material is relatively poor or weak is improved. For example, carbon fiber composite shows small strain at expensive price. Adding proper amount of glass fiber can increase the elongation and reduce the cost. Adding appropriate glass fiber can improve the damping of the material as long as the fiber content and placement order are proper (Yang et al. 2002). Aramid fiber is an organic fiber which can effectively make up the deficiency of the brittleness and toughness of carbon fiber and glass fiber. The selected aramid fiber is Kevlar-29, which is a high-damping fiber, and plays an important role in improving the damping of the material. The performance parameters of various fibers are shown in Table 8.1.

In the composite, the action of the matrix is to support the reinforcing material and to transfer the external load to the reinforcing material in a shear stress form. In addition, the matrix can also protect the reinforcing material from the erosion of the external environment and protect it. Furthermore, the damage tolerance and the practical temperature of the composite material depend mainly on the matrix. The resin matrix has an important influence on the mechanical property of the composite material, especially the most significant influence on transverse performance and compression and shearing performance of the composite material. The resin matrix with excellent toughness can improve the anti-damage ability and the fatigue life of the composite material, and absorb energy. The heat resistance, aging resistance, viscosity and layup property, gel time, shape temperature and pressure time of composite material are determined by resin matrix, so the choice of resin is also very important. E-type 618 epoxy resin produced by the epoxy resin factory of Yueyang Petro-chemical Plant is selected in this experiment, which is characterized by conve-

Table 8.1 Performance parameters of reinforced fiber

Fiber name	Tensile strength/MPa	Elastic modulus/GPa	Elongation/(%)	Density/(g·cm^{-3})	Diameter/μm
Carbon fiber (M40J)	4,400	377	1.2	1.77	6
Kevlar-29	2,800	63	3.6	1.45	12
Glass fiber	4,018	83.3	5.7	2.54	10

8.1 Preparation and Properties of Fiber Reinforced Polymer Composites

nient curing, strong adhesion, and low curing shrinkage. In addition, the cured product has the advantages of good mechanical property, chemical stability, dimensional stability and durability, alkali resistance and acid resistance, solvent resistance and the like. The performance parameters of epoxy resin matrix are as shown in Table 8.2.

The composite material uses carbon fiber, glass fiber, and aramid fiber as reinforcing material, and takes the epoxy resin as base material. The contents of carbon fiber, glass fiber, and aramid fiber are 55, 5, and 5%, respectively, and the content of the epoxy resin is 35%, so that the use requirements for the cable can be met. Since the simple epoxy resin is a linear structure before being cured and cannot be directly applied, curing agent must be added to the resin, and the cross-linking and curing reaction takes place at certain temperature to produce the polymer which can be used. The curing agent added in the experiment is methyl tetrahydrophthalic anhydride. Meanwhile, the accelerator DMP-30 is added to meet the requirement of the molding process, and accelerate the curing. The main proportion of the glue solution is 100 parts of epoxy resin, 90 parts of methyl tetrahydrophthalic anhydride and three parts of DMP-30. The proportion is mass percent.

The hybrid fiber composite material developed is one-dimensional bar continuous fiber composite material, so the mechanical and damping design is based on the design requirements for unidirectional continuous fiber composite materials, and meets the following formula:

Longitudinal elasticity modulus of unidirectional composite material (tensile strength, compressive strength)

$$E = \sum E_i \psi_i + E_m \psi_M \tag{8.1}$$

Mechanical properties mixing ratio of unidirectional composite

$$X^n = \sum X_i^n \psi_i + X_m^n \psi_m \tag{8.2}$$

where E_i and E_m refer to the elasticity modulus of the fiber and matrix; ψ_i and ψ_m refer to the percentage composition of fiber and matrix; X refers to mechanical property in certain direction; i and m refer to the fiber and the matrix component, respectively; n refers to the power of a number, which is $n = 1$ in parallel model and $n = -1$ in series.

Damping theory model

$$\tan\delta_c = \frac{V_{fc} \tan\delta_{fc} E_{fc} + V_{fb} \tan\delta_{fb} E_{fb} + V_{fa} \tan\delta_{fa} E_{fa} + V_m \tan\delta_m E_m}{V_{fc} E_{fc} + V_{fb} E_{fb} + V_{fa} E_{fa} + V_m E_m} \tag{8.3}$$

Table 8.2 Performance parameters of epoxy resin matrix

Tensile yield strength/MPa	Tensile yield strain/(%)	Tensile modulus/GPa	Density/(g·cm^{-3})	Shear modulus/MPa
78	2.97	2.5	1.24	1211

where E_{fa}, E_{fb}, E_{fc}, and E_m refer to the storage modulus of aramid fiber, glass fiber, carbon fiber, and the matrix, respectively; V_{fa}, V_{fb}, V_{fc} and V_m refer to the volume percent of aramid fiber, carbon fiber and glass fiber, and the matrix, respectively; tan δ_{fa}, tan δ_{fb}, and tan δ_c refer to the damping loss factor of aramid fiber, glass fiber, carbon fiber, and the matrix, respectively; tan δ_c and tan δ_m refer to the damping loss factor of the composite material and the matrix, respectively.

The total content of the required fiber is 75%, of which, the contents of carbon fiber, glass fiber, and aramid fiber are 65%, 5%, and 5%, respectively, and the content of the epoxy resin matrix is 25%. The performance parameters of various fibers and matrix are substituted into Formula (8.3), and the calculated theoretical value of the composite material is as shown in Table 8.3.

8.1.2 Performance Test and Analysis of Fiber Reinforced Polymer Bar

As a structural material, the most important performance is mechanical properties. Only when the composite material is placed in the force field, the external force field can bond the filler and the matrix to each other through the interface, and play a synergistic effect. The transfer of force must be carried out through the interface, so the interface becomes one of the key factors that directly affect the overall performance of the composite material (Du et al. 2003).

1. Mechanical Properties

The composite cable is mainly used to replace the cable in the traditional steel cable-stayed bridge, and serve as the main force component of the cable-stayed bridge, which has higher requirements for tensile strength, modulus and impact resistance. Therefore, tensile strength, bending strength and impact toughness of hybrid fiber reinforced composites need to be tested.

According to the testing standard of unidirectional continuous fiber reinforced composites, the mechanical properties of the composites are tested (Li et al. 2012b; Li and Sun 2014). The tensile test shall be carried out according to GB/T 3354-1999 *Test Method for Tensile Properties of Oriented Fiber Reinforced Plastics*, and the sample preparation shall be carried out according to the specified requirements, and the loading test shall be carried out through MTS material universal tester. The impact toughness test method draws on the impact test method for simply supported beam of the fiber reinforced plastics (GB/T 1451-2005), and the testing equipment is a

Table 8.3 Theoretical mechanical parameters of reinforced composites

Material	Tensile strength/MPa	Tensile elasticity modulus/GPa	Density/(g·cm^{-3})
FRP	3,220.4	252.5	1.66

8.1 Preparation and Properties of Fiber Reinforced Polymer Composites

non-metallic material pendulum type impact testing machine. The bending strength test method adopts *Fiber reinforced Plastic Composites-Determination of Flexural Properties* (GB/T 1449-2005), which adopts an unconstrained support, bends at three points, and causes the sample to be damaged at a constant loading rate.

The tested sample is the same batch of pultrusion product, which shows small discreteness. Take five sample mean values and calculate as shown in Table 8.4. Therein, the tensile strength is about 2,600 MPa, and the elasticity modulus is 200 GPa, which is slightly lower than the theoretical value and is related to the molding process and the proportion of fiber and resin. However, the value is higher than 1,860 MPa tensile strength and 200 GPa elasticity modulus of the traditional steel strand and meets the mechanical requirements for cable stretching. Its impact toughness is also higher than that of the common carbon fiber reinforced composite material.

2. Morphology Analysis

The bending and longitudinal shearing sections can best characterize the intensity level of the material interface bonding. Carry out load test on the material mechanics universal testing machine according to national standard GB/T 13096—2008. Hitachi S-570 Japan scanning electron microscope (SEM) is used for morphological analysis. The microscope is 20~20,000 times continuously adjustable, with a resolution of 10 nm at 20 kV.

Select the bending and shearing sections, and cut out a section of about 5 mm thickness and length and width within 1 cm with stainless steel saw blade, for ease of the electron microscope focusing. The opposite face of the fracture section is then repeatedly polished with sandpaper, and it can be placed levelly until it is very smooth. Because the fiber reinforced composite material is non-conducting, it should be subjected to surface metal spraying treatment immediately after cleaning, and the surface metal spraying thickness is about 0.5 mm.

See Fig. 8.1 for the microstructure of the overall bending section of hybrid fiber reinforced composite. Therein, Fig. 8.1a, b are the overall morphology of two sections on the same intelligent cable wire, respectively. There is just a slight difference in magnification times. It can be seen from the figure that the carbon fiber fracture in whole section is relatively neat, and glass fiber and aramid fiber show different degree of pull-out and tearing, respectively. This means that the interface bonding between carbon fiber and the resin is good, and the transferring stress shows good effect. Since

Table 8.4 Mechanical properties of hybrid fiber reinforced composites using theoretical and measuring method

Category	Tensile strength/MPa	Elastic modulus/GPa	Bending strength/GPa	Bending modulus/GPa	Impact toughness/(kJ·m^2)	Density/(g·cm^{-3})
Theoretical data	2,788.2	215.66	–	–	–	1.61
Measured data	2,600	200	1.14	159	145	–

(a) Section 1(×500)   **(b)** Section 2(×50)

Fig. 8.1 Overall bending cross-section SEM micrograph of HFRP

the fibers bear external load together with the resin, the section is relatively neat, and the carbon fiber will first be stressed and fracture because of smaller ultimate tensile strain. However, the ultimate tensile strain of aramid fiber and glass fiber is greater, so the elongation of fracture is obvious. The fiber gratings are in pull-off state, rather than fracturing from the bonding of resin and coating layer, which means the bonding of optical fiber grating and resin is relatively ideal.

The microstructure diagram for the bending section of carbon fiber bundle reinforced composite material is as shown in Fig. 8.2a, b. It can be seen that there are a large number of grooves along the longitudinal direction on the surface of carbon fiber, the surface is rough, and the diameter is about 6 μm. This facilitates interface bonding between carbon fiber and epoxy resin. The carbon fiber bending section and the fiber section are relatively neat, and there is no sign of being pulled out of the epoxy resin. The bonding is tight, and the entire interface bonding is also in good condition.

The microstructure for the bending section of glass fiber bundle reinforced composite material is as shown in Fig. 8.3a, b. The diameter of glass fiber is about 10 μm, and the surface is relatively smooth. The smooth surface is very unfavorable to the bonding between glass fiber and epoxy resin, and the strength is low even in bonding status. The glass fiber is pulled out of the resin, the visible pull-out length is large, and only small amount of resins are bonded on the surface; the whole section is also jagged, and has not shared the external load well with the epoxy resin. This indicates that the glass fiber shows weaker degree of bonding with the epoxy resin under the such process conditions. The interface is broken first, so a continuous whole transferring stress smoothly cannot be formed by the fiber and the resin.

The microstructure for the bending section of aramid fiber reinforced composite material is as shown in Fig. 8.4a, b. The aramid fiber used is Kevlar-29 with 12 μm diameter, and the surface of aramid fiber is uniformly adhered with a layer of resin.

8.1 Preparation and Properties of Fiber Reinforced Polymer Composites

(a) Low times (b) High times

Fig. 8.2 Microstructure of bending section for carbon fiber reinforced composites

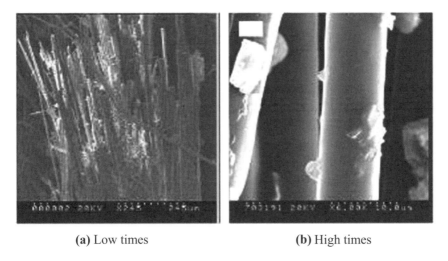

(a) Low times (b) High times

Fig. 8.3 Microstructure of bending section for glass fiber reinforced composites

The aramid fiber is torn into thinner filaments, and a layer of resin can be seen on the surface, which may be the indication of good interface bonding between aramid fiber and the resin. When subjected to external load, the interface can transmit the external force to the reinforced fiber well. Therefore, aramid fiber exhibits a tearing status due to an external force.

As with the bending section, the interface bonding on shearing section can also be reflected well. The material universal tester is used for shear failure of the composite cable wire and then used to conduct SEM scanning electron microscope analysis.

(a) Low times   (b) High times

Fig. 8.4 Microstructure of bending section for aramid fiber reinforced composites

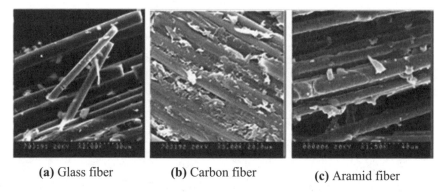

(a) Glass fiber   (b) Carbon fiber   (c) Aramid fiber

Fig. 8.5 Microstructure of shear fracture of HFRP

The microstructure of the shearing section of the hybrid fiber reinforced composite material is as shown in Fig. 8.5. The shearing sections between the three fibers and resin matrix are significantly different. On the section of glass fiber, there is a line of clearly visible fibers, the fiber surface is relatively smooth, and the bonding resin is less, which indicates that the failure of the material mainly occurs at the connection of interface between the fiber and the matrix, and the interface is weakly bonded. On the shearing section of carbon fiber, the fiber and the resin are tightly bonded together, and a few fibers are damaged, which indicates that the interface bonding between carbon fiber and the resin is good. Many resins are uniformly bonded on the surface of aramid fiber, which indicates that the material failure occurs not only between the interfaces but also between the resin matrix, and the interface bonding is good.

In conclusion, no matter the micro-topography analysis on the shearing section or the bending section, the interface bonding between glass fiber and the epoxy resin can be proved to be the worst, aramid fiber takes the second place, and the interface bonding between carbon fiber and the epoxy resin is the best.

3. Damping Performance

Dynamic mechanical analysis technology (DMA) is a shortcut method to evaluate the heat resistance, cold resistance, compatibility and damping efficiency of high polymer. It has been widely used in the performance test of composites, and it is of significant importance for performance evaluation, design optimization, material selection, process development, and quality assurance (Gao 1990).

The load test is carried out with the VA400 viscoelasticity analyzer in the United States, which can provide load and strain amplitude in the material analysis. This dynamic mechanical analysis system is used to study the dynamic viscoelastic properties of fiber composites in three-point bending mode. According to the dynamic mechanical experiment standard established by American Society for Testing Materials "ASTM D2236-1969". Load is applied in the form of three-point bending. According to the test requirements, the sample is a cuboid thin plate: 45 mm (L) × 5 mm (W) × 2 mm (H). Note that the precision of the cut sample shall be controlled within ±0.1 mm.

Under the condition of strain amplitude 0.000 1 m at 20 °C, simulate the low-frequency vibration of the cable and select 10 interval points from 0.1 to 2.0 Hz for testing. As shown in Fig. 8.6, under such test condition, as the load frequency changes, either carbon fiber epoxy resin-vinyl composite material or glass fiber epoxy resin-vinyl composite material, their damping will increase with the increase of frequency. Adams et al. argue that there is a complex relationship between the frequency and the damping loss factor of the material (Qian and Ai 2004). When the other conditions remain unchanged, the damping property of the composite material results from the interaction of the resin, the fiber and the interface between them. The damping property of resin is from its strain hysteresis on stress, which is more significant in the event of smaller frequency. Taking the glass fiber vinyl composite material as an example, the 0.1 Hz damping loss factor is 0.18, and the 2 Hz damping loss factor is up to 0.3.

For the same fiber, either carbon fiber composite material or glass fiber composite material, the fiber vinyl composite material, their damping is greater than the epoxy resin-based fiber composite material. Corresponding to the same test point, the damping loss factor of carbon fiber vinyl composite material is 0.02 greater than that of carbon fiber epoxy resin composite, and the damping loss factor of glass fiber vinyl composite material is 0.01 greater than that of glass fiber epoxy resin composite material. Except for the correlation with damping of fiber and matrix, the damping of the material also may be related to the interface. This may be due to the relatively poor bonding between vinyl and glass fiber interface. Under the condition of external force, it is likely that a part of the energy is absorbed by the micro-crack at the interface. The damping loss factor is relatively large under the same test conditions.

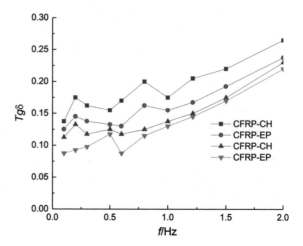

Fig. 8.6 Damping curve of various fiber composites under different frequencies

For the same resin matrix, the damping loss factor of carbon fiber reinforced composite material is slightly greater than that of glass fiber reinforced composite material. Corresponding to the same test point, the damping loss factor of carbon fiber epoxy resin composite material is about 0.02 greater than that of glass fiber epoxy resin composite, and the damping loss factor of carbon fiber vinyl composite material is about 0.04 greater than that of glass fiber vinyl composite material. Although the damping contribution to the composite material of glass fiber is greater than that of carbon fiber under the same volume fraction and test condition, the damping of the composite material is also related to the microstructure of the interface, and the strength of the interface and the micro-cracks will affect the damping. The carbon fiber used in this experiment is a 6 K small fiber strand with 6 μm diameter, while the diameter of glass fiber is 10 μm. The specific surface area, the interface bonding area, and volume are great under the same volume fraction, so the damping loss factor is great under imposed load condition.

Figure 8.7a indicates the process to simulate the operating temperature of the cable under the condition of 0.000 1 m strain amplitude at 0.5 Hz, and select 8 temperature test points from −20 to 60 °C for the scanning test. It is found that in the temperature range of −20 to 60 °C, the damping loss factor of glass fiber composite material decreases with the increase of temperature, and the temperature of vinyl composite material is slightly higher than that of epoxy resin-based composite material with little difference. This may be due to the fact that both the epoxy resin and the vinyl are relatively stable within this temperature range, the distance is relatively far from the damping peak caused by forming phase transition, there is no change in the internal structure, and the viscoelastic difference is not very obvious. In this temperature transition range, the damping loss factor of carbon fiber epoxy resin composite material is about 0.03 higher than that of glass fiber epoxy resin composite material, which may also result from the smaller diameter of carbon fiber. The specific surface area, the interface bonding area, and interface bonding energy

8.1 Preparation and Properties of Fiber Reinforced Polymer Composites

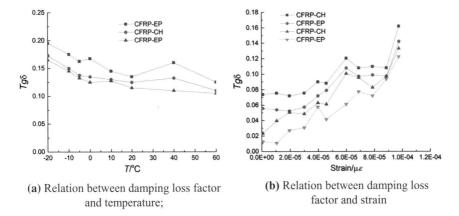

(a) Relation between damping loss factor and temperature;

(b) Relation between damping loss factor and strain

Fig. 8.7 Relationship between damping loss factor and temperature and strain

are great under the same volume fraction, so the damping loss factor of material is great in the event that the interface energy absorbed by the interface is greater under imposed load condition.

Figure 8.7b indicates the process to simulate the strain amplitude caused by imposed load on the cable under 0.5 Hz and 20 °C condition, select 11 test points from the strain amplitude range of 0.000 01~0.000 1 m to carry out the scanning test, and obtain the damping and strain relation diagram. It is found that under the condition of small strain, the damping of the four fiber composites increases gradually with the increase of strain. This is mainly because with the increase of strain, the greater the expansion of the micro-crack inside the material, the more energy absorbed. Therefore, the damping of composite material increases with the increase of strain amplitude in small range. For the same kind of fiber, either carbon fiber composite material or glass fiber composite material, the damping of vinyl fiber composite material is about 0.02 greater than that of epoxy resin-based fiber composite material. The damping of material is also related to the interface microstructure. This may be due to the relatively poor bonding between vinyl and glass fiber interface. Under the condition of external force, it is likely that a part of the energy is absorbed by the micro-crack at the interface. Therefore, the damping loss factor is relatively large under the same test conditions. For the same resin matrix, the damping loss factor of carbon fiber reinforced composite material is slightly greater than that of glass fiber reinforced composite material, and the difference value is between 0.01 and 0.02, which may result from the smaller diameter of carbon fiber. The specific surface area, the interface bonding area, and interface bonding energy are great under the same volume fraction, so the damping loss factor of material is great in the event that the interface energy absorbed by the interface is greater under imposed load condition.

8.1.3 Experiment Study on Anchorage System for Fiber Reinforced Polymer Bar

Since FRP is a crystal material and FRP bars are directional fiber composite materials, its radial and transverse strength ratio are up to 20:1. For the cable of cable-stayed bridge, the traditional prestressing anchor is not suitable for FRP bars under the consideration of premature failure of the anchor zone due to its too low transverse strength. Therefore, a study on the anchor which is suitable for the property of FRP bar (Wang and Huang 2000; Zhang et al. 2000) is an important issue in applying fiber composite materials to stay cables.

1. Experiment Scheme and Process

For the anchorage of single bar, the adhesive type anchor is used, and glass fiber composite bar based on the same matrix is applied to carry out bonding and loading experiment. The composite bar is GFRP supplied by Beijing Glass Fiber Reinforced Plastic Composite Co., Ltd., and its tensile strength guarantee value is 1,000 MPa. It will be subjected to the stretching of MTS material universal tester at 25 MPa/min loading speed, and the constant static load lasts for 1 min each time. The hard aluminum alloy material is selected as a pipe sleeve, and are engraved with threads in the inner part. The bonding medium is Dr-II planting bar glue, and the bonding strength is 35 MPa. The same kind of rods are selected in the experiment. One rod has a smooth surface, and the other is subjected to spiral notch treatment on the surface. As shown in Table 8.5.

The experimental process is as shown in Fig. 8.8. It should be noted that the rods and sleeves are washed multiple times with acetone to ensure that the surface is free of impurities, especially oil scale. The structural adhesive is AB glue, which should be prepared in proportion. Since the glue is very easy to cure at high temperature, it is important to shorten the preparation time of the glue mixing and the rod sleeve

Table 8.5 Test samples

Number	Surface	Anchor length/mm	Rod diameter/mm	Thickness of adhesive layer/mm
1#	Surface scoring	100	11	0.5
2#	Smooth	100	11	0.5
3#	Surface scoring	100	11	1.5
4#	Smooth	100	11	1.5
5#	Surface scoring	100	11	2
6#	Smooth	100	11	2
7#	Surface scoring	100	11	2.5
8#	Surface scoring	150	11	2.5
9#	Surface scoring	250	11	2.5

8.1 Preparation and Properties of Fiber Reinforced Polymer Composites

as far as possible during the experiment, especially in summer. The curing will be natural curing in 72 h.

2. Experimental Results and Analysis

Load test is carried out with MTS material universal tester, and the failure of anchorage system is generally in two forms; one is that FRP bars slip out of the sleeve and the other is that FRP bars are broken in the range of free length. The first is the problem frequently occurred on the adhesive anchor, which is also the most undesirable problem. The results of the experiment are shown in Table 8.6.

When the anchor length and the thickness of the adhesive layer are the same, the average bonding strength of the FRP bars with surface scoring is about three times higher than the average bonding strength of the FRP bars with smooth surface. This is because the bonding strength of bonding anchor is mainly composed of three parts: the chemical adhesive force between the bonding medium and the FRP bar surface, the frictional force on the contact surface of the FRP bar and the bonding medium, and the mechanical lock effect caused by rough surface of the FRP bar. The bonding strength of the smooth FRP bars mainly depends on the chemical adhesive force and the smaller friction force of the adhesive and the rod, while the FRP bars with surface scoring are provided with mechanical bite force that it does not possess originally, so the average bonding strength of the smooth bars is much higher. At the same time, it is also stated that in the whole bonding system, the mechanical bite force with a large specific gravity can have a significant influence on the adhesive property.

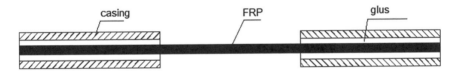

Fig. 8.8 Schematic diagram of adhesive anchor

Table 8.6 Test sample results

Number	Ultimate tensile stress/MPa	Average bond strength/MPa	Failure mode
1#	225	6.18	1
2#	100	2.75	1
3#	400	11	1
4#	125	3.44	1
5#	350	9.6	1
6#	125	3.4	1
7#	300	8.25	1
8#	550	10.08	2
9#	700	7.7	2

Note Failure mode 1—FRP bars are pulled out of the bonding medium; failure mode 2—FRP bars are broken in the free length

With the same adhesive layer thickness, the rod with 150 mm bonding length shows the greatest bonding strength, which means that the bonding length is not the longer the better. The overlength bonding will only cause material waste, but also reduce the bonding strength of the structure, which may be caused by the bubble formation or nonuniform bonding during glue injection as a result of adopting overlength integral pipe sleeve. However, with the increase of the anchorage length, the ultimate tensile strength also increases significantly.

For the same anchorage length, when the adhesive layer is 1.5 mm thick, either the adhesive strength or the ultimate tensile stress will be the best. This is because, when the adhesive layer is too thin, the adhesive cannot realize good bonding, and the binding between the rod piece and the anchor is insufficient; and when the adhesive layer is too thick, there are many flaws in the excessively thick adhesive layer since the adhesive strength is mainly provided by the shearing strength between the adhesive layers, and it is easily deformed and damaged. Therefore, the adhesive layer should neither be too thin nor too thick.

The variation trends of bonding strength and ultimate tensile stress are identical, which is mainly due to the linear relationship between ultimate tensile stress and bonding strength of isometric anchorage.

$$\tau = \frac{\sigma d}{4l} \tag{8.4}$$

where τ is average bonding strength (MPa); σ is ultimate tensile stress (MPa); d is FRP bar diameter (mm); and l is bond length (mm).

8.2 Interface Bonding Analysis of Fiber Bragg Grating Sensors and Composite Materials

The interface bonding effect directly affects the sensing characteristics of the fiber grating. Only when the interface bonding is good, the embedded optical fiber grating can truly pass through the mechanical environment in the external state of the interface reaction, so as to realize the real-time monitoring of the intelligent fiber reinforced composite cable (Li et al. 2012a).

Figure 8.9 indicates the interface bonding micro-topography of bare optical fiber after bending failure. As can be seen from the figure, the pull-off diameter of optical fiber is about 120 μm, which is much smaller than 250 μm diameter of the bare optical fiber. This suggests that the optical fiber breakage is drawn out from the interface between the cladding and the coating layer. The bending failure composite itself and the optical fiber coating layer are still bonded well, and the bending failure cracks even occur between carbon fibers and the resin, which may be due to the fact that, the optical fiber coating layer will be integrated with resin matrix after high-temperature melting, so that the interface between the optical fiber and the composite material are well bonded. A well-bonded interface allows the external load to be successfully

8.2 Interface Bonding Analysis of Fiber Bragg Grating Sensors ...

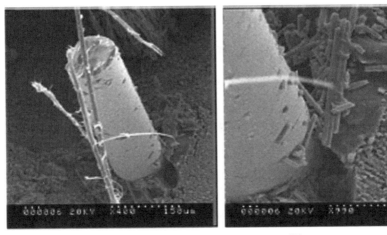

(a) Low times ; (b) High times

Fig. 8.9 Microgram of FBG interface damage

transmitted to the sensing element, and the optical fiber grating can faithfully reflect the external mechanical environment, thus providing reliable guarantee for realizing the intelligent function of the hybrid fiber reinforced composite cable.

Figure 8.10 shows the section pictures of stay cable wire. As shown in the figure, the microsensor is located in the middle of the intelligent cable wire, and the microsensor is well bonded with the composite material itself. Figure 8.10 shows the micro-morphology of the section of the cable wire. Table 8.6 shows the micro-morphology of the section fiber of the cable wire. As shown in the figure, the diameter of the optical fiber is 132 μm, which basically coincides with 125 μm theoretical diameter of the optical fiber; the inner diameter of the packaged steel pipe is about 301 μm, which highly coincide with 0.31 mm design structure size of the microsensor. Figure 8.13 shows the micro-morphology of interface between the microsensor and composite material. As shown in the figure, the interface between optical fiber sensor and composite material is well bonded. A well-bonded interface allows the external load to be successfully transmitted to the sensing element, and the optical fiber grating can faithfully reflect the external mechanical environment, thus providing a reliable guarantee for realizing the intelligent function of the hybrid fiber reinforced composite cable (Figs. 8.11 and 8.12).

8.3 Sensing Characteristics of Smart FRP Rod

In consideration to the strength requirement of carbon fiber, the high-strength and high-modulus M40J produced by Toray Industries Inc. is selected as the main reinforced fiber. In addition, a small amount of high-damping aramid Kevlar-29 and

Fig. 8.10 Cross-section view of smart FRP rod

Fig. 8.11 Cross section SEM of smart FRP rod

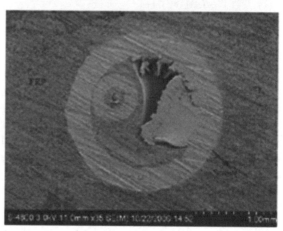

high-strength glass fiber 240Tex are added, so as to increase damping and adjust the ultimate tensile deformation and toughness, and reduce the cost effectively. E-type 618 epoxy resin produced by the epoxy resin factory of Yueyang Petro-chemical Plant is selected as the matrix, which is characterized by convenient curing, strong adhesion, and low curing shrinkage. In addition, the cured product has the advantages of good mechanical property, chemical stability, dimensional stability and durability, alkali resistance and acid resistance and the like. The selected fiber grating is relatively widely used optical FBG. The bandwidth is less than 0.2 nm, and the center wavelength is between 1,530 and 1,550 nm.

Fig. 8.12 Cross section SEM of optical fiber

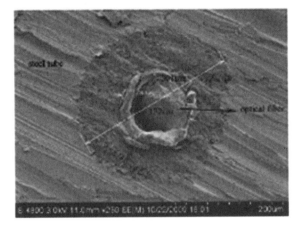

Fig. 8.13 SEM micrograph of FBG-FRP interface

8.3.1 Preparation of Smart FRP Rod

Drawing on the traditional pultrusion technology and equipment, the intelligent fiber composite rod is prepared by adjusting the material selection ratio, and its preparation process is as shown in Fig. 8.14. After resin impregnation with the optical fiber grating placed in the central opening of combined beam plate, the continuous reinforced fiber will be cured and formed in the molding die together with other fibers. In the course of fiber drawing, it is important to note that due to the fragility of the fiber grating, it is necessary to be cautious as far as possible. The traction force cannot be too large, and it should be placed as straight as possible to avoid the measurement error caused by internal bending. Finally, weld and seal the pulled optical fiber along with the jumper.

The epoxy resin is high in viscosity, and the incorporation is great because of high-strength requirement, which made it difficult to manufacture the whole process. The

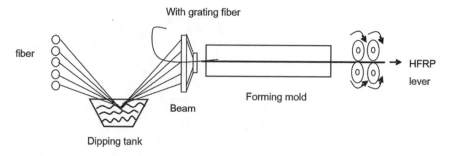

Fig. 8.14 Forming technology of fiber reinforced polymer bar based on optical fiber sensing

die blockage, traction yarn breakage and the like occur frequently. After a plurality of experiments, the concrete parameters of the manufacturing process are finally determined: the traction speed is 0.1 m/min, the molding pressure is 0.4~0.5 MPa, the molding temperature is 170~180 °C, and the mold is a circular die cavity with an inner diameter of 10 mm. The FRP-OFBG intelligent sensing rod produced according to the above-mentioned process and steps is as shown in Fig. 8.15. The black rod is an intelligent sensing rod, and the yellow color on both ends refers to single-mode jumper.

The prepared fiber composite material is a one-dimensional rod and is suitable for pultrusion processes as shown in Fig. 8.16. Attention should be paid to six key factors during the pultrusion process: ① fiber arrangement system, such as yarn frame, felt frame, yarn guide plate, etc.; ② resin dipping tank; ③ pre-forming system and forming mold; ④ control over temperature, speed, traction force, and other specific parameters; ⑤ traction system; and ⑥ cutting system.

After a plurality of experiments, the concrete parameters are finally determined: the traction speed is 0.1 m/min, the molding pressure is 0.4~0.5 MPa, the molding temperature is 170~180 °C, and the mold is a circular die cavity with an inner diameter

Fig. 8.15 FRP-OFBG intelligent sensing rods

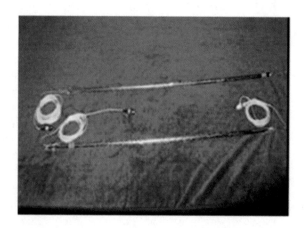

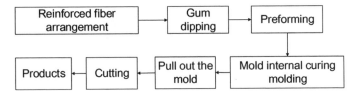

Fig. 8.16 Flow chart of FRP tendon pultrusion process

of 11 mm and 2 mm × 10 mm sheet cavity. The release agent is XTEND802 imported from the United States. The die blockage problem of large incorporation fiber-epoxy resin composite material is solved by manufacturing fiber composite material with such process, and the prepared sample is continuum with smooth surface.

8.3.2 Test and Analysis on Sensing Characteristics of the Smart FPR Rod

Figure 8.17 is the site photograph for sensing characteristics test of fiber intelligent rods (Li et al. 2012a). Three FRP-OFBG intelligent rods are prepared by four FBGs. One of them is embedded with two gratings, and the other two gratings are respectively embedded in one grating. The test methods for each rod piece are the same, that is, the light emitted by the broadband light source BBS (1,525~1,575 nm) is incident on the optical fiber grating through a 3 dB coupler. Under the continuous load of MTS material universal tester, the Bragg central wavelength is shifted, the load light is reflected by FBG, and then is introduced into the spectrum analyzer OSA through the coupler. The movement of the Bragg wavelength can be monitored in the spectrum analyzer. By verifying the relationship between the wavelength and strain of the fiber grating, the coincidence degree between strain sensitivity and the theoretical calculation value is tested and compared.

After the optical fiber is coated, the diameter of the bare optical fiber is 250 μm, the area occupied by the optical fiber is small in comparison with the sectional area of the rod, which may not have great influence on the mechanical property (Du et al. 2008). However, because the optical fiber is single and fragile, attention should be paid to the survival rate of the grating during its molding process. In this experiment, the fiber grating is buried directly during pultrusion, and the molding condition is 0.4~0.5 MPa at 160~180 °C. In addition, the curing deformation of the resin during the molding process can also influence the exposed grating and even fracture it, so it is very necessary to check the survival condition. After pultrusion, there are optical signals passing through the three hybrid fiber composite material rods embedded with the optical fiber grating prepared by the fiber optic spectrometer inspection. Therein, two gratings are multiplexed in the first intelligent rod with two wavelength peaks of 1,547,200 and 1,551,366, while the other two are respectively embedded

Fig. 8.17 Measurement of intelligent HFRP

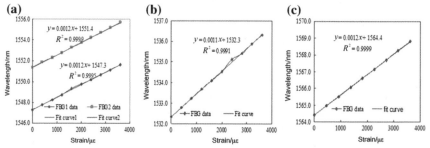

Fig. 8.18 Sensing performance of three fiber composite rods based on optical fiber

with a grating with single peak value of 1,532,342 and 1,564.755, respectively. It means under such implantation molding process, the survival rate of the fiber grating is very high, which is 100%.

The sensing performance of three fiber composite rod based on fiber grating is as shown in Fig. 8.18, in which Fig. 8.18a contains two gratings and Fig. 8.18b, c contain a grating, respectively. From the experimental results, the correlation coefficient between the center wavelength and strain of FBG is over 0.999, there is no hysteresis phenomenon, and a very good linear relationship is formed. It means that FBG is a very ideal strain sensing element. The theoretical strain sensitivity coefficient of one fiber grating is 1.1 pm/$\mu\varepsilon$, and the strain sensitivity coefficient of the other three fiber gratings is 1.2 pm/$\mu\varepsilon$, which is consistent with the theoretical value of 1.21 pm/$\mu\varepsilon$ and 1.22 pm/$\mu\varepsilon$. It also shows that the FBG in the hybrid fiber rod can accurately reflect the external force situation, and such molding process is suitable for developing a fiber composite material with self-monitoring function suitable for the stay cable.

8.4 Summary

This chapter mainly introduces the research on preparation and performance test of the fiber reinforced composite, and carries out experiment on the bonding property of the grating sensor and the composite material. The results demonstrate that the grating sensor can be fully bonded with the interface of the composite material and will not affect the use of material. At the same time, the intelligent fiber composite rod is developed, and the intelligent monitoring technology of fiber composite material based on fiber grating is established through intensive study on its sensing characteristics.

References

Chen K, Zhu Q, Shen W (2016) Experimental search damage detection of glass-fiber composite materials with FBG sensing network. J Exp mech 31(04):475–482. https://doi.org/10.7520/1001-4888-15-189
Du YL, Zhang XS, Zeng KY (2003) Research on a new type of bamboo-glass fiber composite structure. J Shijiazhang Inst Railw Technol 16(1):1–5. 10.13319j.cnki.sjztddxxbzrb.2003.01.001
Du YL, Shao L, Li JZ et al (2008) Study on the intelligent hybrid composites suitable for stayed cable. J Funct Mater 39(2):282–285
Gao J (1990) Polymeric material thermal analysis curve set. Science Press, Beijing
Li JZ, Sun BC (2014) Lower-frequency damping properties of fiber reinforced composite. Chin J Sens Actuators 27(08):1022–1026. https://doi.org/10.3969/j.issn.1004-1699.2014.08.004
Li J, Du YL, Sun BC (2012a) Staying cable wires of fiber bragg grating/fiber-reinforced composite. J Comput 7(9):2148–2191. https://doi.org/10.4304/jcp.7.9.2184-2191
Li JZ, Sun BC, Du YL (2012b) Damping properties of fiber reinforced composite suitable for stayed cable. In: Leng J, BarCohen Y, Lee I, Lu J (eds) Third international conference on smart materials and nanotechnology in engineering, vol 8409, pp 1844–1864. https://doi.org/10.1117/12.920368
Qian X, Ai Y (2004) Damping measures of cables in cable stayed bridge. Ind Tec Eco 23(06):106–108
Wang J (2007) Application of fiber reinforced composites (FRP) in structural strengthening. J Sanming Univ 24(04):361–363+369
Wang B (2014) Properties testing of M40J high modulus carbon fiber. Phys Chem Testing (Physical Volume) 50(07):495–498
Wang Z, Huang D (2000) Studies on application of ERP tendons. CN Muni Eng 01:18–21
Wang T, Zhou G (2017) Research on cable tension monitoring and temperature compensation of bridge cable based on fiber gratings force ring. J CN Fore HW 37(03):112–117. https://doi.org/10.14048/j.issn.1671-2579.2017.03.024
Yang S, Sun K, Hao R (2002) Research on damping properties of hybrid fiber composites. Fiber Compos Mater 6(01):6–10
Zhang B, Benmokrane B, Chennouf A (2000) Prediction of tensile capacity of bond anchorages for FRP tendons, vol 4. https://doi.org/10.1061/(asce)1090-0268(2000)4:2(39)

Chapter 9
Concrete Crack Monitoring Using Fully Distributed Optical Fiber Sensor

As a result of its own and external factors, the concrete structure is easy to produce cracks. The cracks seriously affect the safety of the structure. Therefore, in order to ensure the durability of the structure, it is necessary to monitor the cracks of the concrete structure. Due to the time and position randomness of concrete crack formation, the fully distributed sensor is an ideal way to capture concrete cracks. However, when the fully distributed sensor is used for large-scale measurement, the drift of spatial position of the optical fiber along with the external temperature, strain, and other external environmental changes will occur. Therefore, aiming at the positioning of concrete crack, this chapter focuses on solving the problem of fiber accurate positioning in engineering application. From the basic principle of Brillouin sensing and fiber grating, the locating drift problem of fully distribution optical fiber is discussed, and the concrete crack monitoring technology based on fully distributed optical fiber sensing is established.

9.1 Main Parameters of Fully Distributed Optical Fiber Sensing Technology

Because of the different sensing mechanisms, various types of fully distributed sensing technology have some individual parameters besides some common parameters. Therefore, the fully distributed fiber sensing technology involves many parameters, and this section only introduces some major performance parameters of fully distributed fiber sensing technology.

(1) Sensitivity

The sensor converts the signal to be measured X into an output signal (usually an electrical signal) V_0. The sensitivity S is the ratio of the output signal to the input signal of the sensing system, and the expression is $V_0 = SX$. Ideally, the sensitivity should remain constant throughout the working range, which is independent of environmental factors such as temperature.

(2) Noise

Noise exists in all sensors, even if the random fluctuation of electrons in the resistor will introduce noise (thermal noise). The wider the bandwidth of the sensor, the larger the noise of its output signal. Therefore, the classification of noise is often associated with frequency.

(3) Signal-to-Noise Ratio

The signal-to-noise ratio is the ratio of signal strength to noise intensity output by the sensor.

(4) Resolution

The resolution is the minimum variation of the measured parameter that can be observed. When the variation of the sensor output voltage caused by the measured quantity is equal to the effective noise voltage, the amplitude of the measured parameter variation is defined as the resolution of the sensor.

An important performance parameter in fully distributed fiber-optic sensors is spatial resolution. It means that the measurement system has the ability to distinguish two nearest event points along the sensing fiber. Since the information obtained along the sensing fiber at any time is actually the accumulation signal on a certain segment of the sensing fiber, the information on any infinitely small segment along the sensing fiber cannot be completely distinguished, i.e., the information of all points less than the spatial resolution along the sensing fiber is superposed in time domain. In practical measurements, spatial resolution is generally defined as the spatial length of the measured signal at a rise time of 10%–90% in the transition period (Bao and Chen 2012).

The spatial resolution is mainly determined by the probe optical pulse width of the sensing system, the corresponding time of the photoelectric conversion device, the A/D conversion speed, and the frequency bandwidth of the amplifying circuit.

If the probe optical pulse is rectangular, the pulse width is τ, the group speed of the light in the optical fiber is V_g, and the dispersion of the optical pulse in the sensing optical fiber is ignored. It is considered that if the frequency band of the optical detector and the amplifier is wide enough, the spatial resolution determined by the probe optical pulse is R_{pulse}

$$R_{pulse} = \frac{\tau V_g}{2} \tag{9.1}$$

If the speed of light in the vacuum is c, and the refractive index of the ordinary single-mode fiber core is n, the group velocity of the light in the optical fiber is

$$V_g = \frac{c}{n} = \frac{3 \times 10^8}{1.46} = 2.05 \times 10^8 \text{ (m/s)} \tag{9.2}$$

9.1 Main Parameters of Fully Distributed Optical Fiber Sensing ...

According to Formulas (9.1) and (9.2), the spatial resolution R_{pulse} in the ordinary single-mode optical fiber is expressed as

$$R_{\text{pulse}} \approx \frac{\tau(ns)}{10} \tag{9.3}$$

The spatial resolution $R_{\text{A/D}}$ determined by the A/D conversion speed f can be estimated to

$$R_{\text{A/D}} \approx \frac{100}{f(\text{MHZ})} \tag{9.4}$$

If the bandwidth of the amplifier is B (including the influence of the rise time of the detector), the spatial resolution R_{amp} determined by it can be estimated to

$$R_{\text{amp}} \approx \frac{100}{B(\text{MHZ})} \tag{9.5}$$

The spatial resolution R of the fully distributed optical fiber sensing system may be expressed as

$$R = \max\{R_{\text{pulse}}, R_{\text{A/D}}, R_{\text{amp}}\} \tag{9.6}$$

In Formulas (9.3)–(9.5), R_{pulse}, $R_{\text{A/D}}$ and R_{amp} are measured in meter (m).

(5) Dynamic Range

There are two definition modes for dynamic range: two-way dynamic range and one-way dynamic range. The two-way dynamic range refers to the range of signal power that the detection curve obtained by probe light back and forth in the optical fiber is from the signal-to-noise ratio equal to 1 to the maximum signal-to-noise ratio. The definition of one-way dynamic range is a half of the two-way dynamic range (in dB).

9.2 Brillouin Scattering Principle and Sensing Mechanism in Optical Fiber

9.2.1 Brillouin Scattering in Optical Fiber

Based on the interaction physical mechanism between light and substance in optical fiber, Raman scattering, Brillouin scattering is the inelastic scattering process. Raman scattering is the inelastic light scattering generated by the interaction of the incident light and the optical phonon (Agrawal 2007), while Brillouin scattering is the inelastic scattering phenomenon generated by the interaction between the incident light and the acoustic phonon. Brillouin scattering in optical fiber is divided into spontaneous

Brillouin scattering (Sp-BS) and stimulated Brillouin scattering (SBS). Such two scatterings will be introduced as follows.

1. Spontaneous Brillouin Scattering

The particles (atoms, molecules or ions) constituting the medium will form a continuous elastic mechanical vibration in the medium due to the spontaneous thermal motion. Such mechanical vibration will result in a periodic variation of the density of the medium over time and space, thereby creating a spontaneous acoustic field inside the medium. The acoustic field causes the refractive index of the medium to be periodically modulated and propagated in the medium at sound velocity V_a, which acts like an optical grating (referred to as a sound field grating). The optical wave will be scattered by the action of the sound field grating when it is incident into the medium, and its scattered light will generate frequency drift related to sound velocity due to Doppler Effect. Such scattered light with frequency shift is called spontaneous Brillouin scattered light (Boyd 2003; Zhang 2008).

In the optical fiber, a physical model of spontaneous Brillouin scattering is as shown in Fig. 9.1. Irrespective of the dispersion effect of the optical fiber on the incident light, the angular frequency of the incident light is set as ω, and the moving sound field grating reflects the incident light through the Bragg diffraction. When the sound field grating, and the incident light move in the same direction, the scattered light moves down relative to the incident light frequency due to the Doppler Effect. The scattered light at this time is referred to as Brillouin Stokes light, and the angular frequency is ω_S as shown in Fig. 9.1a. When the sound field grating, and the incident light move in different directions, the scattered light moves up relative to the incident light frequency due to the Doppler Effect. The scattered light at this time is referred to as Brillouin anti-Stokes light, and the angular frequency is ω_{AS} as shown in Fig. 9.1b.

It is assumed that the periodic acoustic wave field caused by the thermal motion of molecule in the incident light field and the optical fiber of the optical fiber are respectively:

$$E(z,t) = E_0 e^{i(\vec{k}\cdot\vec{r}-\omega t)} + \text{c.c.} \tag{9.7}$$

$$\Delta p = \Delta p_0 e^{i(\vec{q}\cdot\vec{r}-\Omega t)} + \text{c.c.} \tag{9.8}$$

where c.c. is the complex conjugate term of the first term in each equation; E_0 is the amplitude of the incident light field; $\vec{k}$ is the wave vector of the incident light; $\vec{r}$ is the displacement; ω is the angular frequency of the incident light wave; Δp_0 is the amplitude of the acoustic wave field; $\vec{q}$ is the wave vector of the sound wave; and Ω is the angular frequency of the sound wave.

The scattered light field in the optical fiber follows the wave equation:

$$\nabla^2 \vec{E} - \frac{n^2}{c^2}\frac{\partial^2 \vec{E}}{\partial t^2} = \frac{4\pi}{c^2}\frac{\partial^2 \vec{P}}{\partial t^2} \tag{9.9}$$

9.2 Brillouin Scattering Principle and Sensing Mechanism in Optical ...

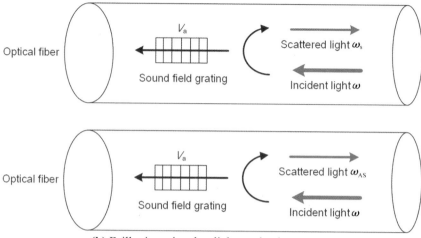

(b) Brillouin anti-stokes light production process

Fig. 9.1 Schematic diagram of Brillouin scattering physical model in fiber

where n is the refractive index of the optical fiber medium; c is the speed of light in the vacuum; and $\vec{P}$ is the additional polarization caused by the fluctuation of the polarization intensity in the medium, which may be expressed as

$$\vec{P} = \frac{\Delta\varepsilon}{4\pi}\vec{E} \tag{9.10}$$

where ε is the dielectric constant of the medium, its change is caused by the fluctuation of the density of the medium, and the change of the density of the medium is caused by the disturbance of the acoustic wave:

$$\Delta\varepsilon = \frac{\partial\varepsilon}{\partial p}\Delta\vec{\rho} \tag{9.11}$$

$$\Delta\rho = \frac{\partial\rho}{\partial p}\Delta\vec{p} \tag{9.12}$$

Substitute Formulas (9.11) and (9.12) into Formula (9.10) and obtain

$$\vec{P} = \frac{1}{4\pi}\left(\frac{\partial\varepsilon}{\partial\rho}\right)\left(\frac{\partial\rho}{\partial p}\right)\Delta\vec{p}\cdot\vec{E} \tag{9.13}$$

Substitute electrostrictive coefficient $\gamma_e = \rho_0\frac{\partial\varepsilon}{\partial\rho}$ and adiabatic compression coefficient $C_s = \frac{1}{\rho_0}\frac{\partial\rho}{\partial P}$ into Formula (9.13),

$$\vec{P} = \frac{1}{4\pi}\gamma_e C_s\Delta\vec{p}\cdot\vec{E} \tag{9.14}$$

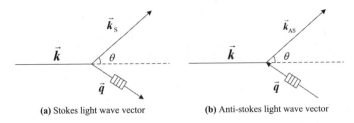

(a) Stokes light wave vector   (b) Anti-stokes light wave vector

Fig. 9.2 Wave-vector relation between Brillouin scattered light and incident and acoustic waves

Integrate Formula (9.14) with Formulas (9.7) and (9.8), and obtain the nonlinear polarization fluctuation equation satisfied by Brillouin scattering in optical fiber (Dong et al. 2009):

$$\nabla^2 \vec{E} - \frac{n^2}{c^2}\frac{\partial^2 E}{\partial t^2} = -\frac{\gamma_e C_s}{c^2}\left[(\omega - \Omega)^2 E_0 \Delta p e^{i(\vec{k}-\vec{q})\cdot\vec{r}-i(\omega-\Omega)t} \right.$$
$$\left. + (\omega + \Omega)_0^2 \Delta p e^{i(\vec{k}+\vec{q})\cdot\vec{r}-i(\omega+\Omega)t} + \text{c.c.} \right. \quad (9.15)$$

The term on the right of Formula (9.15) shows that both the Stokes and anti-Stokes scattering spectrum lines are symmetrically distributed on both sides of the incident angle frequency ω, and the frequency shift of such scattered lights relative to the incident light is equal to the frequency of the sound field Ω, and their frequency shift amount relative to the incident light is referred to as the Brillouin frequency shift.

The relationship between angular frequency ω_S and wave vector $\vec{k}_S$ of Stokes light and the angular frequency ω and wave vector $\vec{k}$ of incident light is

$$\omega_S = \omega - \Omega \quad (9.16)$$
$$\vec{k}_S = \vec{k} - \vec{q} \quad (9.17)$$

The relationship between angular frequency ω_{AS} and wave vector $\vec{k}_{AS}$ of anti-Stokes light and the angular frequency ω and wave vector $\vec{k}$ of incident light is

$$\omega_{AS} = \omega + \Omega \quad (9.18)$$
$$\vec{k}_{AS} = \vec{k} + \vec{q} \quad (9.19)$$

Figure 9.2 simply reflects the wave vector relation between the Stokes, anti-Stokes scattered lights, the incident lights, and the acoustic waves. To better reflect the momentum conservation relation between the incident light, Stokes light, and anti-Stokes light, triangular vector diagram of the three is indicated in Fig. 9.3.

$\omega \approx \omega_S \approx \omega_{AS}$ and $|\vec{k}| \ll |\vec{k}_S| \ll |\vec{k}_{AS}|$ can be considered as a result of $\Omega \ll \omega$ and $|\vec{q}| \ll |\vec{k}|$. Then the following can be obtained from Fig. 9.3.

9.2 Brillouin Scattering Principle and Sensing Mechanism in Optical …

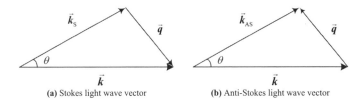

(a) Stokes light wave vector (b) Anti-Stokes light wave vector

Fig. 9.3 Vector conservation relation of Brillouin scattered light

$$|\vec{q}| = 2|\vec{k}|\sin\left(\frac{\theta}{2}\right) \tag{9.20}$$

In addition, the following relationships exist between the angular frequencies and wave vectors of the incident light and the acoustic wave, respectively,

$$\omega = |\vec{k}|\frac{c}{n} \tag{9.21}$$

$$\Omega = |\vec{q}|V_a \tag{9.22}$$

where V_a is the sound velocity in optical fiber medium, and the Brillouin frequency shift can be obtained by Formula (9.20).

$$v_B = \frac{\Omega}{2\pi} = \frac{2nV_a}{\lambda_0}\sin\left(\frac{\theta}{2}\right) \tag{9.23}$$

where λ_0 is the incident light wavelength and θ is the included angle between the scattered light wave vector and the incident light wave vector. It can be seen from Formula (9.23) that the frequency shift of Brillouin scattered light is related to the scattering angle. In a single-mode fiber, propagation modes other than the axial direction are suppressed, so that the Brillouin scattered light is only shown as forward and backward propagation. When the scattering occurs in the forward direction ($\theta = 0$), $v_B = \Omega/2\pi = 0$, that is, no Brillouin scattering occurs; when the scattering occurs in the backward direction ($\theta = \pi$), $v_B = \Omega/2\pi = 2nV_a/\lambda_0$. It can be concluded that the frequency shift of the Brillouin scattering in backward direction is proportional to the effective refractive index of the optical fiber and the acoustic velocity in the optical fiber and is inversely proportional to the wavelength of the incident light. If the refractive index $n = 1.46$, the sound velocity $V_a = 5,945$ m/s, and the incident light wavelength $\lambda_0 = 1,550$ nm of the quartz optical fiber, the Brillouin frequency shift of the quartz optical fiber is about 11.2 GHz.

In practical cases, the acoustic wave is attenuated in the optical fiber medium, so that the Brillouin scattering spectrum has a certain width and is in the form of Lorenz curve,

$$G_B(v) = G\frac{\left(\frac{\Gamma_B}{2}\right)^2}{(v - v_B)^2 + \left(\frac{\Gamma_B}{2}\right)^2} \tag{9.24}$$

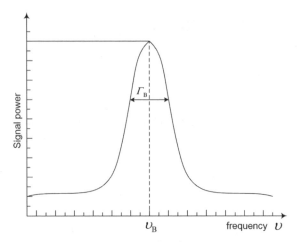

Fig. 9.4 Schematic diagram of spontaneous Brillouin scattering in single-mode fiber

where Γ_B is full width at half maximum (FWHM) of the Brillouin scattering spectrum. Γ_B is related to the acoustic phonon lifetime and the Γ_B in ordinary single-mode optical fiber is usually dozens of megahertz. When $v = v_B$, the signal power is at the Brillouin scattering peak G, and the Brillouin scattering spectrum is as shown in Fig. 9.4.

2. Stimulated Brillouin Scattering

In 1964, stimulated Brillouin scattering was observed for the first time in bulk crystals (Agrawal 2007). The stimulated Brillouin scattering process can be described as the inelastic interaction between incident light waves and Stokes waves through sound waves. Unlike spontaneous Brillouin scattering, the stimulated scattering process is derived from the effect of the strongly induced acoustic field on incident light. When the incident light wave reaches a certain power, the incident light waves generate sound waves by electrostriction, cause periodic modulation of the medium refractive index, and greatly strengthen the sound field satisfying the phase matching, so that most energy of the incident light wave is coupled to the backward transmitted Brillouin scattered light, thereby forming stimulated Brillouin scattering.

In the process of stimulated Brillouin scattering, the incident light can only excite acoustic field transmitted in the same direction, so only the Stokes spectral line with downward frequency is presented in general, and its frequency shift equals to the audio frequency in the medium. From the point of quantum mechanics, this scattering process can be seen as an annihilation of incident light, which produces a Stokes quantum of light and an acoustic phonon.

The frequency and wave vector relation between the incident light field of stimulated Brillouin scattering, the Stokes light, and the acoustic wave field is similar to those in the spontaneous Brillouin scattering process, and there is no need to retelling. The Brillouin amplification process is a nonlinear effect related to stimulated Brillouin scattering, which is an important mechanism for fiber-optic sensing technology.

9.2 Brillouin Scattering Principle and Sensing Mechanism in Optical ...

The stimulated Brillouin scattering process is usually described by the classical three-wave coupling equation. In stable cases, the typical three-wave coupling equation can be simplified as

$$\frac{dI_P}{dz} = -g_B(\Omega)I_P I_S - \alpha I_P \qquad (9.25)$$

$$\frac{dI_S}{dz} = -g_B(\Omega)I_P I_S - \alpha I_S \qquad (9.26)$$

where I_P and I_S are the intensities of the incident light wave and the Stokes scattered light are respectively the intensity of the incident light wave and the Stokes scattered light; α is the loss coefficient of the optical fiber. The Brillouin gain factor $g_B(\Omega)$ has Lorentz spectral pattern, which can be expressed as

$$g_B(\Omega) = g_0 \frac{(\Gamma_B/2)^2}{(\Omega_B - \Omega) + (\Gamma_B/2)^2} \qquad (9.27)$$

where the peak gain factor g_0 can be expressed as

$$g_0 = g_B(\Omega_B) = (2\pi^2 n^7 p_{12}^2)/(c\lambda_0^2 \rho_0 V_a \Gamma_B) \qquad (9.28)$$

where p_{12} is the elasto-optical coefficient; ρ_0 is the material density; $\Gamma_B = 1/\tau_p$ is the Brillouin gain spectrum bandwidth; τ_p is the acoustic phonon lifetime. For ordinary single-mode optical fibers and continuous incident light of 1,550 nm, if the refractive index of the optical fiber is $n = 1.45$, $V_a = 5.96$ km/s, then $g_0 = 5.0 \times 10^{-10}$ m/W. It can be seen from Formula (9.28) that when $|\Omega - \Omega_B| \gg 0$, the Brillouin gain will become very small and the Brillouin scattering at $\Omega = \Omega_B$ has the greatest gain g_0, i.e., only when the frequency difference Ω of the two light fields approaches Ω_B, significant amplification effect for simulated Brillouin amplification effect will be produced. Such amplification effect is applied to realize the sensing by the sensing technology based on stimulated Brillouin scattering.

3. Stimulated Brillouin Scattering Threshold

Threshold is one of the important characteristics of stimulated Brillouin scattering (Smith 1972). Due to the continuous thermal motion of molecular atoms, etc., which make up the optical fiber medium, there is always a different degree of thermotropic acoustic field in the optical fiber. The thermotropic acoustic field causes periodic modulation of the refractive index of the fiber and generates spontaneous Brillouin scattered light when the light is incident into the optical fiber. When the incident light power is gradually increased to a certain extent, the back-transmitted Brillouin scattered light interferes with the incident light, and the refractive index of the optical fiber is periodically modulated to generate a refractive index grating. Such refractive index grating will be further strengthened along with the further strengthening of incident light power, and the backward scattered light on such optical grating will be strengthened continuously, causing most of the incident light to be converted into backward scattered light, and producing stimulated Brillouin scattering. It can be concluded that stimulated Brillouin scattering has obvious threshold characteristics: when the intensity of incident light is small, the power of Brillouin scattered light is linearly related to the power of incident light; however, when the incident light power exceeds a certain value, i.e., the stimulated Brillouin scattering threshold, the power of Brillouin scattered light increases dramatically, producing stimulated Brillouin scattering. The stimulated Brillouin scattering threshold in the optical fiber has different definitions in the literature, which can be described as follows: ① the incident light power in case that incident light power equals to the backscattered light power; ② the incident light power in case that the transmitted light power equals the backward scattered light power; ③ the incident light power in case that the backscattered light power increases rapidly; ④ the incident light power in case that backward scattered light power on the incident end of the optical fiber equals to η times of the incident light power. A commonly used stimulated Brillouin scattering threshold formula is given by (9.29) (Smith 1972).

$$P_{cr} = G \frac{K_P A_{eff}}{g_0 L_{eff}} \tag{9.29}$$

where G is the gain factor of stimulated Brillouin scattering threshold; K_P is the polarization factor ($1 \leq K_P \leq 2$), which is dependent on the polarization state of the incident light and the Brillouin scattered light; A_{eff} is the effective fiber core area; L_{eff} is the effective action length.

$$L_{eff} = [1 - \exp(-\alpha L)]/\alpha \tag{9.30}$$

where L is the fiber length; α is the fiber loss coefficient.

Through theoretical and experimental studies, Sebastien Le Floch et al. think that there are many factors affecting Brillouin scattering threshold in optical fiber. Except for optical fiber length and fiber sectional area, the wavelength of pump light is also included. Therefore, he puts forward that Brillouin scattering threshold coefficient can be expressed as (Li et al. 2003):

$$G \approx \ln\left(\frac{4 A_{eff} f_B \pi^{1/2} B^{3/2}}{g_0 L_{eff} k_B T f_p \Gamma}\right) \tag{9.31}$$

where f_p is the frequency of the pump light; $\Gamma = 1/T_B$ is the acoustic phonon attenuation rate, and $T_B = 10$ ns is the acoustic phonon lifetime; $B = 21$ is the constant related to the dispersion of the optical fiber, k_B is the Boltzmann constant; A_{eff} is the effective mode field area.

9.2.2 Sensing Mechanism Based on Brillouin Scattering

By Formula (9.23), the frequency shift of backward Brillouin scattering in the optical fiber is

$$v_B = 2nV_a/\lambda_0 \qquad (9.32)$$

It can be concluded that the Brillouin frequency shift is proportional to the effective refractive index of the optical fiber and the acoustic velocity in the optical fiber and is inversely proportional to the wavelength of the incident light.

The wave velocity of the known optical fiber can be expressed in the following formula:

$$V_a = \sqrt{\frac{(1-k)E}{(1+k)(1-2k)\rho}} \qquad (9.33)$$

where k is Poisson's ratio; E is Young's modulus; and ρ is the density of the optical fiber medium. The refractive index n and these parameters are functions of temperature and stress. They are denoted as $n(\varepsilon, T)$, $E(\varepsilon, T)$, $k(\varepsilon, T)$ and $\rho(\varepsilon, T)$, respectively, and substituted into Formula (9.32) to obtain the Brillouin frequency shift amount.

$$v_B(\varepsilon, T) = \frac{2n(\varepsilon, T)}{\lambda_0} \sqrt{\frac{[1-k(\varepsilon, T)]E(\varepsilon, T)}{[1+k(\varepsilon, T)][1-2k(\varepsilon, T)]\rho(\varepsilon, T)}} \qquad (9.34)$$

1. Relationship Between Brillouin Frequency Shift and Strain

Under the constant temperature condition, when the strain of optical fiber changes, the interaction potential between atoms inside the optical fiber changes, resulting in the variation of Young's modulus and Poisson's ratio, which allows the change of the refractive index, thereby influencing the variation of Brillouin frequency shift amount.

If the reference temperature is T_0, Formula (9.34) can be written as

$$v_B(\varepsilon, T_0) = \frac{2n(\varepsilon, T_0)}{\lambda_0} \sqrt{\frac{[1-k(\varepsilon, T_0)]E(\varepsilon, T_0)}{[1+k(\varepsilon, T_0)][1-2k(\varepsilon, T_0)]\rho(\varepsilon, T_0)}} \qquad (9.35)$$

Since the constituent components of the optical fiber are mainly the brittle material SiO_2, its tensile strain is smaller. In the case of micro-strain, conduct Taylor expansion for Formula (9.35) at $\varepsilon = 0$, ignore more than one order of higher order entries and obtain

$$v_B(\varepsilon, T_0) \approx v_B(0, T_0)\left[1 + \Delta\varepsilon \frac{\partial v_B(\varepsilon, T_0)}{\partial \varepsilon}\bigg|_{\varepsilon=0}\right]$$
$$= v_B(0, T_0)[1 + \Delta\varepsilon(\Delta n_\varepsilon + \Delta k_\varepsilon + \Delta E_\varepsilon + \Delta\rho_\varepsilon)] \quad (9.36)$$

At room temperature, taking the typical value of each parameter: $\lambda = 1{,}550$ nm, $\Delta n = -0.22$, $\Delta k = 1.49$, $\Delta E = 2.88$, $\Delta\rho = 0.33$, the variation of Brillouin frequency shift along with stress may be expressed as

$$v_B(T_0, \varepsilon) \approx v_B(T_0, 0)(1 + 4.48\Delta\varepsilon) \quad (9.37)$$

Formula (9.37) shows that the Brillouin frequency shift is proportional to the optical fiber strain. When incident light having a wavelength of 1,550 nm is incident on an ordinary single-mode quartz optical fiber at a constant temperature, the strain varies 100 μm and the corresponding Brillouin frequency shifts is about 4.5 MHz.

2. Relationship Between Brillouin Frequency Shift and Temperature

When the optical fiber is relaxed, i.e., strain $\varepsilon = 0$, the following can be obtained by Formula (9.35):

$$v_B(0, T_0) = \frac{2n(0, T_0)}{\lambda_0}\sqrt{\frac{[1 - k(0, T_0)]E(0, T_0)}{[1 + k(0, T_0)][1 - 2k(0, T_0)]\rho(0, T_0)}} \quad (9.38)$$

When the temperature of optical fiber changes, the thermal expansion effect and thermal optical effect respectively cause the change of optical fiber density and refractive index. Meanwhile, the free energy of optical fiber will change with temperature, so that Young's modulus, Poisson's ratio, and other physical quantities of the optical fiber also change with temperature. When the temperature changes within a small range, it is assumed that the temperature change is ΔT, conduct Taylor expansion for Formula (9.38), ignore more than one order of higher order number entries, and obtain

$$v_B(0, T_0) \approx v_B(0, T_0)\left[1 + \Delta T \frac{\partial v_B(0, T_0)}{\partial \varepsilon}\bigg|_{T=T_0}\right]$$
$$= v_B(0, T_0)[1 + \Delta T(\Delta n_T + \Delta k_T + \Delta E_T + \Delta\rho_T)] \quad (9.39)$$

At room temperature ($T = 20$ °C), when the wavelength of incident light for ordinary single-mode optical fiber is 1,550 nm, the corresponding relation between Brillouin frequency shift and temperature change is

$$v_B(T, 0) \approx v_B(T_0, 0)[1 + 1.18 \times 10^{-4}\Delta T] \quad (9.40)$$

According to Formula (9.40), when the ordinary single-mode optical fiber is relaxed at room temperature $T = 20$ °C, the wavelength of incident light is 1,550 nm.

Every 1 °C temperature rise will increase the corresponding Brillouin frequency shift by 1.2 MHz.

Above analysis shows that the variation of Brillouin frequency shift Δv_B varies linearly with the variation of optical fiber temperature and strain, which can generally be expressed as

$$\Delta v_B = C_{v,T} \Delta T + C_{v,\varepsilon} \Delta \varepsilon \tag{9.41}$$

where $C_{v,T}$ and $C_{v,\varepsilon}$ are the temperature coefficient and strain coefficient for the variation of Brillouin frequency shift respectively. When the incident light wavelength is 1,553.8 nm, $C_{v,T} = 1.1$ MHz/°C, $C_{v,\varepsilon} = 0.0483$ MHz/$\mu\varepsilon$.

3. Corresponding Relation Between Brillouin Scattering Power and Temperature and Strain

The change of ambient temperature and strain will not only change the Brillouin frequency shift in the optical fiber but also change the power of Brillouin scattered light. Tests conducted by T. R. Parker et al. indicate that the Brillouin scattered light power in the optical fiber has the following corresponding relation with the strain and temperature of the optical fiber:

$$\frac{100 \Delta P_B}{P_B(\varepsilon, T)} = C_{P,\varepsilon} \Delta \varepsilon + C_{P,T} \Delta T \tag{9.42}$$

where ΔP_B is the variation amount of the Brillouin power; $C_{P,\varepsilon}$ and $C_{P,T}$ are the temperature coefficient and the strain coefficient for the variation of the Brillouin scattered light power respectively. According to the test statistics, when an incident light having a wavelength of 1,550 nm is incident on an ordinary single-mode optical fiber, two coefficient values related to strain and temperature are, respectively, $C_{P,\varepsilon} = -(7.7 \pm 1.4) \times 10^{-5}$ %/$\mu\varepsilon$ and $C_{P,T} = -(0.36 \pm 0.06)$ %/°C.

The spontaneous Brillouin scattered light signal is weak, and the insertion loss, welding loss, end face reflection, and other problems may also exist in the optical fiber, which can cause the change of scattered light power, resulting in inaccurate measurement of Brillouin signal power. Thus, in practical applications, Landau–Placzek ratio (LPR), i.e., the ratio of Rayleigh scattering power to spontaneous Brillouin scattering power, is often used for sensing (Li et al. 2003). The method of introducing Rayleigh scattering optical power can compensate the error caused by optical fiber loss, so as to ensure more accurate measurement results.

If the variation of LPR is ΔP_B^{LPR}, its relationship with the variation of temperature and strain should be

$$\Delta P_B^{LPR} = C_{P,T} \Delta T + C_{P,\varepsilon} \Delta \varepsilon \tag{9.43}$$

The matrix equation of Formula (9.44) can be obtained according to Formulas (9.41) and (9.43):

$$\begin{bmatrix} \Delta P_B^{LPR} \\ \Delta v_B \end{bmatrix} = \begin{bmatrix} C_{P,T} & C_{P,\varepsilon} \\ C_{v,T} & C_{v,\varepsilon} \end{bmatrix} \begin{bmatrix} \Delta T \\ \Delta \varepsilon \end{bmatrix} \quad (9.44)$$

When $C_{v,\varepsilon} C_{P,T} \neq C_{v,T} C_{P,\varepsilon}$, temperature and strain can be determined simultaneously according to the variation of Brillouin frequency shift and the variation of LPR.

$$\begin{bmatrix} \Delta T \\ \Delta \varepsilon \end{bmatrix} = \frac{1}{C_{P,T} C_{v,\varepsilon} - C_{P,\varepsilon} C_{v,T}} \begin{bmatrix} C_{v,\varepsilon} & -C_{P,\varepsilon} \\ -C_{v,T} & C_{P,T} \end{bmatrix} \begin{bmatrix} \Delta P_B^{LPR} \\ \Delta v_B \end{bmatrix} \quad (9.45)$$

Thus, the measurement error of the temperature and strain is estimated to be

$$\delta T = \frac{|C_{P,\varepsilon}| \delta v_B + |C_{v,\varepsilon}| \delta P_B}{|C_{v,T} C_{P,\varepsilon} - C_{v,\varepsilon} C_{P,T}|} \quad (9.46)$$

$$\delta \varepsilon = \frac{|C_{P,T}| \delta v_B + |C_{v,T}| \delta P_B}{|C_{v,T} C_{P,\varepsilon} - C_{v,\varepsilon} C_{P,T}|} \quad (9.47)$$

Among them, δT, $\delta \varepsilon$, δv_B, and δP_B are the root mean square errors of temperature strain, Brillouin frequency shift, and LPR, respectively.

9.3 FBG-Based Positioning Method for BOTDA Sensing

9.3.1 Traditional Positioning Method for Fully Distributed Optical Fiber Sensing

The full-distributed optical fiber sensing method based on Brillouin sensing has the outstanding advantages of full-scale continuity and long-distance measurement. But due to the double sensitivity of optical fiber to temperature and strain, the length of optical fiber will change when temperature, strain or any other external parameter changes, and the spatial position obtained by the Brillouin sensing time domain test (BOTDR/A) method will also change. As a result, when the same position of the structure is in different temperature environments, the spatial position from the time domain method is completely different, so that the measured event of the structure cannot be accurately positioned.

With respect to the optical fiber positioning method, there are two kinds of general methods: the first method is to set a section of optical fiber without position change in the different ambient temperatures, strain and the like as a reference position, and the reference position of optical fiber is used to obtain the position of measured event by the distance between measured event and reference position. In engineering application, however, due to that: ① the environmental is severe; ② the length variation cannot be determined, and the comparison and finding of the reference section are difficult and complex; ③ the optical fiber for large-scale engineering structure is prearranged as reference section. Therefore, these methods have great challenges in practical engineering application. The second method is the resistance heating method at the reference position in the structure. The disadvantage of the method is that: ① power supply is required on site; ② it is difficult to arrange heating point for long, large and hidden structure. Up to now, we have not found a method that can realize the accurate positioning of measured event, which seriously restricts the application of fully distributed sensing in the practical engineering.

9.3.2 Description of FBG-Based Positioning Method

According to Loranger et al. (2015), it is known that FBGs can be used to improve signal strength, and therefore the sensitivity in distributed temperature and strain sensing using frequency domain Rayleigh scatter. According to the above study, we demonstrated the use of both FBG sensors as positioning tool. Figure 9.5 schematically depicts the hybrid sensor network employing FBGs as the location indicator we proposed. The system can be separated into two parts: the FBG indicator itself and Brillouin distributed sensor (Li et al. 2016a). The positioning mechanism is based on higher signal strength from FBGs than Brillouin scattering signal. For instance, after the ambient temperature and strain vary, the measured points of applied strain or temperature via the BOTDA varied from L_5 to L_6. The actual location of applied strain or temperature in this structure is unable to be discriminated without FBG indicator due to the length variations of optical fiber from ambient temperature or strain. However, if FBG would be integrated in this sensing fiber, when the pump light passes through FBG, the back-reflected peak of FBG occurs. Thus, the reflected peak can be used as an indicator of actual location, the so-called FBG-based location indicator. The locations of FBG_1 and FBG_2 are respectively L_1 and L_3 before ambient temperature varied. Moreover, after the ambient temperature and strain changed, the location of FBG_1 (L_1) varied from L_1 to L_2 and that of FBG_2 varied from L_3 to L_4. Assuming the distance from FBG_1 and FBG_2 to the applied strain event remains constant due to the short distance, the location of applied strain can be calculated by the value of $L_4 - L_6$ or $L_6 - L_2$.

To experimentally demonstrate the feasibility of FBG indicator to identify the location of strain or temperature event, the strain-stretched experiment was carried out. The specimen with 3 FBGs in sensing fiber was used in the test. Corning 28E fiber was used as the distributed sensing fiber in this work. FBGs were directly written into this sensing fiber without hurting the fiber, which is favorable and convenient for the engineering application. And FBG points were written based on the wavelength of FBGs along sensing fiber. We refer to them as in-line FBGs fiber for short. Their parameters were listed in Table 9.1, and their configurations are shown in Fig. 9.6. The BOTDA positioning setup used is depicted in Figs. 9.7 and 9.8. A heated oven was used in this experiment to shift the location of sensing fiber, which simulated positioning shift due to the ambient temperature variations. The temperature inside

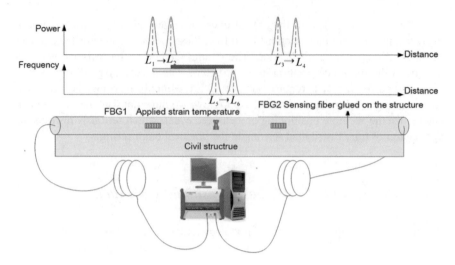

Fig. 9.5 Mechanism of FBG-based positioning method © [2002] IEEE. Reprinted, with permission, from Li et al. (2016b)

Table 9.1 FBG parameters in corning 28E fiber © [2002] IEEE

Wavelength (nm)	Reflectivity (%)	Bandwidth (nm)	FBG interval (m)
1514.80, 1519.79, 1524.8	90.98, 90.68, 89.50	0.22, 0.23, 0.22	8.015, 8.01

Reprinted, with permission, from Li et al. (2016b)

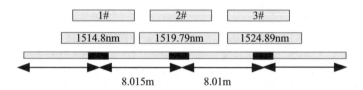

Fig. 9.6 The configuration of in-line FBGs sensing fiber © [2002] IEEE. Reprinted, with permission, from Li et al. (2016b)

the oven is measured by a thermocouple, which varies with a step shape from 5.1 to 45.8 °C. The NBX-7020 BOTDA was employed to test the SBS signal. The low-wavelength end of in-line FBGs fiber was connected with the pump connector of the instrument. Two segments of the optical fiber were statically strained by a displacement driving device. Tension segment 1 and 2 were strained with 1,000, 1,600, 2,300, 2,800, 3,300, and 3,800 με, respectively. Spatial resolutions of 100, 50, 20, 10, 5, and 2 cm were used in these measurements. Sampling interval, sampling time, and refractive index were respectively 1 cm and 2×10^{15} and 1.46 in this work.

9.3 FBG-Based Positioning Method for BOTDA Sensing

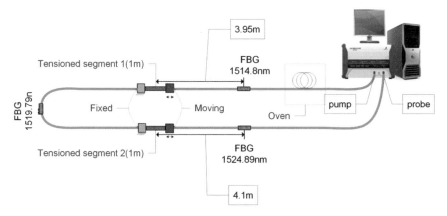

Fig. 9.7 Schematic of experimental setup to position the strain stimulus © [2002] IEEE. Reprinted, with permission, from Li et al. (2016b)

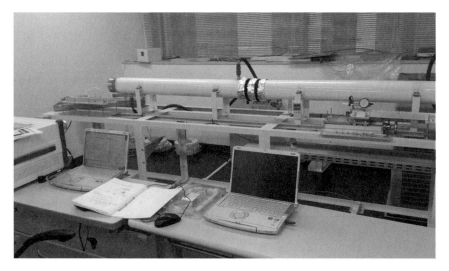

Fig. 9.8 Experiment photo © [2002] IEEE. Reprinted, with permission, from Li et al. (2016b)

9.3.3 Results and Discussion

Table 9.2 illustrates the positions of FBGs in different temperatures (Li et al. 2016b). It is found that the position of FBGs depends on the ambient temperature. The position of FBG_1 varies from 1,052.538, 1,060.577, 1,068.627 m to 1,052.836, 1,060.885, 1,068.924 m in the range of 5.1~45.8 °C. These results are also as shown in Fig. 9.9. Three back-reflected peaks of 3 FBGs are revealed at first sight, and the location of each peak matches with the actual location of each FBG. The positioning shift is about 7.5 cm/10 °C for 1 km sensing bare fiber to our knowledge, which is strongly

Table 9.2 Position of FBG in different temperatures (0 µε) © [2002] IEEE

Case	FBGs positioning data (m)			FBG interval (m)		Temperature (°C)
	FBG_1	FBG_2	FBG_3	FBG_1-FBG_2	FBG_3-FBG_2	
Ruler	–	–	–	8.015	8.01	25.6
BOTDA	1,052.538	1,060.577	1,068.627	8.039	8.05	5.1
BOTDA	1,052.713	1,060.762	1,068.811	8.049	8.049	25.8
BOTDA	1,052.836	1,060.885	1,068.924	8.049	8.039	45.8

Reprinted, with permission, from Li et al. (2016b)

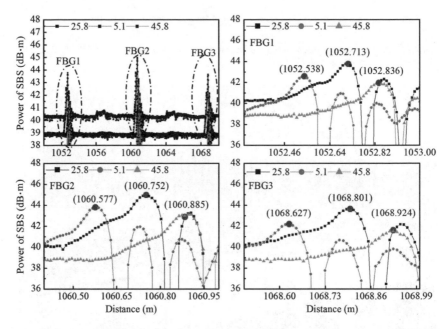

Fig. 9.9 Amplitude of BOTDA signal along the distance in different temperatures © [2002] IEEE. Reprinted, with permission, from Li et al. (2016b)

dependent of optical fiber jacket, applied strain, and the length of sensing fiber as well as ambient temperature. Hence, how to position the strain or temperature events precisely is extremely important due to employing several tens kilometers of fiber for structure health monitoring. The positioning error of the FBGs interval from the NBX-7020 BOTDA is about 3~4 cm compared with the actual data using a ruler (Sun et al. 2017). Figure 9.10 shows that the repeatability of the power spectrum of the reflected light is excellent in different measurements, and the maximum error of FBG positioning is 2 cm. Therefore, it is possible to employ the reflected peak of FBG as location indicator.

9.3 FBG-Based Positioning Method for BOTDA Sensing 311

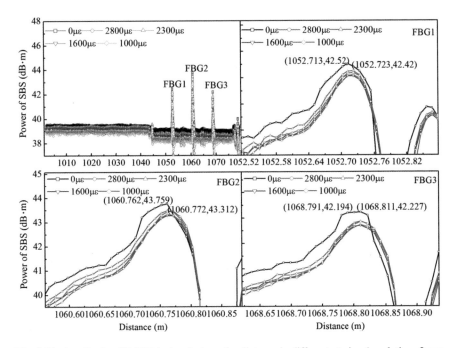

Fig. 9.10 Amplitude of BOTDA signal along the distance in different strains (resolution: 2 cm; temperature: 25.8 °C) © [2002] IEEE. Reprinted, with permission, from Li et al. (2016b)

To prove the feasibility of the technique of positioning strain events, both segments of 1 m-length fiber were statically strained (Li et al. 2017). There have both occurred strain events, as shown in Figs. 9.11 and 9.12. From Fig. 9.12, two segments were strained, and the stressed segment position from Brillouin frequency spectrum and FBG position from the SBS power spectrum is dependent of the ambient temperature, and the distances begin from segment 1 to FBG_1 and segment 2 to FBG_2, respectively. The stretched length from BOTDA measurement (one end of segment 1 and segment 2 to the other end) is approximately 1 m, as shown in Fig. 9.12, which is in good agreement with the actual stretched length (1 m). Therefore, based on FBG positioning method proposed in this paper, the relative distance between strain event and FBG could be calculated accurately according to the $L_{segment1}$-L_{FBG1} or L_{FBG3}-$L_{segment2}$. Figure 9.13 depicts the positioning error with a resolution of 2, 5, 10, 20, 50 and 100 cm. The positioning error is less than 20 cm, as illustrated in Fig. 9.13. Additionally, the small black dots are data points, and the large circle dots indicate the data mean M in Fig. 9.13. For each case, the small black dots on the left show original data points, those in the middle show SD, and those on the right show SE. For each case, we assume that 95% confidence interval includes the true value. The error for the strained distance is approximately 20 cm compared to the true value (m) as shown in Fig. 9.13. Therefore, it is inferred with more confidence that the positioning error is about 20 cm for a segment fiber and approximately 10 cm

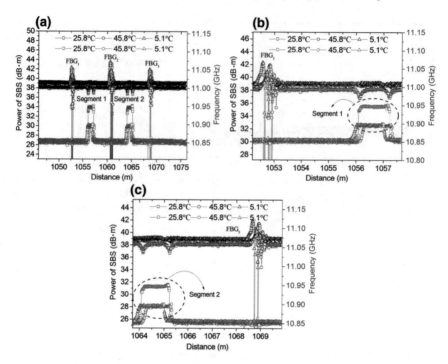

Fig. 9.11 Brillouin frequency shift and amplitude of BOTDA signal along the distance in different strains (spatial resolution: 2 cm, the data was measured under 5.1 °C, 25.8 °C and 0 με; 45.8 °C and 1,000 με (50 MHz)) © [2002] IEEE. Reprinted, with permission, from Li et al. (2016b)

for just one point. The SD values over spatial resolution of 20, 50, and 100 cm are more than those of 2, 5, and 10 cm.

In order to further investigate the positioning accuracy, positioning ability for strain events is analyzed. Based on the above analysis, the positioning errors over 20, 50, and 100 cm of spatial resolution are focused on. The position of strain events from FBG indicator via BOTDA system is less than 10 cm in different temperatures. The actual interval between FBG_1 or FBG_3 and the strain events are respectively 3.95 and 4.10 cm (marked by the dotted black and blue line), as shown in Figs. 9.14, 9.15, and 9.16. Figure 9.12 shows that the positioning error has different temperatures. It is determined that the positioning error is independent of the temperature. Means with error bars for three cases: $n = 4$, $n = 6$ and $n = 4$. The column denotes the data mean M. The bars at the top of column show standard deviation (SD). M and SD are the same for every case. $2 \times$ SD error bars could encompass roughly 9% of the sample. The positioning errors are respectively about 10, 7, and 10 cm over a spatial resolution of 20, 50, and 100 cm, as shown in Fig. 9.17, which is agreement with the error of the strained length (approximately 20 cm for a segment of fiber, corresponding to half of 20 cm for a point positioning). As a result, it is confidently determined that

9.3 FBG-Based Positioning Method for BOTDA Sensing

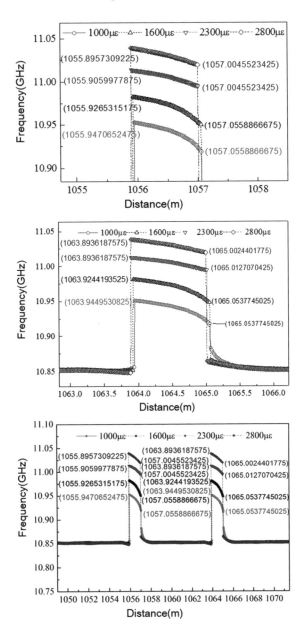

Fig. 9.12 Brillouin frequency shift along the distance in different strains (resolution: 100 cm; temperature: 5.1 °C) © [2002] IEEE. Reprinted, with permission, from Li et al. (2016b)

the positioning error is independent on the temperature and approximately 10 cm, which depends on the ability of BOTDA instrument.

We have successfully proposed and demonstrated experimentally for the first time the ability of FBG indicator to precisely determine the position of strain or temperature events. Results indicate that the positioning accuracy is independent of the

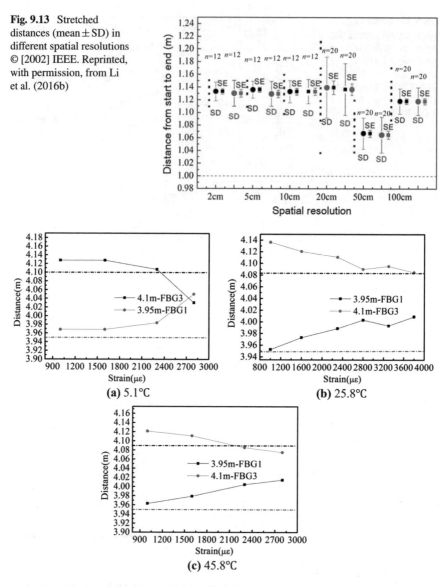

Fig. 9.13 Stretched distances (mean ± SD) in different spatial resolutions © [2002] IEEE. Reprinted, with permission, from Li et al. (2016b)

Fig. 9.14 Distance of applied strain from FBG_1 and FBG_3 in different strains at a spatial resolution of 20 cm © [2002] IEEE. Reprinted, with permission, from Li et al. (2016b)

ambient temperature and spatial resolution. The positioning error is approximately 10 cm. Further improvements in positioning accuracy to deal with case when the length of applied strain is less than the spatial resolution in BOTDA system are the subject of ongoing research.

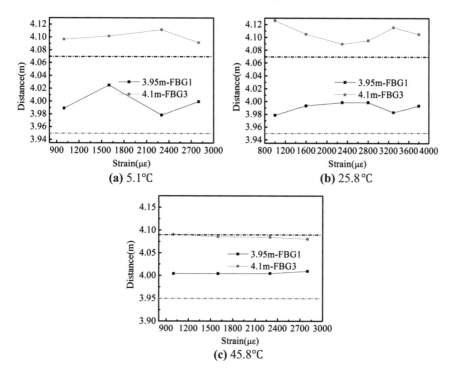

Fig. 9.15 Distance of applied strain from FBG$_1$ and FBG$_3$ in different strains at a spatial resolution of 50 cm © [2002] IEEE. Reprinted, with permission, from Li et al. (2016b)

Table 9.3 FBG performance parameters

Grating sensor	Performance parameters		
	Wavelength (nm)	Bandwidth (nm)	Reflectivity (%)
FBG$_1$	1,544.74	0.22	73.39
FBG$_2$	1,529.91	0.23	91.54

9.4 Concrete Crack Monitoring Using Fully Distributed Optical Fiber Sensing Technology

9.4.1 Tests

1. Test Material

The optical fibers used in the test are strain-sensing optical cables tightly packed with single-core single-mode (G652B) (Sun et al. 2007). Two Bragg grating sensors are used as positioning indicators, which are named FBG$_1$ and FBG$_2$, respectively. Therein, the performance parameters of the grating are shown in Table 9.3.

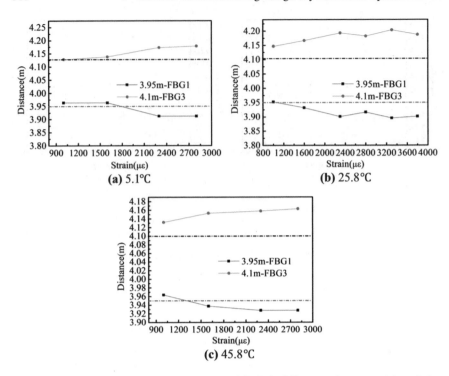

Fig. 9.16 Distance of applied strain from FBG$_1$ and FBG$_3$ in different strains at a spatial resolution of 100 cm © [2002] IEEE. Reprinted, with permission, from Li et al. (2016b)

2. Test Program

The sensing optical fiber is adhered to the upper surface of the concrete beam, and two pasting modes are adopted to lay the optical fiber. One is full-mounted pasting, the other is two-point-mounted pasting with point-to-point interval of 10 cm, the pasted length of the optical fiber is 1.5 m, and the arrangement of the optical fiber is as shown in Fig. 9.18. The sensing fiber distribution is as shown in Fig. 9.19.

3. Test Procedure

Figure 9.20 is a schematic diagram of the test device. In the experiment process, the beam is fixed on the steel frame with bolts and channel steel, and the jack and the pressure sensor are placed in the proper position under the beam; the surface of the beam is ground and leveled with the coarse sandpaper and cleaned with acetone; setting-out is conducted, and the optical fiber is adhered to the surface of the beam with AB glue (modified acrylic resin glue). After the completion of the optical fiber adhesion, the test site is as shown in Fig. 9.21. In order to ensure that the initial stress of the optical fiber of the pasting section is consistent, both ends of the optical fiber pasting section are hung with the same counterweight, and the tensile stress is applied; after curing for 24 h, the sensing optical fiber is connected to BOTDA test

9.4 Concrete Crack Monitoring Using Fully Distributed Optical Fiber ... 317

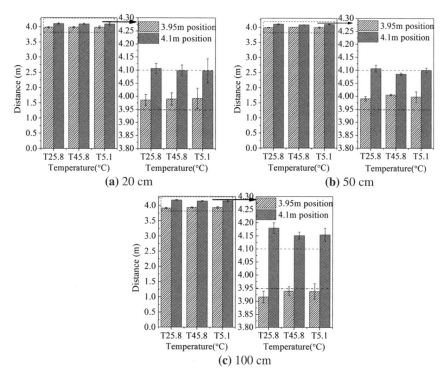

Fig. 9.17 Positioning distance (mean±SD) in different spatial resolutions © [2002] IEEE. Reprinted, with permission, from Li et al. (2016b)

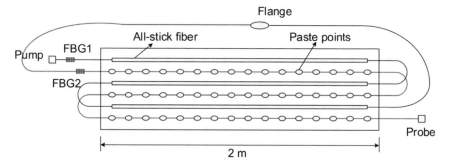

Fig. 9.18 Optical sensing fiber glued on concrete beam

instrument according to Fig. 9.20; step-by-step loading is conducted with QLD32 screw jack, and the loading range is 0~8.5 kN. LH-S10-2T pressure sensor and LH-PTC reading displayer are adopted in the force measuring system to observe the appearance time and specific position of the crack on the surface of the beam, and the acquisition instrument is used to test and record the data. The spatial resolution

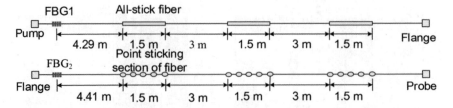

Fig. 9.19 Location sensing fiber glued on concrete beam along optical fiber

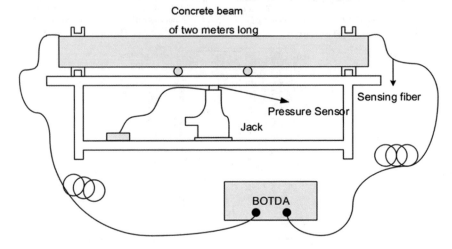

Fig. 9.20 Diagram of test device for distributed sensor glued on concrete beam

of BOTDA test instrument is 10 cm, the sampling rate is 2×10^{15}, and the sampling point is 5 cm.

9.4.2 Results and Discussion

9.4.2.1 Result Analysis for the Fiber Grating Positioning Test

Figure 9.22 is a time domain spectrum for Brillouin scattering power of the sensing optical fiber. The rectangular shadow in the graph shows the reflection spectrum of FBG_1 and FBG_2 in the stimulated Brillouin power time domain. From the figure, it can be seen that the intervals of the two gratings in the power time domain spectrum are 3.131 and 70.585 m, respectively. The figure also indicates the unconstrained fluctuation along the variation of the optical fiber power, which is due to the bending caused by optical fiber winding. In addition, since there are many methods for optical fiber connection, and the optical fiber at the joint of the flange is subjected to serious

9.4 Concrete Crack Monitoring Using Fully Distributed Optical Fiber ...

Fig. 9.21 Photo of experimental system

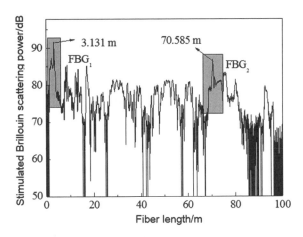

Fig. 9.22 Power spectrum of Brillouin scattering power for sensing fiber

optical loss, there is a significant attenuation of the optical energy, and the shock of the signal can be improved by straightening the optical fiber and not using the flange application.

Figure 9.23 is the distributed strain of the optical fiber when the load is not applied. It is evident from the figure that the six optical fiber sections have the initial stress, the first three sections are full-attached optical fiber sections, and the last three segments are two-point attached optical fiber sections. Based on BOTDA test, it is not possible to determine the specific position of both ends on the structure due to the limitation

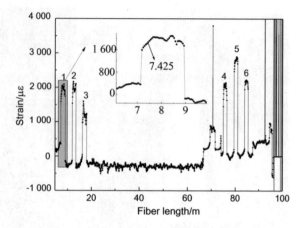

Fig. 9.23 Distribution of strain for sensing fiber at 0 kN load

of the spatial resolution. For example, according to an enlarged view of the first section of the adhesive optical fiber section, if no FBG is used, the specific position of the adhesive optical fiber in the distributed strain diagram will be uncertain. If the corresponding position of the drastic strain variation is regarded as the starting point of the beam, the position is 7.187 m. However, the actual position of the beginning of the structure is 7.421 m. After the FBG is used as indicator, it can be clearly seen that the distance from the beginning position to FBG is 4.29 m. At this time, the position of the structure can be accurately determined according to the spatial position of FBG, that is, 7.425 m in the drawing. The experimental results show that the positioning error of the structure can be up to 30 cm by using the traditional time domain analysis, while this further proves the superiority of using the proposed positioning method integrating FBG and BOTDA. In other words, accurate positioning can be realized by using Brillouin fully distributed positioning method based on an optical fiber grating.

9.4.2.2 Concrete Crack Identification Based on Fully Mounted Optical Fiber

Along with the step-by-step loading of the jack, the cracks on the beam surface gradually occur and develop. Finally, nine cracks are visible to the naked eye appear on the beam, and all the cracks are transverse cracks (see Fig. 9.24). When it is loaded to 4 kN, five cracks ①, ②, ③, ④, ⑤ occur on the surface of the beam. When it is loaded to 5 kN, the other two cracks ⑥ and ⑦ appear, and cracks ⑧ and ⑨ occur under 8 kN load.

Figure 9.24 is the Brillouin frequency shift along the full-mounted sensing optical fiber under loading condition. Therein, Fig. 9.25b~d are enlarged views of three adhesive optical fiber sections in Fig. 9.25a. As can be seen from the figure, there are sharp frequency shift peaks at the loaded fiber section due to the increase in loading. It

9.4 Concrete Crack Monitoring Using Fully Distributed Optical Fiber ...

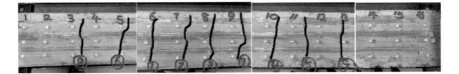

Fig. 9.24 Cracks occurred on the surface of concrete beam

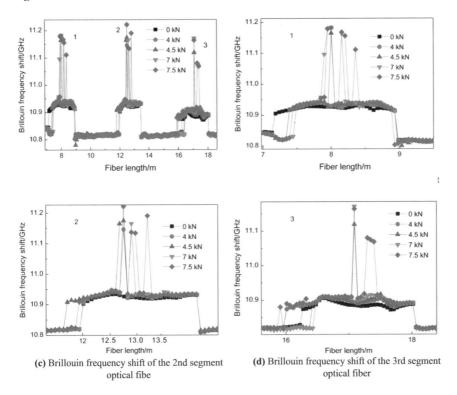

(c) Brillouin frequency shift of the 2nd segment optical fibe

(d) Brillouin frequency shift of the 3rd segment optical fiber

Fig. 9.25 Brillouin frequency shift of full-glued distributed sensor along the optical fiber

is the outcome of the sharp strain increase due to the crack, correspondingly resulting in the dramatic frequency. In Fig. 9.25b, when it is loaded to 4.5 kN, the first sharp peak appears, the second sharp peak appears when it is loaded to 7 kN, and the third sharp peak appears when it is loaded to 7.5 kN. In figure c and d, the number of sharp peaks in the Brillouin frequency shift gradually increases with load. However, the sequence of cracks observed in Fig. 9.25 lags behind the actual occurrence of cracks, and the identifiable cracks are less than the actual cracks.

According to the Brillouin frequency shift along the full-mounted sensing optical fiber, and in combination with Brillouin fully distributed positioning method based on an optical fiber grating, the measured specific distribution of the cracks on the surface is as shown in Fig. 9.26. The black curve is the actual position of the crack,

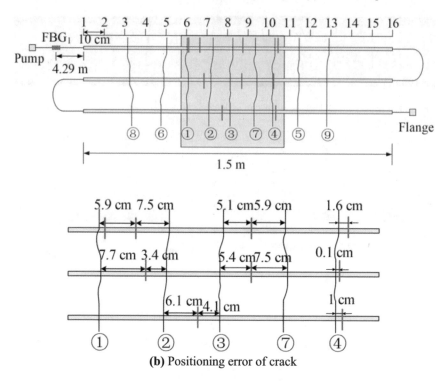

Fig. 9.26 Crack location of fully viscous sensing fiber

and the red short line represents the process to determine the crack position on the basis of fiber grating positioning method. The crack tested by the Brillouin distributed positioning method based on the fiber grating mainly focuses on the shaded part in the graph. Figure 9.26b is an enlarged view of the shaded part. Using the Brillouin distributed positioning method based on the fiber grating, the positioning for several cracks ①, ②, ③, ④, and ⑦ can be realized, and the error is within 8 cm. However, the positioning method has not identified the location of all the cracks. Possibly due to the fact that the stress is released as a result of the optical fiber degumming at the cracks, the lower spatial resolution of the instrument, and too dense cracks appear, the crack event cannot be identified by the instrument.

9.4.2.3 Concrete Crack Identification Based on Two-Point Mounted Optical Fiber Sensor

Figure 9.27 is Brillouin frequency shift along the point-mounted sensing optical fiber under loading condition. Figure 9.27b~d is the enlarged views of three adhesive optical fiber sections in Fig. 9.27a, respectively. As for the point-mounted optical fiber, Brillouin frequency shift of 3 optical fiber sections is independent on the increase

9.4 Concrete Crack Monitoring Using Fully Distributed Optical Fiber ...

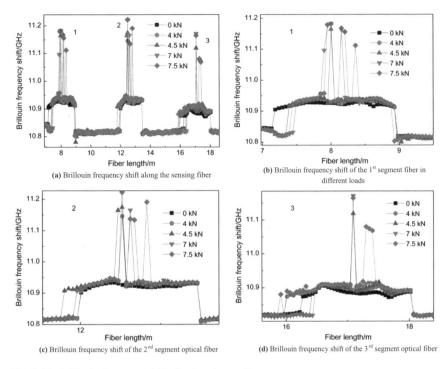

Fig. 9.27 Brillouin frequency shift of point viscous fiber

of loading. Unlike the full-mounted optical fiber, there is no sharp peak in Brillouin frequency shift of the point-mounted optical fiber after the crack occurs. However, since the optical fiber in the occurrence area of cracks is subjected to tensile stress, the frequency shift of a fiber section increases. As the load increases progressively, cascade steps appear on both sides of the platform with frequency shift increase as shown in Fig. 9.27b and c. It is reflected that the cracks on the beam increase as the load increases, which conforms to the actual conditions.

Figure 9.28 is the schematic diagram of the crack positioning of the point-mounted adhesive optical fiber. According to Figs. 9.27 and 9.28, it can be seen that for the point-mounted adhesive optical fiber, it is not possible to position each crack. Since the crack is dense, only the area where cracks are generated can be observed. However, when it is loaded to 4 kN, obvious grooves appear in the center of three point-mounted adhesive optical fiber sections. At this time, cracks ①~⑤ actually appear. When it is gradually loaded to 5 kN, the frequency at the groove increases obviously. At this time, cracks ⑥ and ⑦ appear on the surface of the beam, crack ⑦ is located just between cracks ③ and ④. As a result of the occurrence of crack ⑦, the optical fiber between the two pasting points (8 and 9) is subjected to tensile stress, resulting in the increase of Brillouin frequency. It is concluded that the groove position appeared in the frequency spectrum shall be the position of crack ⑦.

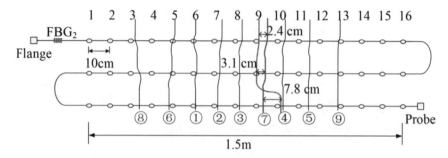

Fig. 9.28 Location of cracks for point-bonding distributed sensor

Fig. 9.29 Heating photo of distributed sensor near location of No. 7 crack

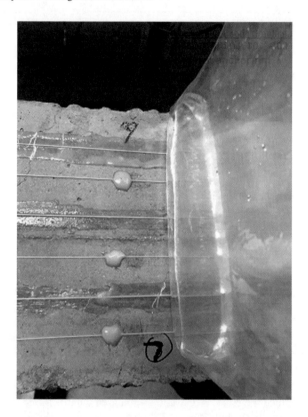

In order to verify that the speculation is correct, place the hot water bag at crack ⑦ (see Fig. 9.29), and observe the Brillouin frequency shift change of the sensing fiber. Figure 9.30 is the stimulated Brillouin spectrum at crack ⑦. In this figure, the circle part is the frequency spectrum at the groove. As the temperature increases, the frequency on the groove increases obviously, thereby confirming that the speculation is correct, and the position of the groove in the frequency spectrum is the position of crack ⑦.

9.4 Concrete Crack Monitoring Using Fully Distributed Optical Fiber ...

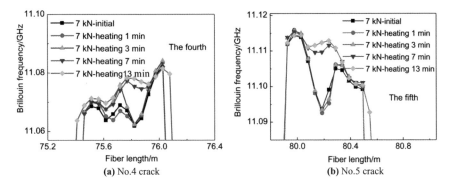

Fig. 9.30 Brillouin frequency shift at location of No. 4 and No. 5 cracks in different times

Crack ⑦ was positioned with Brillouin fully distributed positioning method based on fiber grating. According to Fig. 9.30, it can be seen that the blue curve represents the actual position of crack ⑦, and the red curve is the position of the crack positioned according to the test result. The results show that the positioning errors of crack ⑦ with FBG_2 on the three sensing fiber sections are 2.4, 3.1, and 7.8 cm, respectively.

9.5 Summary

Based on the principle of full Brillouin sensing and optical fiber grating, this chapter discusses the problem of positioning drift of fully distribute optical fiber sensor in detail. Based on the principle of unchanged spatial relative position, the principle and method of optical fiber grating positioning are described. Two methods of full-mounted and point-mounted optical fiber are employed to accurately capture the number of cracks and the process of crack generation and expansion of concrete structure. The crack monitoring and crack positioning of concrete structure are realized, and the positioning error is controlled within 8 cm.

References

Agrawal GP (2007) Nonlinear fiber optics. Academic Press, The Salt City
Bao X, Chen L (2012) Recent progress in distributed fiber optic sensors. Sensors 12(7). https://doi.org/10.3390/s120708601
Boyd RW (2003) Nonlinear optics. Academic Press, Beijing
Dong Y, Bao X, Li W (2009) Differential Brillouin gain for improving the temperature accuracy and spatial resolution in a long-distance distributed fiber sensor. Appl Opt 48(22):4297–4301. https://doi.org/10.1364/AO.48.004297
Li J, Furuta T, Goto H et al (2003) Theoretical evaluation of hydrogen storage capacity in pure carbon nanostructures. Chem Phys 119(4). https://doi.org/10.1063/1.1582831

Li JZ, Sun BC, Du YL (2016a) A fully-distributed fiber-optic sensing test method and patent for invention that can be accurately positioned. China patent ZL2,014,104,645,710, 26 Nov 2014

Li JZ, Xu LX, Kishida K (2016b) FBG-based positioning method for BOTDA sensing. IEEE Sens J 16(13):5236–5242. https://doi.org/10.1109/JSEN.2016.2556748

Li JZ, Zhao DS, Hou YM et al (2017) Power coupling characteristics between FBG and backscattering signals. Int J Mod Phys B 31(7):1741014. https://doi.org/10.1142/S0217979217410144

Loranger S, Gagné M, Lambin Iezzi V et al (2015) Rayleigh scatter based order of magnitude increase in distributed temperature and strain sensing by simple UV exposure of optical fibre. Sci Rep 5. https://doi.org/10.1038/srep11177

Smith RG (1972) Optical power handling capacity of low loss optical fibers as determined by stimulated Raman and Brillouin scattering. Appl Opt 11(11):2489–2494. https://doi.org/10.1364/AO.11.002489

Sun BC, Xu H, Li JZ et al (2007) New technology study about crack monitoring based on optical net. Chin J Sens Actuators 20(7):1672–1675

Sun BC, Hou YM, Li F et al (2017) Coupling characteristics between fiber grating and stimulated Brillouin signal. Chin Opt 10(4):484–490. https://doi.org/10.3788/CO.20171004.0484

Zhang M (2008) Laser light scattering spectroscopy. Science Press, Beijing

Chapter 10
Engineering Applications of Optical Fiber Sensing Technology

The optical fiber sensing technology combines optical waveguide technology with optical fiber technology, which has the characteristics of strong anti-electromagnetic interference ability, high measurement accuracy, long-term stability, perception and transmission integration, multi-sensor reuse and distributed sensing, and plays an increasingly important role in the health monitoring of large-scale engineering structures (Bremer et al. 2016; Giurgiutiu 2016; Hong et al. 2017). The author of this book has done lots of works in the application of optical fiber sensing technology and established a long-term health monitoring and alarm system of structure based on a number of optical fiber sensing, and alarm system have been established and applied successfully in Wuhu Yangtze River Bridge, Liaohe Bridge on Qinhuangdao-Shenyang passenger dedicated line, Hemaxi Grand Bridge, Shanxi Xinyuan Highway Xiaogou Grand Bridge and the high and steep slope of Shuohuang Railway.

10.1 Long-Term Health Monitoring and Alarm System for Wuhu Yangtze River Bridge

The long-term health monitoring and alarm system for Wuhu Yangtze River Bridge is a complete set of comprehensive long-term health monitoring system of large bridge structure, making the environmental monitoring, bridge structure performance, and train running safety monitoring as the main body composed of multiparameters, multi-sensors and related conditioning equipment, multi-station and two-level computer local area network. The strain monitoring employs the white light interference fiber strain sensor and its corresponding signal demodulation system.

10.1.1 Brief Introduction to Wuhu Yangtze River Bridge

Wuhu Yangtze River Bridge is the first large-span road-rail bridge constructed with plate-truss composite structure in China. The railway bridge is 10,521 m long and the highway bridge is 5,681 m long. Due to navigation clearance requirements and flight restrictions near the airport, the main part of the bridge employs three-span continuous low tower and cable-stayed continuous steel truss beam structure, the span of cable-stayed bridge is arranged as 180+312+180 m, the tower is 33.2 m high above the highway, the ratio of side span to main span is 0.577, ratio of tower height to main span is 0.11, the outermost cable inclination angle is about 15°, these ratios are significantly different from the conventional cable-stayed bridge, the structural stress characteristics is between the general cable-stayed bridge and the continuous steel truss bridge. The highway load acts on the concrete bridge deck, and the concrete bridge deck is combined with the upper chord of main truss through M22 shear key studs to participate in the stress of main truss. Due to the above structural characteristics, the stress state of bridge structure is more complex.

Wuhu Yangtze River Bridge (see Fig. 10.1) was opened to traffic in 2,000, representing the highest level of bridge design and construction in China at that time. To ensure the safety of bridge operation, verify and improve the design theory of large-scale bridge structure, improve the detection, maintenance, and management level of the bridge, a long-term health monitoring and alarm system based on optical fiber sensing are established.

Fig. 10.1 Yangtze River Bridge in Wuhu

10.1.2 General Overview of the Long-Term Health Monitoring and Alarm System

The long-term health monitoring and alarm system are mainly composed of sensor system, data acquisition system, data transmission system, data processing and management system, safety evaluation system, and information display system (see Fig. 10.2).

In the network structure system of the monitoring system, the data acquisition terminals includes eight dynamic data acquisition terminals (data acquisition substations), one train axle load monitoring terminal, one driving safety (train derailment coefficient) monitoring terminal based on embedded system. the whole system integrates 49 vibration displacement sensors, 10 acceleration sensors, 50 dynamic strain measuring points, 6 deflection measuring points, 4 beam end displacement measuring points, 6 driving speed measuring points, 24 temperature measuring points, 12 train axle monitoring signals 8 driving safety monitoring signals, etc. and a total of 219 long-term monitoring measuring points.

The monitor data is output from data acquisition terminals on the bridge and transmitted to the main monitor center under the bridge by the optical fiber converted by the optical fiber transceiver. The main monitoring center is composed of three units: machine room, control room, and large screen terminal. The machine room consists of two database servers, a network page server and a plurality of data browsing user operation terminals; the control room consists of a television monitoring screen and

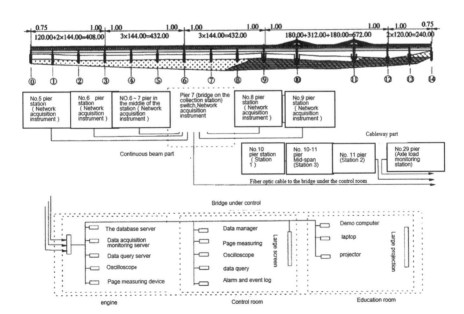

Fig. 10.2 Topology diagram of Wuhu Yangtze River Bridge long-term health monitoring system (unit: m)

a number of user terminals, which can be used to browse and query the monitoring results at any time, and can display the bridge dynamic response waveform when the train crosses the bridge in real time. The large screen projection terminal can be used to synchronously browse, query, and demonstrate the monitoring results and display the real-time dynamic waveform.

The long-term health monitoring and alarm system for Wuhu Yangtze River Bridge have the following functions.

(1) In the real-time monitoring process, the monitoring system can automatically send out grading alarm information: under the effect of live load (running vehicles), when the maximum measured value (or minimum value) changes within the normal range, the alarm information is green light; when the maximum value (or minimum) overrun, that is, the system automatically alarms (or yellow light, or red light means the highest alarm level) beyond its normal range.

(2) The monitoring system automatically records the changes of main measured parameters (such as the mid-span deflection of the bridge, the longitudinal displacement of the beam end, the stress of the main truss member and the natural vibration frequency of the structure) at any time or the environment temperature under the action of dead load (i.e., there is no vehicle on the bridge at that time), and gives the change curves of main measured parameters at any time or the environment temperature.

(3) The current value of main measure parameter of the bridge structure under the action of dead load can be automatically compared with the reference value under the same environmental temperature in the past. Once the value of a certain parameter is found to be abrupt (i.e., the current value of a certain parameter is larger than 1.2 times of the reference value under the same condition in the past), the system automatically sends a warning signal to remind the maintenance management department to pay attention to the sudden change, and the reason for it in the parameter is timely ascertained according to the monitoring data results to analyze and judge the possible damage in the structure.

10.1.3 Strain Monitoring System Based on Optical Fiber Sensing

10.1.3.1 Composition of Strain Monitoring Subsystem

The strain monitoring subsystem of Wuhu Yangtze River Bridge long-term health monitoring and alarm system is mainly composed of white light interference optical fiber strain sensor, Bus System optical fiber signal demodulator, and network acquisition controller (Dai et al. 2007; Zhang et al. 2008).

The strain sensor is a surface-mounted (Cui et al. 1999; Liu and Zhang 2000) white light interference type with the measuring precision of 1 $\mu\varepsilon$ and measuring range of $\pm 2,000$ $\mu\varepsilon$. The sensor consists of three parts: optical fiber sensing head,

10.1 Long-Term Health Monitoring and Alarm System for Wuhu Yangtze River Bridge

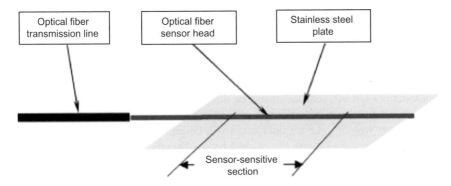

Fig. 10.3 Schematic diagram of surface-mounted white light interference type fiber strain sensor

stainless steel plate, and optical fiber transmission line (see Figs. 10.3 and 10.4). The optical fiber sensing head encapsulates the optical fiber sensing sensitive element in a stainless steel pipe with a diameter of 1 mm and a length of 57 mm; the stainless steel plate is 25 mm long, 6 mm wide, and 0.3 mm thick, optical fiber sensing head and stainless steel plate firmly welded together; the optical fiber transmission line is an armored optical cable with a diameter of 4 mm and length of 5 m.

The optical fiber sensing signal acquisition employs the Bus System optical fiber signal demodulation instrument, and completes the acquisition control, storage, transmission and management of the strain signal through the network data acquisition instrument. The strain monitoring mode is a real-time trigger acquisition mode, a trigger command sent by a main monitoring center is received through an RS-232 interface, and train load trigger is the real-time trigger condition, namely, when the train reaches a test area, the system triggers a network data acquisition instrument, the intelligent controller Rabbit 3,000 automatically starts a Bus System to acquire strain signals and output analog voltage signals, and then the analog voltage signals are transmitted to the main monitoring center through a communication network after A/D conversion through Rabbit 3000 serial port for real-time display and storage.

Fig. 10.4 Surface paste type white light interference optical strain sensor

10.1.3.2 Installation Process of Optical Fiber Strain Sensor

The surface-mounted white light interference optical fiber strain sensor can be directly attached to the surface of steel structure by using a bonding mode, but the adhesive is easy to age and fail after long-term work in wild, so that the sensor is degummed or fall off; sensor can also be directly welded on the surface of the steel beam, but the welding heat affected area is larger (depth is generally greater than 2 mm), after welding, the residual stress will be generated inside the steel girder, which may affect the fatigue life of steel beam. Therefore, the traditional bonding or welding method is not suitable for long-term strain monitoring of Wuhu Yangtze River Bridge.

To fix the surface adhesive white light interference type optical fiber strain sensor firmly on the steel beam, a set of installation scheme of bonding, pressing plate, bolt fastening is designed. The scheme comprise the following steps: weld six bolts with a diameter of 6 mm on a steel beam, fix a pressure plate on the steel beam in a bolt connection mode, and place an optical fiber sensor between the steel beam and the pressure plate, so that synchronous deformation of the optical fiber sensor and the steel beam can be ensured, as well as the and long-term reliability of the installation effect of sensor. The key of this scheme is to weld the bolt firmly on the steel beam; the traditional welding method will produce a large heat affected area, and then affect the fatigue life of steel beam. The stud welding machine produced by British Tyler Stud Welding System Co., Ltd. provides the possibility for the realization of the scheme. The stud welding machine is to install special bolts in the stud welding gun, to tighten the welding gun on the pre-welding parts, after the welding gun switch is buckled, the welding gun will shoot the stud, using the principle of high voltage discharge of capacitance, and the stud can be firmly welded on the welded parts within 4 ms. The main advantage of stud welding method is that the welding strength is high and the strength of the parent material can be reached. The heat affected zone is small (less than 0.3 mm), and the service life of steel plate with a thickness greater than 10 mm is hardly affected.

The specific installation process is as follows:

(1) Design and manufacture a stainless steel pressure plate of an optical fiber strain sensor at the size of $60 \times 24 \times 6$ mm, and with six $\phi 6$ bolt through-holes uniformly distributed;
(2) Weld six studs on the steel beam according to the position of bolt holes of the stainless steel pressure plate by adopting a stud welding method;
(3) Attach the optical fiber strain sensor on the surface of steel structure by adopting an adhesive;
(4) Install the pressure plate on the stud, apply sealant between the steel beam and the pressure plate, apply anti-loosening glue between the nut and the stud, set the standard nut on the stud, and tighten the nut to firmly press the sensor and the pressure plate on steel beam.

The purpose of applying sealant between the steel plate and the pressure plate is to prevent the installation position of optical fiber strain sensor from being corroded after long-time operation, and the purpose of applying anti-loosening glue between the nut and the stud is to prevent the nut from loosening after long-term operation of the bolt.

The optical fiber strain sensor installed according to that method can ensure the synchronous deformation of sensor and steel beam and prevent the corrosion of the steel plate at the installation position and the loosening of nut, thereby ensuring the long-term stable and reliable operation of sensor.

10.1.3.3 Strain Monitoring Data Analysis and Alarm Software

To realize the real-time analysis and alarm function of monitoring data, a set of data analysis and alarm software system is developed by using LabVIEW software platform. The software system is divided into 4 functional modules: real-time monitoring module, data query module, data analysis module, and report output module.

The real-time monitoring module can display the strain time history curve, the maximum value and the temperature of strain measuring point of all main truss members in real time in the same window, display the information of train speed, wheelbase, axle load, axle number, etc. in real time, and carry out classification alarm according to the monitoring value.

The data inquiry and analysis module can select different inquiry modes on the control board, analyze necessary data according to requirement, and export data in a text mode; the printing function outputs the data currently queried to the printing device in the form of report, and the graphic saving function can save the current graphic as a bitmap file.

The real-time monitoring and data processing interface are as shown in Fig. 10.5.

Since the installation and operation of Wuhu Yangtze River Bridge long-term health monitoring and alarm system in 2003, a large number of real-time monitoring data have been obtained, which plays an important role in ensuring the scientific

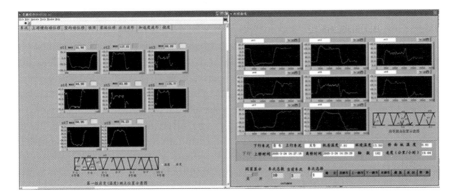

Fig. 10.5 Real-time monitoring interface

maintenance and safe operation of the bridge. The strain monitoring results show that the optical fiber strain sensor not only can reflect the relative increment of steel beam stress under external load accurately but also realize the absolute measurement of stress change caused by the change of stress state throughout the operation. The cumulative stress value of stress change in the process of structural stress state changing can be measured, which provides an important basis for effective evaluation of bridge stress state.

10.2 Monitoring System of Liaohe Bridge on Qinhuangdao-Shenyang Passenger Dedicated Line

10.2.1 Brief Introduction to Liaohe Bridge

Qinhuangdao-Shenyang passenger dedicated line is the first high-speed railway designed by China, with a design speed of 200~250 km/h. To ensure the comfort of passengers and the smoothness of track when the train passes through at high speed, it is necessary to ensure that the bridge has enough rigidity and integrity. Therefore, the bridge of Qinhuangdao-Shenyang passenger dedicated line employs integral box girder, among which Liaohe Bridge employs 32 m double-line prestressed concrete box girder with the largest span (Wei and Du 2004).

Liaohe Bridge on Qinhuangdao-Shenyang passenger dedicated line (see Fig. 10.6) is located at about 600 m upstream of Liaohe Bridge of Beijing–Shenyang highway, with the center mileage DK337+633.79, and the total length of 2,433.59 m is located at +0.0‰ ramp and curve, straight line and transition curve with radius of 5,500 and 6,000 m. The lower structure is bored pile foundation, ear-wall abutment, round end plate pier, rectangular pier cap. The beam is 74 holes double-line whole-hole single-cell-box prestressed concrete box simple beam with equal altitude, length of 32 m, height of 2.6 m high, roof width of 12.40 m, bottom width of 6.0~6.4 m, beam length of 32.60 m, each hole weighs about 754 t, TQF-1 waterproof layer is set at bridge deck, JHPZ 5,000 kN basin-type rubber bearing is employed.

Liaohe large bridge is designed as a single box girder with equal height and single box and single chamber, the web plate at the end of the girder is locally widened, and the bottom plate is locally thickened; and partition wall is arranged at the support seat at the both ends; the upper web plate of beam is provided with an exhaust hole to reduce the temperature difference between the inside and the outside of the box; pavement and beam body top surface are perfused as a whole, and consider to design the beam top as 12.4 m wide for the installation of sidewalk railings. Box girder manufacturing employs the segmental precast, and the precast beam section is divided into five sections, the sections at both ends are 5.35 m long, while the middle section is 6.5 m long, with four wet joints having a width of 0.6 m. The beam section is positioned on the bridge-assembling machine by transportation and

Fig. 10.6 Liaohe Bridge

The method does not require large-scale lift equipment, can build bridge at high speed, and ensure that construction quality of the box girder. However, compared with the traditional construction technology, it is necessary to further monitor and verify the construction technology, construction quality, and operation performance so as to ensure the construction safety, operation safety and improve the construction technology and design theory of double-line prestressed concrete box girder.

10.2.2 Application of Optical Fiber Sensor in Concrete Hydration Heat Testing

Liaohe Bridge employs double-line whole-hole prestressed concrete box beam with maximum span of 32 m. During pouring and hardening, the temperature change caused by hydration heat in this double-line whole-hole concrete is very obvious. To master the reasonable construction method, ensure the construction quality and prevent the cracking of concrete caused by excessive temperature difference between inside and outside, the embedded white light interference optical fiber temperature sensor and BGK 3,700 temperature sensor are used to monitor the temperature of concrete hydration process.

According to the prefabricated condition of box girder segment and the simulation analysis result, the section of 48-hole and 67-hole beam ends at 0.9 m away from the beam ends is selected for temperature monitoring, and the temperature sensors are respectively arranged in the web plate, bottom plate and top plate of the section. Each section is equipped with 14 temperature sensors. Continuous real-time monitoring was carried out within 7 days after April 25 and April 30, 2001, until the temperature change within the concrete was basically consistent with the changes in ambient temperature. The layout of the measuring points is as shown in Fig. 10.7.

Compared with finite element calculation and the measured data of the sensors, the maximum temperature, and the history curve are in good agreement (see Fig. 10.8). Therefore, it is feasible to monitor the hydration heat temperature by using the embed-

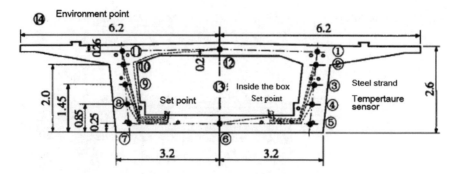

Fig. 10.7 Measuring point arrangement of hydration heat temperature (unit: m)

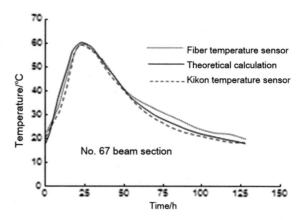

Fig. 10.8 Comparison of typical hydration heat test results using different sensors

ded white light interference optical fiber temperature sensor, and the test results can be used for construction control.

The results show that the vertical temperature difference of box beam is nonlinear and the horizontal temperature difference is nearly linear; the maximum temperature difference of each measuring point is 0~11 °C, the maximum temperature difference between the structure and the surface is about 20 °C, which is close to the empirical control value of 0~20 °C, so the temperature stress analysis should be carried out and combined with the concrete strength analysis over time to determine the temperature difference control value.

It has been proved that there is no surface cracking in the prefabricated beam section. The measured and calculated results show that the maximum temperature of concrete occurs at around 30 h, indicating that the hydration rate of high strength concrete is fast, the heating rate is about 2 °C/h and cooling rate is 0.3 °C/h on average, and the maximum cooling rate is 0.55 °C/h; the duration of high temperature (0.8 times the maximum temperature) is about 24 h, while the duration of temperature higher than environment temperature (15 °C) is about 75 h.

10.2.3 Application of Optical Fiber Sensor in Construction Quality Monitoring of Concrete Bridge

As the long-span double-line whole-hole prestressed concrete box girder is first applied to railway bridge construction, its manufacturing process is not proven enough, its design theory is not perfect, it is, therefore, necessary to conduct in-depth research on its manufacturing process and mechanical properties, so the embedded white light interference optical fiber strain sensor is used to monitor its construction and long-term monitoring. The main monitoring contents include the development rule of shrinkage and creep with time, the length of prestressing transmission, the distribution of prestress in the top and bottom plates of mid-span section, the transverse tensile stress of end block, and the local stress of entrance hole, etc.

According to the precast box beam segment and simulation analysis results, select precast concrete beam across section top, bottom, web, respectively, for fiber-optic strain sensor and differential strain sensor parity arrangement, measuring point arrangement as shown in Fig. 10.9.

The monitoring is mainly conducted at the concrete main beam tensioning stage (see Fig. 10.10) and bridge long-term operation stage (see Fig. 10.11); it shows the tension stress curve consistency is higher through the comparison of finite element calculation and measured data. Therefore, it is feasible to use embedded white light interference optical fiber strain sensor for monitoring, and the monitoring results can provide the basis for construction control.

10.2.4 Application of Optical Fiber Strain Sensor in Dynamic Monitoring of Concrete Box Beam

By using the white light interference optical fiber strain sensor embedded in the construction monitoring, some strain gauges and white light interference optical

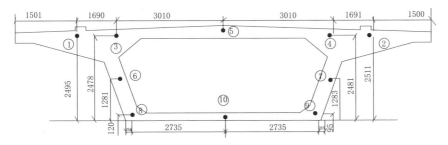

Fig. 10.9 Main beam cross-section of strain measuring point placement of Liaohe Bridge (unit: mm)

Fig. 10.10 Comparison of measured stress for typical comparison of typical test results

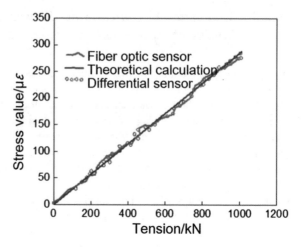

Fig. 10.11 Points in main beam tensioning stage comparison of typical test results during long-term operation (6 months)

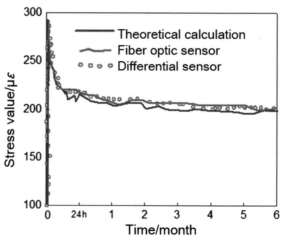

fiber strain sensors are pasted on the surface of box girder, the dynamic contrast monitoring of driving performance on Liaohe Bridge is carried out.

Measuring points are arranged on the longitudinal axis of 72-hole span section (see Fig. 10.12), measuring points 1~4 are located on the top plate surface in the box, 5–8 are located on the bottom plate surface in the box, 9~10 are the original embedded optical fiber strain sensor, measuring points position and sensor type are as shown in Table 10.1, and the dynamic monitoring instruments and equipment used are as shown in Table 10.2.

From the monitoring results, it can be seen that the response of box beam structure is small when the train passes through, while the stress of each strain measuring point is corresponding to the live load, and the direction is clear. The maximum strain is about 28 $\mu\varepsilon$ when the bottom plate of mid-span beam is pulled, that of top plate is about 11 $\mu\varepsilon$; the elastic modulus is calculated by 40 GPa, and the corresponding

10.2 Monitoring System of Liaohe Bridge on Qinhuangdao-Shenyang …

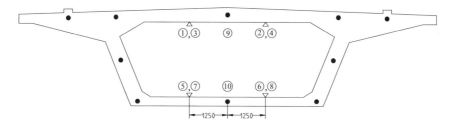

Fig. 10.12 Installed location and number of optical fiber sensors at the 72-holen medium-cross-section (unit: mm)

Table 10.1 Measuring point positions and sensor types

Measuring point number	Acquisition instrument channel number	Type	Location
1	1	Strain gauges	Roof
2	2	Strain gauges	Roof
3	3	Fiber strain sensor	Roof
4	4	Fiber strain sensor	Roof
5	5	Strain gauges	Floor
6	6	Strain gauges	Floor
7	7	Fiber strain sensor	Floor
8	8	Fiber strain sensor	Floor
9	9	Fiber strain sensor	Floor
10	10	Fiber strain sensor	Floor

stress is 0.96 and 0.44 MPa respectively, which indicates that the stress of beam body in the train is smaller, and the maximum dynamic strain is as shown in Table 10.3.

Figure 10.13 shows the strain time history curve of each working condition of each measuring point, and the following conclusions can be obtained after analysis.

(1) The strain history curves of the strain gauge and the fiber strain sensor are consistent, and the fiber strain sensor has no response lag, the peak time of strain gauge is the same as that of the fiber strain sensor, which indicates that the dynamic response characteristics of fiber strain sensor can meet the monitoring requirements.

(2) The dynamic strain monitoring results of strain gauges and optical fiber strain sensors are in good agreement, but there are some differences. The reason is that the absolute amount of dynamic strain is very small, and some relative errors will be caused by the factors such as patch technology, environmental temperature, etc.

(3) The dynamic response of the embedded optical fiber strain sensor and the surface-mounted optical fiber strain sensor are consistent. The strain value of the sensor is inconvenient to compare directly due to the different section height

Table 10.2 Dynamic test equipment and models

Serial number	Equipment name	Specifications model	Origin	Unit	Quantity	Use
1	Fiber strain test system	BUS System 8 channels	Canada	Set	1	Dynamic strain test
2	Dynamic data acquisition system	7V15	China	Set	1	Dynamic strain test
3	Automatic storage and processing system for large capacity data	INV303/306	China	Set	1	Dynamic strain test

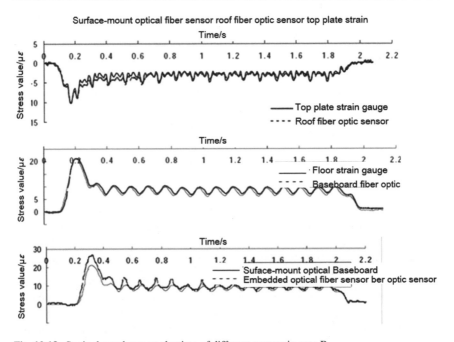

Fig. 10.13 Strain dependence on the time of different sensors in case B

position of the sensor, but the strain near the bottom surface of the beam body (embedded optical fiber strain sensor) is larger than the strain on the upper surface of the bottom plate (adhesive optical fiber strain sensor).

10.2 Monitoring System of Liaohe Bridge on Qinhuangdao-Shenyang ... 341

Table 10.3 Maximum dynamic strain of cross-section in beam

Working part number	Measuring point number	Strain value/$\mu\varepsilon$	Position	Sensor type
11	1	−11.2	Top edge of the roof	Strain gauges
	3	−11.3		Fiber-optic sensor
	6	22.6	Floor	Strain gauges
	7	23.0		Surface-mounted optical fiber sensor
	10	27.9		Embedded optical fiber sensor
13	1	−10.1	Top edge of the roof	Strain gauges
	3	−10.1		Fiber-optic sensor
	6	21.0	Floor	Strain gauges
	7	21.2		Surface-mounted optical fiber sensor
	10	26.9		Embedded optical fiber sensor
14	1	−11.0	Top edge of the roof	Strain gauges
	3	−10.7		Fiber-optic sensor
	6	21.4	Floor	Strain gauges
	7	22.4		Surface-mounted optical fiber sensor
	10	27.1		Embedded optical fiber sensor
16	1	−10.7	Top edge of the roof	Strain gauges
	3	−10.2		Fiber-optic sensor
	6	21.3	Top edge of the roof	Strain gauges
	7	19.7		Surface-mounted optical fiber sensor
	10	22.8		Embedded optical fiber sensor

10.3 Long-Term Health Monitoring System for Xinyuan Highway Xiaogou Grand Bridge

10.3.1 Brief Introduction to Xiaogou Grand Bridge

Xiaogou Grand Bridge is located on K106+802.061 of the newly built Guangwu-Yuanping Highway, it is a double bridge type with separate left and right bridge, medial strip width of 1.5 m, anti-collision guardrail on both sides of 2×0.5 m, and net width of single bridge carriageway of 11 m. Bridge span composition at left line: (55 m + 5 × 100 m + 55 m) (main bridge) + (7 × 30 m) (approach bridge); bridge span composition at right line: 30 m (approach bridge) + (55 m + 5 × 100 m + 55 m) (main bridge) + (7 × 30 m) (approach bridge). The main bridge of the bridge employs the rigid frame-continuous composite structure system, the upper structure for single box single chamber bidirectional prestressing (longitudinal and vertical) variable cross-section box beam, box beam root height of 5.3 m, span box height of 2.0 m, box girder roof width of 12.0 m, floor width of 6.0 m, flange plate cantilever length of 3.0 m. The lower structure of the main bridge employs reinforced concrete thin wall hollow pier and bored pile foundation. Design load grade: automobile-over level 20, trailer-120, basic seismic intensity is VII.

10.3.2 Composition of Long-Term Health Monitoring System

The system is mainly composed of three parts: field automatic measurement system, remote control center system, and analysis center system. The composition of the whole system is as shown in Fig. 10.14. The main functions of the system are as follows:

1. On-site Automatic Measuring System

   The system is mainly used to collect data of strain, temperature, deflection deformation of bridge site and changes in vehicles. The single-chip microcomputer running under the TCP/IP protocol is used to collect and transmit data, and the different sensor groups are started regularly to collect, preprocess and transmit data. The system block diagram is as shown in Fig. 10.15.

2. Remote Control Center System

   The system is mainly used to store and count the collected data and send the system control commands. The system employs the LabVIEW (Zhang 2004) of NI Company to receive the data sent by the lower computer through TCP/IP, alarm when abnormal situation occurs, and carry out the original data processing and analysis. It can display the change trend of bridge structure strain, deflection deformation, and other parameters in the form of report graphics, and transmit data to the analysis center. The system block diagram is as shown in Fig. 10.16.

10.3 Long-Term Health Monitoring System for Xinyuan Highway Xiaogou ... 343

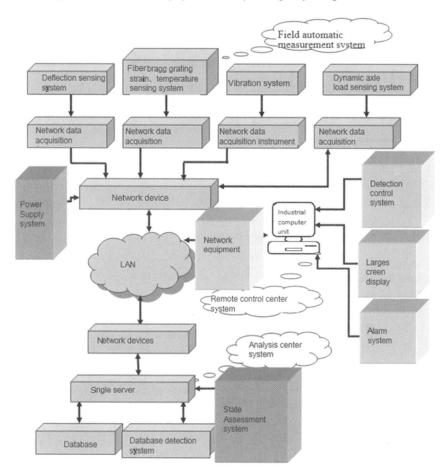

Fig. 10.14 Block diagram of remote control center system

3. Analysis Center System

The system mainly analyzes and judges the original data collected by the system and the processed data, evaluates the bridge operation status during monitoring period, and provides suggestions for maintenance and repair of the bridge. The system is the most important part of the whole health monitoring system, which needs to use scientific methods to establish a mathematical model reflecting the health status of bridge. The system block diagram is as shown in Fig. 10.17.

By using the health monitoring system, the real-time online monitoring of the strain, deflection, temperature, vibration, and dynamic axle load of the 7×100 continuous rigid frame bridge of Xiaogou Bridge main body is realized, and the strain monitoring system, deflection monitoring system, vibration monitoring system, temperature monitoring system and dynamic axle load monitoring system are

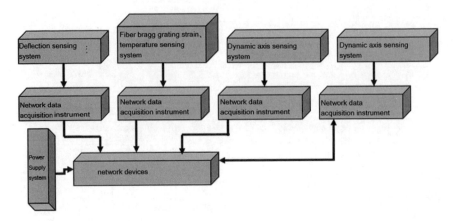

Fig. 10.15 Block diagram of remote control center system

Fig. 10.16 Block diagram of remote control center system

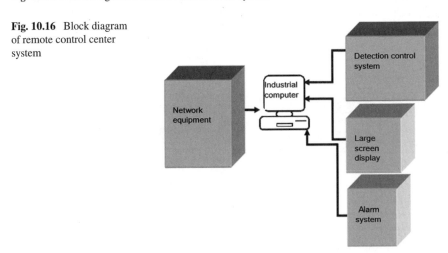

established, so that the working characteristics of Xiaogou Bridge during normal operation can be comprehensively and scientifically analyzed. The strain monitoring system employs the fiber grating strain sensor and fiber grating sensor network demodulator and other monitoring equipment.

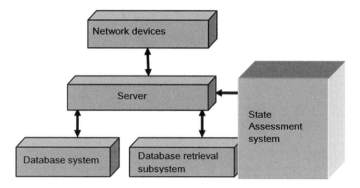

Fig. 10.17 Block diagram analysis center system

10.3.3 Strain Monitoring System Based on Fiber Bragg Grating

10.3.3.1 Selection of Strain Monitoring Section and Layout of Measuring Points

The strain monitoring section of Xinyuan Highway Xiaogou Grand Bridge is mainly selected in the middle span of box beam and root section of box girder. Considering that the right bridge is in the direction of heavy vehicle and the left bridge is in the direction of empty vehicle, 13 monitoring sections are selected for the right bridge, and 3 monitoring sections are selected for the left bridge, and a total of 16 monitoring sections (see Fig. 10.18) are selected. Each monitoring section has a strain measuring point in the middle and four corners of the upper and lower bottom of the box beam, and a total of 6 measuring points in each section (see Fig. 10.19), with a total of 96 strain measuring points.

10.3.3.2 Design of Strain Acquisition and Transmission System

The strain monitoring system employs the fiber grating strain sensor and fiber grating sensor network demodulator and other monitoring equipment, but there are still some technical problems such as data communication interface, remote control, and integration with the whole monitoring system. For this purpose, a software and hardware system for remote control of optical fiber strain acquisition is designed and developed. FBG strain signal demodulator controlled by Rabbit 3000 is used for strain signal acquisition, and the data is output through an Ethernet port and transmitted to a monitoring center through a communication network for real-time display and storage (see Fig. 10.20).

Two data acquisition and communication stations are arranged in the Xiaogou Grand Bridge and are connected with the remote control and analysis center arranged

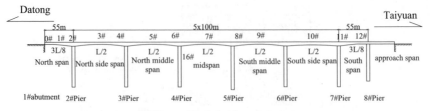

Right amplitude bridge strain test

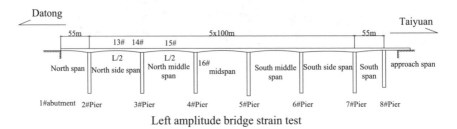

Left amplitude bridge strain test

Fig. 10.18 Strain monitoring point along Xinyuan Highway Bridge

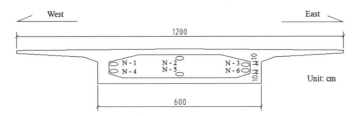

Fig. 10.19 Strain monitoring point distribution in cross-section of Xinyuan Highway Bridge

in the office building of Xinyuan Highway Construction Co., Ltd. through a data optical terminal machine and an optical cable. The communication systems are built with devices as per the Standard 802.3 to ensure that all system-wide devices are interconnected.

The data output by the sensor is transmit to the data acquisition instrument through an analog-to-digital conversion card (A/D), the network interface of the data acquisition instrument is connected with the optical interface of the data optical terminal machine through an optical fiber converter, the data of electric signal is converted into an optical signal for long-distance transmission through data optical terminal machine. The receiving end converts the optical signal into an electrical signal and then inputs it to a remote control analysis center.

The remote control and analysis center consists of several high-performance computers and servers and is equipped with relevant software systems, which can realize the display, processing, and analysis of collected data, control, management and maintenance of the system, and analyze and evaluate the safety state of whole bridge structure.

10.3 Long-Term Health Monitoring System for Xinyuan Highway Xiaogou …

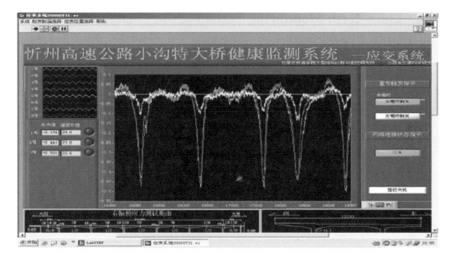

Fig. 10.20 Display interface of strain monitoring for Xiaogou Grand Bridge

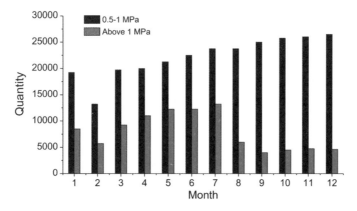

Fig. 10.21 Histogram of maximum profit at north right span of Xiaogou Grand Bridge

10.3.4 *Effect Analysis of Strain Monitoring*

The long-term health monitoring system of Xinyuan Highway Xiaogou Grand Bridge was built and put into use in 2006. The overall operation of the system is stable and normal, and a large number of real-time monitoring data are obtained, which plays an important role in ensuring the scientific maintenance and safe operation of the bridge. The statistical results of strain monitoring show that the measured stress value of each monitoring section is far less than the design maximum stress control value, which is less than half of the design load stress value (see Fig. 10.21), indicating that the bridge structure is in good stress state and can meet the normal operation requirements.

10.4 Long-Term Monitoring System of Shuohuang Railway High-Steep Slope

Shuohuang Railway is the second largest channel to transport coal from west to east in China, passing through Taihang Mountain, Heng Mountain, Yunzhong Mountain, and other mountains. The terrain is complex with high mountains and deep valley. There are potential geological hazards such as landslide or collapse due to the existence of numerous high embankments and high-steep slopes. Therefore, a long-term monitoring system for high-steep slope of Shuohuang railway based on optical fiber tilt sensor is established.

10.4.1 Brief Introduction to Monitoring Section

High-steep slope monitoring section is Shuohuang Railway K3+000–K3+410 section, located in Shenchi County, Shanxi Province (see Fig. 10.22, in the north side of the railway line. It is highway deep excavation slope from east to west with maximum-height about 30 m. The monitoring area is a temperate continental monsoon climate having significant seasonal changes, long and cold in winter, dry and windy in spring, mild in summer without the intense heat, cool and rainy in autumn; the west wind prevails in the area for many years with an annual average wind speed of 4.1 m/s; the average temperature in the area for many years is 4.6 °C, the lowest temperature is -28.2 °C, the average annual freezing time is 160 days, the maximum freezing depth is 1.5 m; the average annual precipitation in the area is 490 mm, the average maximum monthly precipitation is 130 mm.

Shuohuang Railway k3+000~k3+410 section is located in the loess hilly area with gentle slope and gully development. Most are meso-Cenozoic strata including new

Fig. 10.22 Monitored area topography of Shuohuang Railway K3+000–K3+410 section

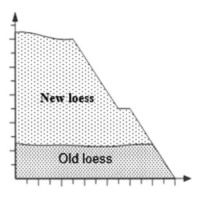

Fig. 10.23 Engineering geological profile (I)

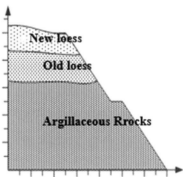

Fig. 10.24 Engineering geological profile (II)

loess, old loess, and argillaceous rock from shallow to deep. Figure 10.22 shows the two typical engineering geological profiles of slope in the monitoring area. The slope is partly protected by mortar rubble covering while others are unprotected with sparse vegetation covering (Figs. 10.23, 10.24).

10.4.2 Monitoring Scheme for Slope Deformation

The variation of the deep horizontal displacement of the slope is an important information reflecting the stable state of the high and steep slope, and is a key index for predicting the occurrence of the landslide. Therefore, the monitoring of deformation of high-steep slope is focused on the monitoring of its deep horizontal displacement. The overall monitoring scheme is shown in Fig. 10.25, it mainly consists of inclinometer (including FBG tilt sensor), FBG demodulator, GPRS communication module, and solar power supply. The signal sensed by the FBG inclination sensor installed inside the inclinometer is collected and demodulated by a demodulator and then transmitted to the monitoring center through GPRS module.

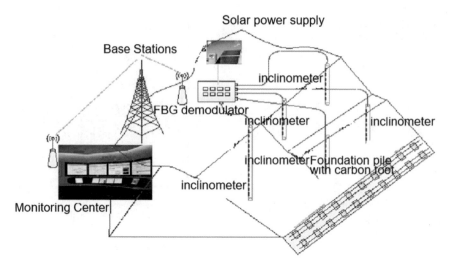

Fig. 10.25 Deep slip monitoring system

The inclinometer device based on fiber grating inclination angle sensor is the core component of the technical scheme. It mainly consists of an optical fiber inclination angle sensor, a connecting part, and an inclined tube. The optical fiber inclination sensor mainly reflects and characterizes the slip condition of deep slope through the drift of central wavelength of the optical fiber grating; the connecting part should ensure the effective connection of the sensing probe and the inclined tube connecting head. Through reasonable design, the connecting part can quickly realize the connection between the sensing probe and the inclined tube, and ensure the internal seal, so as to facilitate the monitoring of deep displacement of the slope; the inclined tube is made of ABS materials, which can better couple the inclined tube with the loess layer of the slope and facilitate the displacement transmission. The entire fiber dip angle inclinometer is as shown in Fig. 10.25.

When the slope slides, the displacement generated in the deep part can cause inclination change (deflection change) of the inclined tube, so that the inclined tube generates a deflection angle, the optical fiber inclination sensor packaged in the inclination measuring device can sense the deflection angle, the central wavelength of optical fiber grating will drift, the demodulation system on the ground can demodulate the drift amount of central wavelength of the optical fiber inclination sensor, and then the inclination angle of the position can be calculated and converted into horizontal displacement to realize the judgment of sliding off the deep part of the slope through the relationship curve of horizontal displacement amount along the depth change.

10.4.3 Monitoring Points Layout and Monitoring Equipments Installation

The layout of sensors in K3+000~K3+410 high-steep slope monitoring section on Shuohuang Railway as shown in Fig. 10.26, slope of deep horizontal displacement measuring holes are arranged in the first-level platform and slope top of A–A' section, the first-level platform and slope top of B–B' section and slope top of D–D' section, the measuring hole is 1 m away from the front edge of the slope. Figure 10.27 shows the first-level platform and top slope hole location. The details of each monitoring section are as shown in Table 10.4. In the actual field installation process, in addition to the five sets of optical fiber inclinometer, two sets of traditional electrical inclinometer are also installed to verify the monitoring effect based on optical fiber tilt sensor inclinometer scheme.

By using the drilling machine to drill the hole with diameter of 110 mm to measure the deep horizontal displacement, and the non-perpendicularity of the hole is not less than 2°. After the drilling is completed, the inclinometer based on the fiber grating inclination sensor with diameter of 70 mm is installed in a timely manner, and the gap between the measuring hole and the inclinometer is filled with fine sand. All transmission lines are protected by steel pipes to the fiber grating demodulator connected to the site monitoring station.

Fig. 10.26 Monitoring point of Shuo Huang Railway K3+000~K3+410 section

Fig. 10.27 Hole position of first-level platform

Table 10.4 Monitoring for specific situation on the section

Monitoring section	Monitoring content	Number and depth of holes	Number of sensors
A-A´(About K3+200)	Deep horizontal displacement	14 m One 20 m One	Tilt sensor 8 Tilt sensor 10
B-B´(About K3+300)	Deep horizontal displacement	13.5 m One 17.5 m One	Light/electric tilt sensor of the 8 Light/electric tilt sensor of the 10
D-D´(About K3+400)	Deep horizontal displacement	25 m One	13 tilt sensors

10.4.4 System Operation and Monitoring Results

Since the long-term monitoring system for high-steep slope of Shuohuang Railway was built and put into use in August 2017, the system has been running smoothly and normally, and a large number of real-time monitoring data have been obtained. The monitoring results show that (see Figs. 10.28 and 10.29): the change rule of displacement–depth curve measured by the inclinometer based on fiber grating inclination sensor is basically consistent with that measured by the traditional electrical inclinometer, the overall monitoring data change is small, and the maximum deformation is less than 2 mm. Compared with the traditional electrical inclinometer, the monitoring data of traditional electrical inclinometer fluctuates greatly (which may be the reason of electromagnetic interference of electric locomotive), this shows that the

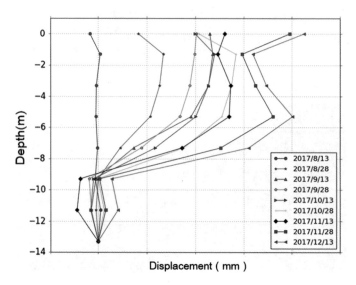

Fig. 10.28 Displacement from FBG sensor versus time at slope position

10.4 Long-Term Monitoring System of Shuohuang Railway High-Steep Slope

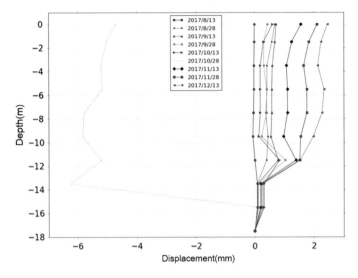

Fig. 10.29 Displacement from electrical sensor versus time at slope position

inclinometer based on fiber grating inclination sensor has strong anti-electromagnetic interference ability and good long-term stability, and is more suitable for long-term monitoring of electrified railway facilities under high current and strong magnetic field environment.

The monitoring results in recent months show that the overall deformation of slope is small, indicating that the Shuohuang Railway K3+000–K3+410 high and steep slope section stability is good, no slope slip or other geological disasters occurring.

10.5 Summary

This chapter introduces the application of long-term health monitoring and alarm system based on optical fiber sensing in Wuhu Yangtze River Bridge, Liaohe Bridge on Qinhuangdao-Shenyang Passenger Dedicated Line, Hemaxi Grand Bridge, Shanxi Xinyuan Highway Xiaogou Grand Bridge and the high-steep slope of Shuohuang Railway. Each monitoring system works well and provides strong support for the safe operation of different engineering structures. Therefore, the author believes that the long-term monitoring technology of project structure based on optical fiber sensing will make great achievements in the field of civil engineering.

References

Bremer K, Wollweber M, Weigand F et al (2016) Fibre optic sensors for the structural health monitoring of building structures. Proc Technol 26 (supplement C):524–529. https://doi.org/10.1016/j.protcy.2016.08.065

Cui F, Yuan W, Shi J (1999) Application of optimal sensor placement algorithms for health monitoring of bridges. J Tongji Univ 27(2):165–169

Dai JY, Zhang WT, Sun BC (2007) Applications of fiber optic sensors in concrete structural health monitoring. In: Proceedings of SPIE—advanced sensor systems and applications III. The International Society for Optical Engineering

Giurgiutiu V (2016) Structural health monitoring of aerospace composites. In: Victor G (ed) Structural health monitoring of aerospace composites, vol 31. Academic Press, Oxford, pp 249–296. https://doi.org/10.1016/B978-0-12-409605-9.00007-6

Hong C, Zhang Y, Li G et al (2017) Recent progress of using Brillouin distributed fiber optic sensors for geotechnical health monitoring. Sens Actuators A 258 (Supplement C):131–145. https://doi.org/10.1016/j.sna.2017.03.017

Liu F, Zhang L (2000) Advances in optimal placement of actuators and sensors. Mech Compos 30(04):506–516

Wei B, Du Y (2004) On the prospects for the application of optical fiber sensors to bridge structures. Nat Defence Traffic Eng Technol 2(1):8–11. https://doi.org/10.13219/j.gjgyat.2004.01.003

Zhang K (2004) LabVIEW virtual instrument engineering design and development. National Defense Industry Press, Beijing

Zhang WT, Dai JY, Sun BC (2008) Experiment on dynamic response of fiber optic Fabry-Perot sensors and its application in structural health monitoring. In: Proceedings of the world forum on smart materials and smart structures technology, SMSST 07, p 504